Illustrative Mathematics

# ACCELERATED 6

## Units 5–9

STUDENT EDITION

Volume 2

ISBN 978-1-26-432852-9

Accel6_8

*Making Connections*



*Voting*



*Designing a Course*

# Illustrative Mathematics

## Unit 5

STUDENT EDITION

Book 2

# Lesson 1: Proportional Relationships and Equations

Let's write equations describing proportional relationships.

## 1.1: Number Talk: Division

Find each quotient mentally.

$645 \div 100$

$645 \div 50$

$48.6 \div 30$

$48.6 \div x$

## 1.2: Feeding a Crowd, Revisited

1. A recipe says that 2 cups of dry rice will serve 6 people. Complete the table as you answer the questions. Be prepared to explain your reasoning.

   a. How many people will 1 cup of rice serve?

   b. How many people will 3 cups of rice serve? 12 cups? 43 cups?

   c. How many people will $x$ cups of rice serve?

| cups of dry rice | number of people |
|---|---|
| 1 | |
| 2 | 6 |
| 3 | |
| 12 | |
| 43 | |
| $x$ | |

2. A recipe says that 6 spring rolls will serve 3 people. Complete the table as you answer the questions. Be prepared to explain your reasoning.

   a. How many people will 1 spring roll serve?

   b. How many people will 10 spring rolls serve? 16 spring rolls? 25 spring rolls?

   c. How many people will $n$ spring rolls serve?

| number of spring rolls | number of people |
|---|---|
| 1 | |
| 6 | 3 |
| 10 | |
| 16 | |
| 25 | |
| $n$ | |

3. How was completing this table different from the previous table? How was it the same?

## 1.3: Denver to Chicago

A plane flew at a constant speed between Denver and Chicago. It took the plane 1.5 hours to fly 915 miles.

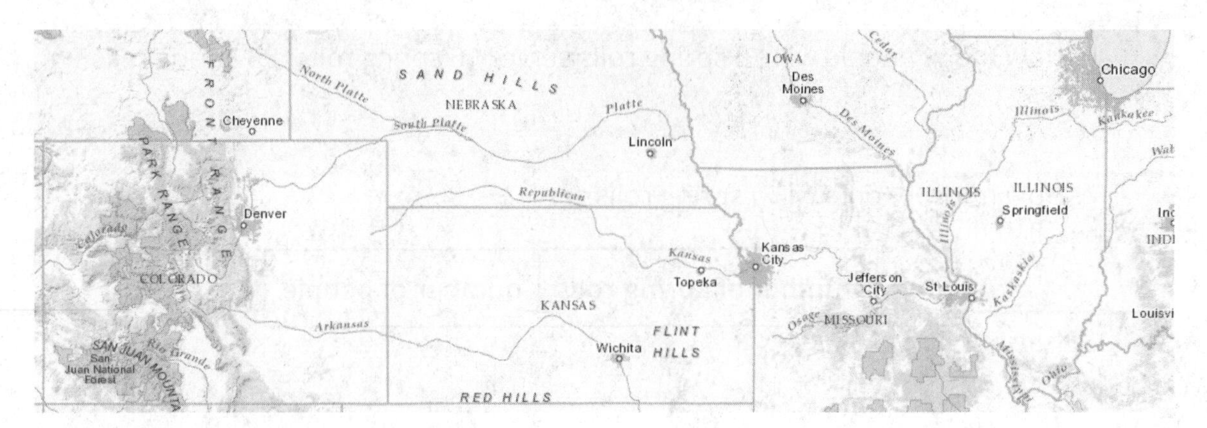


1. Complete the table.

| time (hours) | distance (miles) | speed (miles per hour) |
|---|---|---|
| 1 | | |
| 1.5 | 915 | |
| 2 | | |
| 2.5 | | |
| $t$ | | |

2. How far does the plane fly in one hour?

3. How far would the plane fly in $t$ hours at this speed?

4. If $d$ represents the distance that the plane flies at this speed for $t$ hours, write an equation that relates $t$ and $d$.

5. How far would the plane fly in 3 hours at this speed? in 3.5 hours? Explain or show your reasoning.

### Are you ready for more?

A rocky planet orbits Proxima Centauri, a star that is about 1.3 parsecs from Earth. This planet is the closest planet outside of our solar system.

1. How long does it take light from Proxima Centauri to reach Earth? (A parsec is about 3.26 light years. A light year is the distance light travels in one year.)

2. There are two twins. One twin leaves on a spaceship to explore the planet near Proxima Centauri traveling at 90% of the speed of light, while the other twin stays home on Earth. How much does the twin on Earth age while the other twin travels to Proxima Centauri? (Do you think the answer would be the same for the other twin? Consider researching "The Twin Paradox" to learn more.)

## 1.4: Revisiting Bread Dough

A bakery uses 8 tablespoons of honey for every 10 cups of flour to make bread dough. Some days they bake bigger batches and some days they bake smaller batches, but they always use the same ratio of honey to flour.

1. Complete the table.

2. If $f$ is the cups of flour needed for $h$ tablespoons of honey, write an equation that relates $f$ and $h$.

3. How much flour is needed for 15 tablespoons of honey? 17 tablespoons? Explain or show your reasoning.

| honey (tbsp) | flour (c) |
|---|---|
| 1 | |
| 8 | 10 |
| 16 | |
| 20 | |
| $h$ | |

## Lesson 1 Summary

The table shows the amount of red paint and blue paint needed to make a certain shade of purple paint, called Venusian Sunset.

Note that "parts" can be *any* unit for volume. If we mix 3 cups of red with 12 cups of blue, you will get the same shade as if we mix 3 teaspoons of red with 12 teaspoons of blue.

| red paint (parts) | blue paint (parts) |
|---|---|
| 3 | 12 |
| 1 | 4 |
| 7 | 28 |
| $\frac{1}{4}$ | 1 |
| $r$ | $4r$ |

The last row in the table says that if we know the amount of red paint needed, $r$, we can always multiply it by 4 to find the amount of blue paint needed, $b$, to mix with it to make Venusian Sunset. We can say this more succinctly with the equation $b = 4r$. So the amount of blue paint is proportional to the amount of red paint and the constant of proportionality is 4.

We can also look at this relationship the other way around.

If we know the amount of blue paint needed, $b$, we can always multiply it by $\frac{1}{4}$ to find the amount of red paint needed, $r$, to mix with it to make Venusian Sunset. So $r = \frac{1}{4}b$. The amount of blue paint is proportional to the amount of red paint and the constant of proportionality $\frac{1}{4}$.

| blue paint (parts) | red paint (parts) |
|---|---|
| 12 | 3 |
| 4 | 1 |
| 28 | 7 |
| 1 | $\frac{1}{4}$ |
| $b$ | $\frac{1}{4}b$ |

In general, when $y$ is proportional to $x$, we can always multiply $x$ by the same number $k$—the constant of proportionality—to get $y$. We can write this much more succinctly with the equation $y = kx$.

## Lesson 1 Practice Problems

1. A certain ceiling is made up of tiles. Every square meter of ceiling requires 10.75 tiles. Fill in the table with the missing values.

| square meters of ceiling | number of tiles |
|---|---|
| 1 | |
| 10 | |
| | 100 |
| $a$ | |

2. On a flight from New York to London, an airplane travels at a constant speed. An equation relating the distance traveled in miles, $d$, to the number of hours flying, $t$, is $t = \frac{1}{500}d$. How long will it take the airplane to travel 800 miles?

3. Each table represents a proportional relationship. For each, find the constant of proportionality, and write an equation that represents the relationship.

| $s$ | $P$ |
|---|---|
| 2 | 8 |
| 3 | 12 |
| 5 | 20 |
| 10 | 40 |

Constant of proportionality:

Equation: $P =$

| $d$ | $C$ |
|---|---|
| 2 | 6.28 |
| 3 | 9.42 |
| 5 | 15.7 |
| 10 | 31.4 |

Constant of proportionality:

Equation: $C =$

4. Diego bought 12 mini muffins for $4.20.

   a. At this rate, how much would Diego pay for 4 mini muffins?

   b. How many mini muffins could Diego buy with $3.00? Explain or show your reasoning. If you get stuck, consider using the table.

| **number of mini muffins** | **price in dollars** |
| --- | --- |
| 12 | 4.20 |
| | |
| | |
| | |

(From Unit 2, Lesson 9.)

5. It takes $1\frac{1}{4}$ minutes to fill a 3-gallon bucket of water with a hose. At this rate, how long does it take to fill a 50-gallon tub? If you get stuck, consider using a table.

(From Unit 2, Lesson 10.)

# Lesson 2: Two Equations for Each Relationship

Let's investigate the equations that represent proportional relationships.

## 2.1: Missing Figures

Here are the second and fourth figures in a pattern.

?

figure 1

figure 2

?

figure 3

figure 4

1. What do you think the first and third figures in the pattern look like?

2. Describe the 10th figure in the pattern.

## 2.2: Meters and Centimeters

There are 100 centimeters (cm) in every meter (m).

| length (m) | length (cm) |
|---|---|
| 1 | 100 |
| 0.94 | |
| 1.67 | |
| 57.24 | |
| $x$ | |

| length (cm) | length (m) |
|---|---|
| 100 | 1 |
| 250 | |
| 78.2 | |
| 123.9 | |
| $y$ | |

1. Complete each of the tables.
2. For each table, find the constant of proportionality.
3. What is the relationship between these constants of proportionality?
4. For each table, write an equation for the proportional relationship. Let $x$ represent a length measured in meters and $y$ represent the same length measured in centimeters.

### Are you ready for more?

1. How many cubic centimeters are there in a cubic meter?
2. How do you convert cubic centimeters to cubic meters?
3. How do you convert the other way?

## 2.3: Filling a Water Cooler

It took Priya 5 minutes to fill a cooler with 8 gallons of water from a faucet that was flowing at a steady rate. Let $w$ be the number of gallons of water in the cooler after $t$ minutes.

1. Which of the following equations represent the relationship between $w$ and $t$? Select **all** that apply.

   a. $w = 1.6t$

   b. $w = 0.625t$

   c. $t = 1.6w$

   d. $t = 0.625w$

2. What does 1.6 tell you about the situation?

3. What does 0.625 tell you about the situation?

4. Priya changed the rate at which water flowed through the faucet. Write an equation that represents the relationship of $w$ and $t$ when it takes 3 minutes to fill the cooler with 1 gallon of water.

5. Was the cooler filling faster before or after Priya changed the rate of water flow? Explain how you know.

## 2.4: Feeding Shrimp

At an aquarium, a shrimp is fed $\frac{1}{5}$ gram of food each feeding and is fed 3 times each day.

1. How much food does a shrimp get fed in one day?

2. Complete the table to show how many grams of food the shrimp is fed over different numbers of days.

| number of days | food in grams |
|---|---|
| 1 | |
| 7 | |
| 30 | |

3. What is the constant of proportionality? What does it tell us about the situation?

4. If we switched the columns in the table, what would be the constant of proportionality? Explain your reasoning.

5. Use $d$ for number of days and $f$ for amount of food in grams that a shrimp eats to write *two* equations that represent the relationship between $d$ and $f$.

6. If a tank has 10 shrimp in it, how much food is added to the tank each day?

7. If the aquarium manager has 300 grams of shrimp food for this tank of 10 shrimp, how many days will it last? Explain or show your reasoning.

## Lesson 2 Summary

If Kiran rode his bike at a constant 10 miles per hour, his distance in miles, $d$, is proportional to the number of hours, $t$, that he rode. We can write the equation

$$d = 10t$$

With this equation, it is easy to find the distance Kiran rode when we know how long it took because we can just multiply the time by 10.

We can rewrite the equation:

$$d = 10t$$

$$\left(\frac{1}{10}\right)d = t$$

$$t = \left(\frac{1}{10}\right)d$$

This version of the equation tells us that the amount of time he rode is proportional to the distance he traveled, and the constant of proportionality is $\frac{1}{10}$. That form is easier to use when we know his distance and want to find how long it took because we can just multiply the distance by $\frac{1}{10}$.

When two quantities $x$ and $y$ are in a proportional relationship, we can write the equation

$$y = kx$$

and say, "$y$ is proportional to $x$." In this case, the number $k$ is the corresponding constant of proportionality. We can also write the equation

$$x = \frac{1}{k}y$$

and say, "$x$ is proportional to $y$." In this case, the number $\frac{1}{k}$ is the corresponding constant of proportionality. Each one can be useful depending on the information we have and the quantity we are trying to figure out.

## Lesson 2 Practice Problems

1. The table represents the relationship between a length measured in meters and the same length measured in kilometers.

   a. Complete the table.

   b. Write an equation for converting the number of meters to kilometers. Use $x$ for number of meters and $y$ for number of kilometers.

| meters | kilometers |
|---|---|
| 1,000 | 1 |
| 3,500 | |
| 500 | |
| 75 | |
| 1 | |
| $x$ | |

2. Concrete building blocks weigh 28 pounds each. Using $b$ for the number of concrete blocks and $w$ for the weight, write two equations that relate the two variables. One equation should begin with $w =$ and the other should begin with $b =$.

3. A store sells rope by the meter. The equation $p = 0.8L$ represents the price $p$ (in dollars) of a piece of nylon rope that is $L$ meters long.

   a. How much does the nylon rope cost per meter?

   b. How long is a piece of nylon rope that costs $1.00?

4. The table represents a proportional relationship. Find the constant of proportionality and write an equation to represent the relationship.

| $a$ | $y$ |
|---|---|
| 2 | $\frac{2}{3}$ |
| 3 | 1 |
| 10 | $\frac{10}{3}$ |
| 12 | 4 |

Constant of proportionality: __________

Equation: $y =$

(From Unit 5, Lesson 1.)

5. Jada walks at a speed of 3 miles per hour. Elena walks at a speed of 2.8 miles per hour. If they both begin walking along a walking trail at the same time, how much farther will Jada walk after 3 hours? Explain your reasoning.

(From Unit 2, Lesson 18.)

# Lesson 3: Using Equations to Solve Problems

Let's use equations to solve problems involving proportional relationships.

## 3.1: Number Talk: Quotients with Decimal Points

Without calculating, order the quotients of these expressions from least to greatest.

$42.6 \div 0.07$

$42.6 \div 70$

$42.6 \div 0.7$

$426 \div 70$

Place the decimal point in the appropriate location in the quotient: $42.6 \div 7 = 608571$

Use this answer to find the quotient of *one* of the previous expressions.

## 3.2: Concert Ticket Sales

A performer expects to sell 5,000 tickets for an upcoming concert. They want to make a total of $311,000 in sales from these tickets.

1. Assuming that all tickets have the same price, what is the price for one ticket?

2. How much will they make if they sell 7,000 tickets?

3. How much will they make if they sell 10,000 tickets? 50,000? 120,000? a million? $x$ tickets?

4. If they make $404,300, how many tickets have they sold?

5. How many tickets will they have to sell to make $5,000,000?

## 3.3: Recycling

Aluminum cans can be recycled instead of being thrown in the garbage. The weight of 10 aluminum cans is 0.16 kilograms. The aluminum in 10 cans that are recycled has a value of $0.14.

1. If a family threw away 2.4 kg of aluminum in a month, how many cans did they throw away? Explain or show your reasoning.

2. What would be the recycled value of those same cans? Explain or show your reasoning.

3. Write an equation to represent the number of cans $c$ given their weight $w$.

4. Write an equation to represent the recycled value $r$ of $c$ cans.

5. Write an equation to represent the recycled value $r$ of $w$ kilograms of aluminum.

### Are you ready for more?

The EPA estimated that in 2013, the average amount of garbage produced in the United States was 4.4 pounds per person per day. At that rate, how long would it take your family to produce a ton of garbage? (A ton is 2,000 pounds.)

## Lesson 3 Summary

Remember that if there is a proportional relationship between two quantities, their relationship can be represented by an equation of the form $y = kx$. Sometimes writing an equation is the easiest way to solve a problem.

For example, we know that Denali, the highest mountain peak in North America, is 20,300 feet above sea level. How many miles is that? There are 5,280 feet in 1 mile. This relationship can be represented by the equation

$$f = 5{,}280m$$

where $f$ represents a distance measured in feet and $m$ represents the same distance measured miles. Since we know Denali is 20,310 feet above sea level, we can write

$$20{,}310 = 5{,}280m$$

So $m = \frac{20{,}310}{5{,}280}$, which is approximately 3.85 miles.

# Lesson 3 Practice Problems

1. A car is traveling down a highway at a constant speed, described by the equation $d = 65t$, where $d$ represents the distance, in miles, that the car travels at this speed in $t$ hours.

   a. What does the 65 tell us in this situation?

   b. How many miles does the car travel in 1.5 hours?

   c. How long does it take the car to travel 26 miles at this speed?

2. Elena has some bottles of water that each holds 17 fluid ounces.

   a. Write an equation that relates the number of bottles of water ($b$) to the total volume of water ($w$) in fluid ounces.

   b. How much water is in 51 bottles?

   c. How many bottles does it take to hold 51 fluid ounces of water?

3. There are about 1.61 kilometers in 1 mile. Let $x$ represent a distance measured in kilometers and $y$ represent the same distance measured in miles. Write two equations that relate a distance measured in kilometers and the same distance measured in miles.

   (From Unit 5, Lesson 2.)

4. In Canadian coins, 16 quarters is equal in value to 2 toonies.

| number of quarters | number of toonies |
|---|---|
| 1 | |
| 16 | 2 |
| 20 | |
| 24 | |

a. Complete the table.

b. What does the value next to 1 mean in this situation?

(From Unit 5, Lesson 1.)

5. Each table represents a proportional relationship. For each table:

a. Fill in the missing parts of the table.

b. Draw a circle around the constant of proportionality.

| $x$ | $y$ |
|---|---|
| 2 | 10 |
| | 15 |
| 7 | |
| 1 | |

| $a$ | $b$ |
|---|---|
| 12 | 3 |
| 20 | |
| | 10 |
| 1 | |

| $m$ | $n$ |
|---|---|
| 5 | 3 |
| 10 | |
| | 18 |
| 1 | |

(From Unit 5, Lesson 1.)

6. Write a multiplication equation that corresponds to each division equation.

a. $10 \div 5 = ?$

b. $4.5 \div 3 = ?$

c. $\frac{1}{2} \div 4 = ?$

(From Unit 3, Lesson 2.)

# Lesson 4: Comparing Relationships with Tables

Let's explore how proportional relationships are different from other relationships.

## 4.1: Adjusting a Recipe

A lemonade recipe calls for the juice of 5 lemons, 2 cups of water, and 2 tablespoons of honey.

Invent four new versions of this lemonade recipe:

1. One that would make more lemonade but taste the same as the original recipe.

2. One that would make less lemonade but taste the same as the original recipe.

3. One that would have a stronger lemon taste than the original recipe.

4. One that would have a weaker lemon taste than the original recipe.

## 4.2: Visiting the State Park

Entrance to a state park costs $6 per vehicle, plus $2 per person in the vehicle.

1. How much would it cost for a car with 2 people to enter the park? 4 people? 10 people? Record your answers in the table.

| number of people in vehicle | total entrance cost in dollars |
|---|---|
| 2 | |
| 4 | |
| 10 | |

2. For each row in the table, if each person in the vehicle splits the entrance cost equally, how much will each person pay?

3. How might you determine the entrance cost for a bus with 50 people?

4. Is the relationship between the number of people and the total entrance cost a proportional relationship? Explain how you know.

### Are you ready for more?

What equation could you use to find the total entrance cost for a vehicle with any number of people?

## 4.3: Running Laps

Han and Clare were running laps around the track. The coach recorded their times at the end of laps 2, 4, 6, and 8.

Han's run:

| distance (laps) | time (minutes) | minutes per lap |
|---|---|---|
| 2 | 4 | |
| 4 | 9 | |
| 6 | 15 | |
| 8 | 23 | |

Clare's run:

| distance (laps) | time (minutes) | minutes per lap |
|---|---|---|
| 2 | 5 | |
| 4 | 10 | |
| 6 | 15 | |
| 8 | 20 | |

1. Is Han running at a constant pace? Is Clare? How do you know?

2. Write an equation for the relationship between distance and time for anyone who is running at a constant pace.

### Lesson 4 Summary

Here are the prices for some smoothies at two different smoothie shops:

Smoothie Shop A

| smoothie size (oz) | price ($) | dollars per ounce |
|---|---|---|
| 8 | 6 | 0.75 |
| 12 | 9 | 0.75 |
| 16 | 12 | 0.75 |
| $s$ | $0.75s$ | 0.75 |

Smoothie Shop B

| smoothie size (oz) | price ($) | dollars per ounce |
|---|---|---|
| 8 | 6 | 0.75 |
| 12 | 8 | 0.67 |
| 16 | 10 | 0.625 |
| $s$ | ??? | ??? |

For Smoothie Shop A, smoothies cost $0.75 per ounce no matter which size we buy. There could be a proportional relationship between smoothie size and the price of the smoothie. An equation representing this relationship is

$$p = 0.75s$$

where $s$ represents size in ounces and $p$ represents price in dollars. (The relationship could still not be proportional, if there were a different size on the menu that did not have the same price per ounce.)

For Smoothie Shop B, the cost per ounce is different for each size. Here the relationship between smoothie size and price is definitely *not* proportional.

In general, two quantities in a proportional relationship will always have the same quotient. When we see some values for two related quantities in a table and we get the same quotient when we divide them, that means they might be in a proportional relationship—but if we can't see all of the possible pairs, we can't be completely sure. However, if we know the relationship can be represented by an equation is of the form $y = kx$, then we are sure it is proportional.

## Lesson 4 Practice Problems

1. Decide whether each table could represent a proportional relationship. If the relationship could be proportional, what would the constant of proportionality be?

   a. How loud a sound is depending on how far away you are.

| distance to listener (ft) | sound level (dB) |
|---|---|
| 5 | 85 |
| 10 | 79 |
| 20 | 73 |
| 40 | 67 |

   b. The cost of fountain drinks at Hot Dog Hut.

| volume (fluid ounces) | cost ($) |
|---|---|
| 16 | $1.49 |
| 20 | $1.59 |
| 30 | $1.89 |

2. A taxi service charges $1.00 for the first $\frac{1}{10}$ mile then $0.10 for each additional $\frac{1}{10}$ mile after that.

Fill in the table with the missing information then determine if this relationship between distance traveled and price of the trip is a proportional relationship.

| distance traveled (mi) | price (dollars) |
|---|---|
| $\frac{9}{10}$ | |
| 2 | |
| $3\frac{1}{10}$ | |
| 10 | |

3. A rabbit and turtle are in a race. Is the relationship between distance traveled and time proportional for either one? If so, write an equation that represents the relationship.

Turtle's run:

| distance (meters) | time (minutes) |
|---|---|
| 108 | 2 |
| 405 | 7.5 |
| 540 | 10 |
| 1,768.5 | 32.75 |

Rabbit's run:

| distance (meters) | time (minutes) |
|---|---|
| 800 | 1 |
| 900 | 5 |
| 1,107.5 | 20 |
| 1,524 | 32.5 |

4. For each table, answer: What is the constant of proportionality?

| a | b |
|---|---|
| 2 | 14 |
| 5 | 35 |
| 9 | 63 |
| $\frac{1}{3}$ | $\frac{7}{3}$ |

| a | b |
|---|---|
| 3 | 360 |
| 5 | 600 |
| 8 | 960 |
| 12 | 1440 |

| a | b |
|---|---|
| 75 | 3 |
| 200 | 8 |
| 1525 | 61 |
| 10 | 0.4 |

| a | b |
|---|---|
| 4 | 10 |
| 6 | 15 |
| 22 | 55 |
| 3 | $7\frac{1}{2}$ |

(From Unit 5, Lesson 1.)

5. Here is a table that shows the ratio of flour to water in an art paste. Complete the table with values in equivalent ratios.

| cups of flour | cups of water |
|---|---|
| 1 | $\frac{1}{2}$ |
| 4 | |
| | 3 |
| $\frac{1}{2}$ | |

(From Unit 2, Lesson 9.)

# Lesson 5: Comparing Relationships with Equations

Let's develop methods for deciding if a relationship is proportional.

## 5.1: Notice and Wonder: Patterns with Rectangles

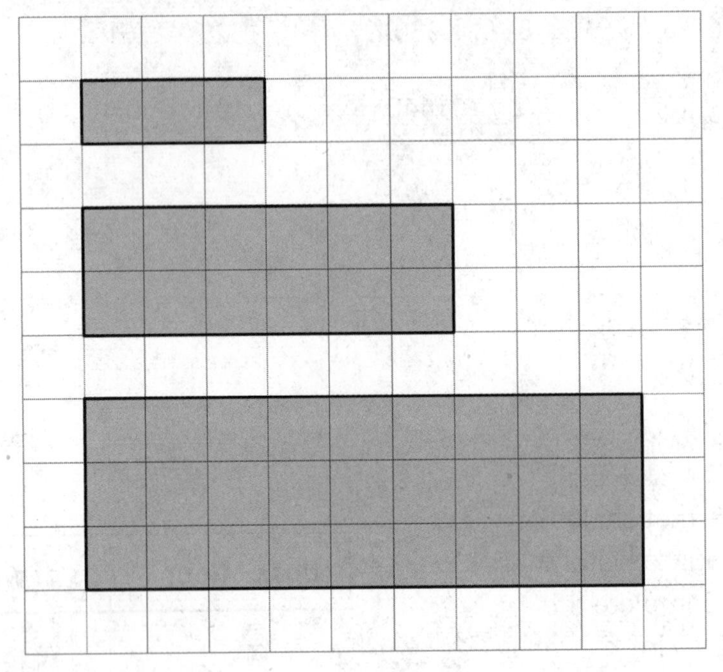

Do you see a pattern? What predictions can you make about future rectangles in the set if your pattern continues?

## 5.2: More Conversions

The other day you worked with converting meters, centimeters, and millimeters. Here are some more unit conversions.

1. Use the equation $F = \frac{9}{5}C + 32$, where $F$ represents degrees Fahrenheit and $C$ represents degrees Celsius, to complete the table.

| temperature (°C) | temperature (°F) |
|---|---|
| 20 | |
| 4 | |
| 175 | |

2. Use the equation $c = 2.54n$, where $c$ represents the length in centimeters and $n$ represents the length in inches, to complete the table.

| length (in) | length (cm) |
|---|---|
| 10 | |
| 8 | |
| $3\frac{1}{2}$ | |

3. Are these proportional relationships? Explain why or why not.

## 5.3: Total Edge Length, Surface Area, and Volume

Here are some cubes with different side lengths. Complete each table. Be prepared to explain your reasoning.

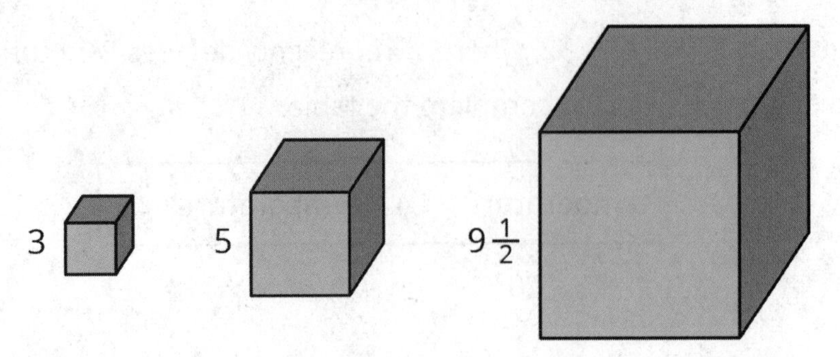


1. How long is the total edge length of each cube?

| side length | total edge length |
|---|---|
| 3 | |
| 5 | |
| $9\frac{1}{2}$ | |
| $s$ | |

2. What is the surface area of each cube?

| side length | surface area |
|---|---|
| 3 | |
| 5 | |
| $9\frac{1}{2}$ | |
| $s$ | |

3. What is the volume of each cube?

| side length | volume |
|---|---|
| 3 | |
| 5 | |
| $9\frac{1}{2}$ | |
| $s$ | |

4. Which of these relationships is proportional? Explain how you know.

5. Write equations for the total edge length $E$, total surface area $A$, and volume $V$ of a cube with side length $s$.

**Are you ready for more?**

1. A rectangular solid has a square base with side length $\ell$, height 8, and volume $V$. Is the relationship between $\ell$ and $V$ a proportional relationship?

2. A different rectangular solid has length $\ell$, width 10, height 5, and volume $V$. Is the relationship between $\ell$ and $V$ a proportional relationship?

3. Why is the relationship between the side length and the volume proportional in one situation and not the other?

## 5.4: All Kinds of Equations

Here are six different equations.

$y = 4 + x$

$y = 4x$

$y = \frac{4}{x}$

$y = \frac{x}{4}$

$y = 4^x$

$y = x^4$

$y = 4 + x$

| $x$ | $y$ | $\frac{y}{x}$ |
| --- | --- | --- |
| 2 | | |
| 3 | | |
| 4 | | |
| 5 | | |

$y = 4x$

| $x$ | $y$ | $\frac{y}{x}$ |
| --- | --- | --- |
| 2 | | |
| 3 | | |
| 4 | | |
| 5 | | |

$y = \frac{4}{x}$

| $x$ | $y$ | $\frac{y}{x}$ |
| --- | --- | --- |
| 2 | | |
| 3 | | |
| 4 | | |
| 5 | | |

$y = \frac{x}{4}$

| $x$ | $y$ | $\frac{y}{x}$ |
| --- | --- | --- |
| 2 | | |
| 3 | | |
| 4 | | |
| 5 | | |

$y = 4^x$

| $x$ | $y$ | $\frac{y}{x}$ |
| --- | --- | --- |
| 2 | | |
| 3 | | |
| 4 | | |
| 5 | | |

$y = x^4$

| $x$ | $y$ | $\frac{y}{x}$ |
| --- | --- | --- |
| 2 | | |
| 3 | | |
| 4 | | |
| 5 | | |

1. Predict which of these equations represent a proportional relationship.

2. Complete each table using the equation that represents the relationship.

3. Do these results change your answer to the first question? Explain your reasoning.

4. What do the equations of the proportional relationships have in common?

## Lesson 5 Summary

If two quantities are in a proportional relationship, then their quotient is always the same. This table represents different values of $a$ and $b$, two quantities that are in a proportional relationship.

| $a$ | $b$ | $\frac{b}{a}$ |
|---|---|---|
| 20 | 100 | 5 |
| 3 | 15 | 5 |
| 11 | 55 | 5 |
| 1 | 5 | 5 |

Notice that the quotient of $b$ and $a$ is always 5. To write this as an equation, we could say $\frac{b}{a} = 5$. If this is true, then $b = 5a$. (This doesn't work if $a = 0$, but it works otherwise.)

If quantity $y$ is proportional to quantity $x$, we will always see this pattern: $\frac{y}{x}$ will always have the same value. This value is the constant of proportionality, which we often refer to as $k$. We can represent this relationship with the equation $\frac{y}{x} = k$ (as long as $x$ is not 0) or $y = kx$.

Note that if an equation cannot be written in this form, then it does not represent a proportional relationship.

## Lesson 5 Practice Problems

1. The relationship between a distance in yards ($y$) and the same distance in miles ($m$) is described by the equation $y = 1760m$.

   a. Find measurements in yards and miles for distances by completing the table.

| distance measured in miles | distance measured in yards |
|---|---|
| 1 | |
| 5 | |
| | 3,520 |
| | 17,600 |

   b. Is there a proportional relationship between a measurement in yards and a measurement in miles for the same distance? Explain why or why not.

2. Decide whether or not each equation represents a proportional relationship.

   a. The remaining length ($L$) of 120-inch rope after $x$ inches have been cut off: $120 - x = L$

   b. The total cost ($t$) after 8% sales tax is added to an item's price ($p$): $1.08p = t$

   c. The number of marbles each sister gets ($x$) when $m$ marbles are shared equally among four sisters: $x = \frac{m}{4}$

   d. The volume ($V$) of a rectangular prism whose height is 12 cm and base is a square with side lengths $s$ cm: $V = 12s^2$

3. a. Use the equation $y = \frac{5}{2}x$ to complete the table.
Is $y$ proportional to $x$ and $y$? Explain why or why not.

| $x$ | $y$ |
|---|---|
| 2 | |
| 3 | |
| 6 | |

b. Use the equation $y = 3.2x + 5$ to complete the table.
Is $y$ proportional to $x$ and $y$? Explain why or why not.

| $x$ | $y$ |
|---|---|
| 1 | |
| 2 | |
| 4 | |

4. To transmit information on the internet, large files are broken into packets of smaller sizes. Each packet has 1,500 bytes of information. An equation relating packets to bytes of information is given by $b = 1{,}500p$ where $p$ represents the number of packets and $b$ represents the number of bytes of information.

a. How many packets would be needed to transmit 30,000 bytes of information?

b. How much information could be transmitted in 30,000 packets?

c. Each byte contains 8 bits of information. Write an equation to represent the relationship between the number of packets and the number of bits.

(From Unit 5, Lesson 3.)

# Lesson 6: Solving Problems about Proportional Relationships

Let's solve problems about proportional relationships.

## 6.1: What Do You Want to Know?

Consider the problem: A person is running a distance race at a constant rate. What time will they finish the race?

What information would you need to be able to solve the problem?

## 6.2: Info Gap: Biking and Rain

Your teacher will give you either a *problem card* or a *data card*. Do not show or read your card to your partner.

If your teacher gives you the *problem card*:

1. Silently read your card and think about what information you need to be able to answer the question.
2. Ask your partner for the specific information that you need.
3. Explain how you are using the information to solve the problem.

   Continue to ask questions until you have enough information to solve the problem.
4. Share the *problem card* and solve the problem independently.
5. Read the *data card* and discuss your reasoning.

If your teacher gives you the *data card*:

1. Silently read your card.
2. Ask your partner *"What specific information do you need?"* and wait for them to *ask* for information.

   If your partner asks for information that is not on the card, do not do the calculations for them. Tell them you don't have that information.
3. Before sharing the information, ask "*Why do you need that information?*" Listen to your partner's reasoning and ask clarifying questions.
4. Read the *problem card* and solve the problem independently.
5. Share the *data card* and discuss your reasoning.

Pause here so your teacher can review your work. Ask your teacher for a new set of cards and repeat the activity, trading roles with your partner.

## 6.3: Moderating Comments

A company is hiring people to read through all the comments posted on their website to make sure they are appropriate. Four people applied for the job and were given one day to show how quickly they could check comments.

- Person 1 worked for 210 minutes and checked a total of 50,000 comments.
- Person 2 worked for 200 minutes and checked 1,325 comments every 5 minutes.
- Person 3 worked for 120 minutes, at a rate represented by $c = 331t$, where $c$ is the number of comments checked and $t$ is the time in minutes.
- Person 4 worked for 150 minutes, at a rate represented by $t = \left(\frac{3}{800}\right) c$.

1. Order the people from greatest to least in terms of total number of comments checked.

2. Order the people from greatest to least in terms of how fast they checked the comments.

**Are you ready for more?**

1. Write equations for each job applicant that allow you to easily decide who is working the fastest.

2. Make a table that allows you to easily compare how many comments the four job applicants can check.

## Lesson 6 Summary

Whenever we have a situation involving constant rates, we are likely to have a proportional relationship between quantities of interest.

- When a bird is flying at a constant speed, then there is a proportional relationship between the flying time and distance flown.
- If water is filling a tub at a constant rate, then there is a proportional relationship between the amount of water in the tub and the time the tub has been filling up.
- If an aardvark is eating termites at a constant rate, then there is proportional relationship between the number of termites the aardvark has eaten and the time since it started eating.

Sometimes we are presented with a situation, and it is not so clear whether a proportional relationship is a good model. How can we decide if a proportional relationship is a good representation of a particular situation?

- If you aren't sure where to start, look at the quotients of corresponding values. If they are not always the same, then the relationship is definitely not a proportional relationship.
- If you can see that there is a single value that we always multiply one quantity by to get the other quantity, it is definitely a proportional relationship.

After establishing that it is a proportional relationship, setting up an equation is often the most efficient way to solve problems related to the situation.

# Lesson 6 Practice Problems

1. For each situation, explain whether you think the relationship is proportional or not. Explain your reasoning.

   a. The weight of a stack of standard 8.5x11 copier paper vs. number of sheets of paper.

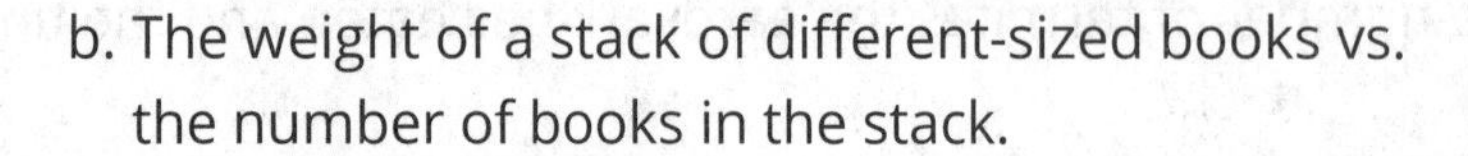

   b. The weight of a stack of different-sized books vs. the number of books in the stack.

2. Every package of a certain toy also includes 2 batteries.

   a. Are the number of toys and number of batteries in a proportional relationship? If so, what are the two constants of proportionality? If not, explain your reasoning.

   b. Use $t$ for the number of toys and $b$ for the number of batteries to write two equations relating the two variables.

   $b =$ $\qquad\qquad$ $t =$

3. Lin and her brother were born on the same date in different years. Lin was 5 years old when her brother was 2.

   a. Find their ages in different years by filling in the table.

| Lin's age | Her brother's age |
|---|---|
| 5 | 2 |
| 6 | |
| 15 | |
| | 25 |

   b. Is there a proportional relationship between Lin's age and her brother's age? Explain your reasoning.

4. A student argues that $y = \frac{x}{9}$ does not represent a proportional relationship between $x$ and $y$ because we need to multiply one variable by the same constant to get the other one and not divide it by a constant. Do you agree or disagree with this student?

(From Unit 5, Lesson 5.)

5. In one version of a trail mix, there are 3 cups of peanuts mixed with 2 cups of raisins. In another version of trail mix, there are 4.5 cups of peanuts mixed with 3 cups of raisins. Are the ratios equivalent for the two mixes? Explain your reasoning.

(From Unit 5, Lesson 1.)

# Lesson 7: Graphs of Proportional Relationships

Let's see how graphs of proportional relationships differ from graphs of other relationships.

## 7.1: Notice These Points

1. Plot the points $(0, 10), (1, 8), (2, 6), (3, 4), (4, 2)$.

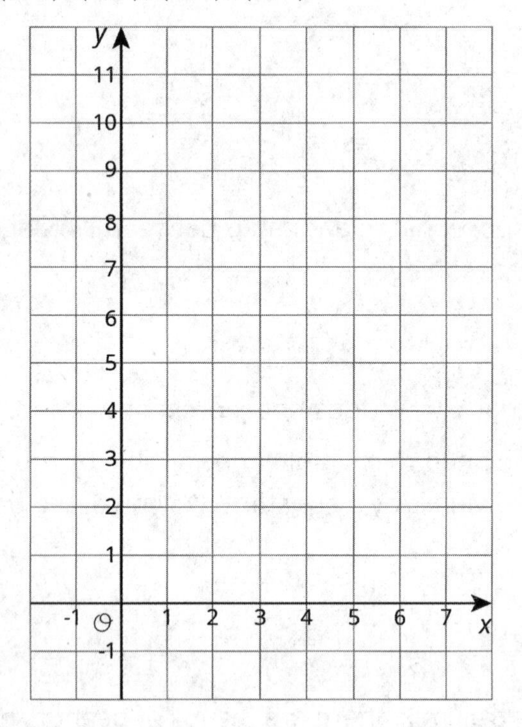


2. What do you notice about the graph?

## 7.2: T-shirts for Sale

Some T-shirts cost $8 each.

| $x$ | $y$ |
|---|---|
| 1 | 8 |
| 2 | 16 |
| 3 | 24 |
| 4 | 32 |
| 5 | 40 |
| 6 | 48 |

1. Use the table to answer these questions.

    a. What does $x$ represent?

    b. What does $y$ represent?

    c. Is there a proportional relationship between $x$ and $y$?

2. Plot the pairs in the table on the **coordinate plane**.

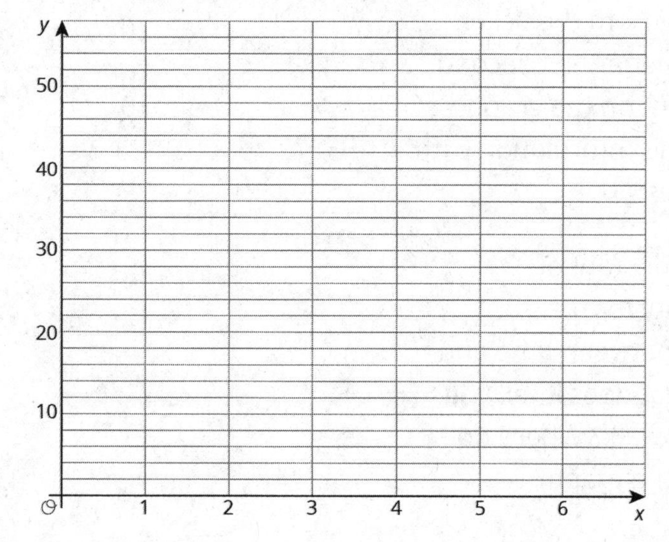


3. What do you notice about the graph?

## 7.3: Tyler's Walk

Tyler was at the amusement park. He walked at a steady pace from the ticket booth to the bumper cars.

1. The point on the graph shows his arrival at the bumper cars. What do the coordinates of the point tell us about the situation?

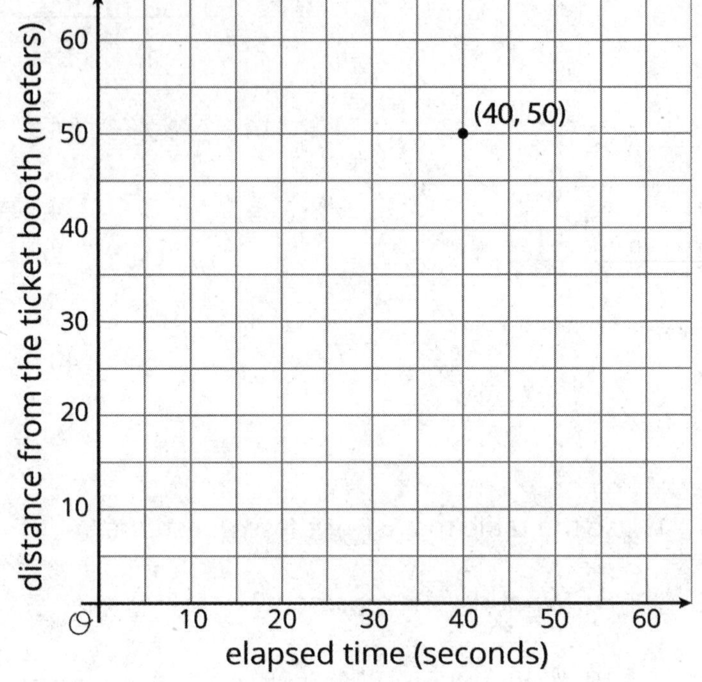


2. The table representing Tyler's walk shows other values of time and distance. Complete the table. Next, plot the pairs of values on the grid.

| time (seconds) | distance (meters) |
|---|---|
| 0 | 0 |
| 20 | 25 |
| 30 | 37.5 |
| 40 | 50 |
| 1 | |

3. What does the point $(0, 0)$ mean in this situation?

4. How far away from the ticket booth was Tyler after 1 second? Label the point on the graph that shows this information with its coordinates.

5. What is the constant of proportionality for the relationship between time and distance? What does it tell you about Tyler's walk? Where do you see it in the graph?

### Are you ready for more?

If Tyler wanted to get to the bumper cars in half the time, how would the graph representing his walk change? How would the table change? What about the constant of proportionality?

### Lesson 7 Summary

One way to represent a proportional relationship is with a graph. Here is a graph that represents different amounts that fit the situation, "Blueberries cost $6 per pound."

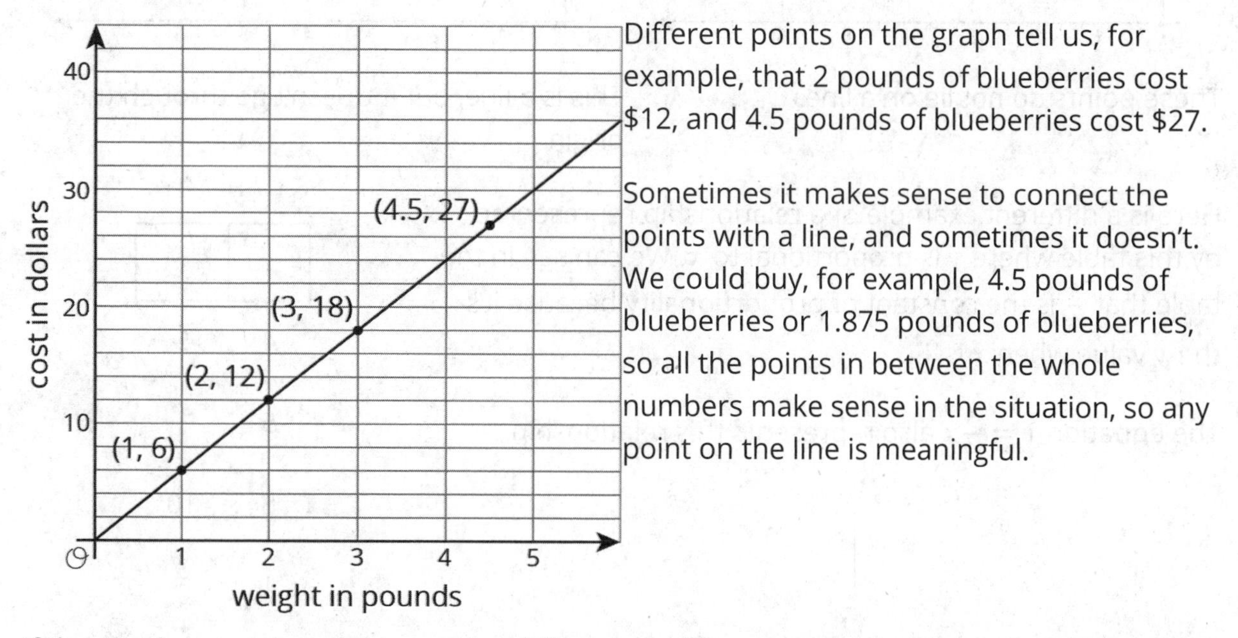


Different points on the graph tell us, for example, that 2 pounds of blueberries cost $12, and 4.5 pounds of blueberries cost $27.

Sometimes it makes sense to connect the points with a line, and sometimes it doesn't. We could buy, for example, 4.5 pounds of blueberries or 1.875 pounds of blueberries, so all the points in between the whole numbers make sense in the situation, so any point on the line is meaningful.

If the graph represented the cost for different *numbers of sandwiches* (instead of pounds of blueberries), it might not make sense to connect the points with a line, because it is often not possible to buy 4.5 sandwiches or 1.875 sandwiches. Even if only points make sense in the situation, though, sometimes we connect them with a line anyway to make the relationship easier to see.

Graphs that represent proportional relationships all have a few things in common:

- There are points that satisfy the relationship lie on a straight line.
- The line that they lie on passes through the **origin**, $(0, 0)$.

Here are some graphs that do *not* represent proportional relationships:

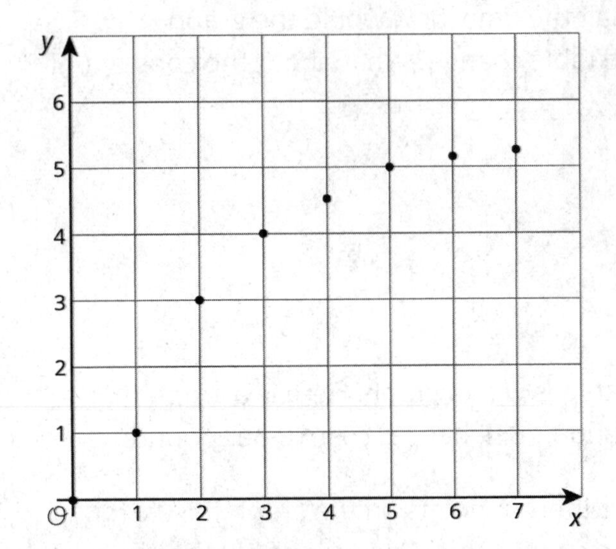


These points do not lie on a line.

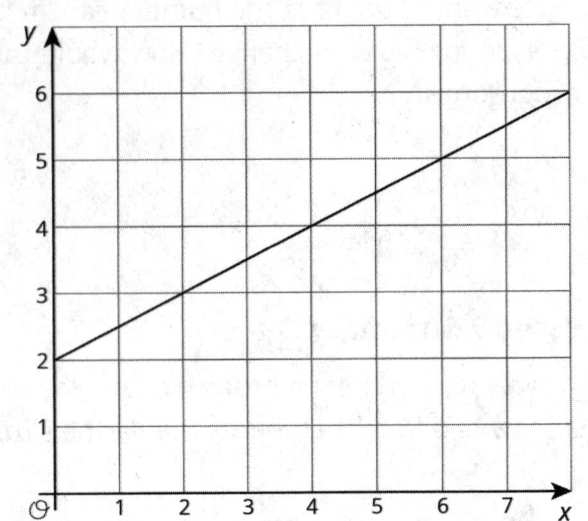


This is a line, but it doesn't go through the origin.

Here is a different example of a relationship represented by this table where $y$ is proportional to $x$. We can see in the table that $\frac{5}{4}$ is the constant of proportionality because it's the $y$ value when $x$ is 1.

The equation $y = \frac{5}{4}x$ also represents this relationship.

| $x$ | $y$ |
|---|---|
| 4 | 5 |
| 5 | $\frac{25}{4}$ |
| 8 | 10 |
| 1 | $\frac{5}{4}$ |

Here is the graph of this relationship.

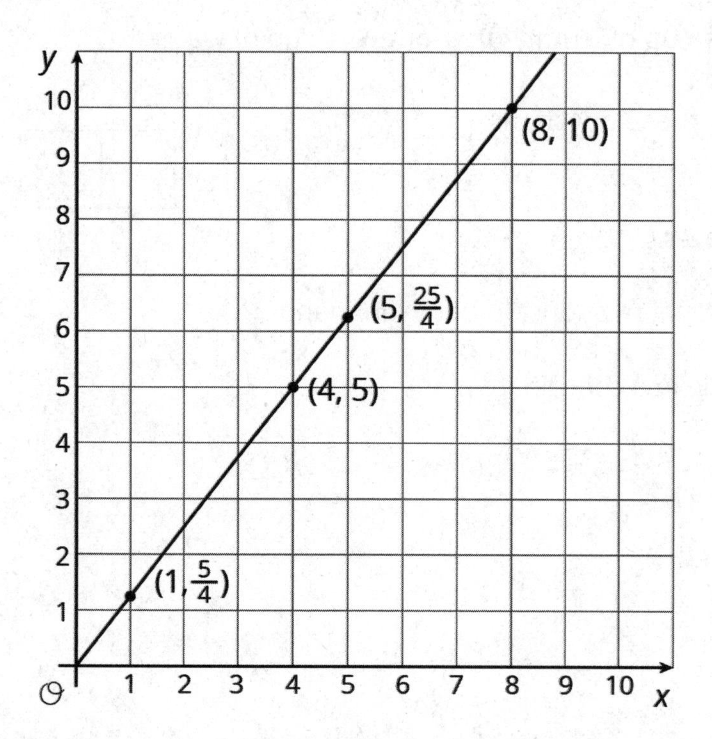


If $y$ represents the distance in feet that a snail crawls in $x$ minutes, then the point $(4, 5)$ tells us that the snail can crawl 5 feet in 4 minutes.

If $y$ represents the cups of yogurt and $x$ represents the teaspoons of cinnamon in a recipe for fruit dip, then the point $(4, 5)$ tells us that you can mix 4 teaspoons of cinnamon with 5 cups of yogurt to make this fruit dip.

We can find the constant of proportionality by looking at the graph, because $\frac{5}{4}$ is the $y$-coordinate of the point on the graph where the $x$-coordinate is 1. This could mean the snail is traveling $\frac{5}{4}$ feet per minute or that the recipe calls for $1\frac{1}{4}$ cups of yogurt for every teaspoon of cinnamon.

In general, when $y$ is proportional to $x$, the corresponding constant of proportionality is the $y$-value when $x = 1$.

## Lesson 7 Practice Problems

1. A lemonade recipe calls for $\frac{1}{4}$ cup of lemon juice for every cup of water.

    a. Use the table to answer these questions.

    i. What does $x$ represent?

    ii. What does $y$ represent?

    iii. Is there a proportional relationship between $x$ and $y$?

    b. Plot the pairs in the table in a coordinate plane.

| $x$ | $y$ |
|---|---|
| 1 | $\frac{1}{4}$ |
| 2 | $\frac{1}{2}$ |
| 3 | $\frac{3}{4}$ |
| 4 | 1 |

2. There is a proportional relationship between the number of months a person has had a streaming movie subscription and the total amount of money they have paid for the subscription. The cost for 6 months is $47.94. The point $(6, 47.94)$ is shown on the graph below.

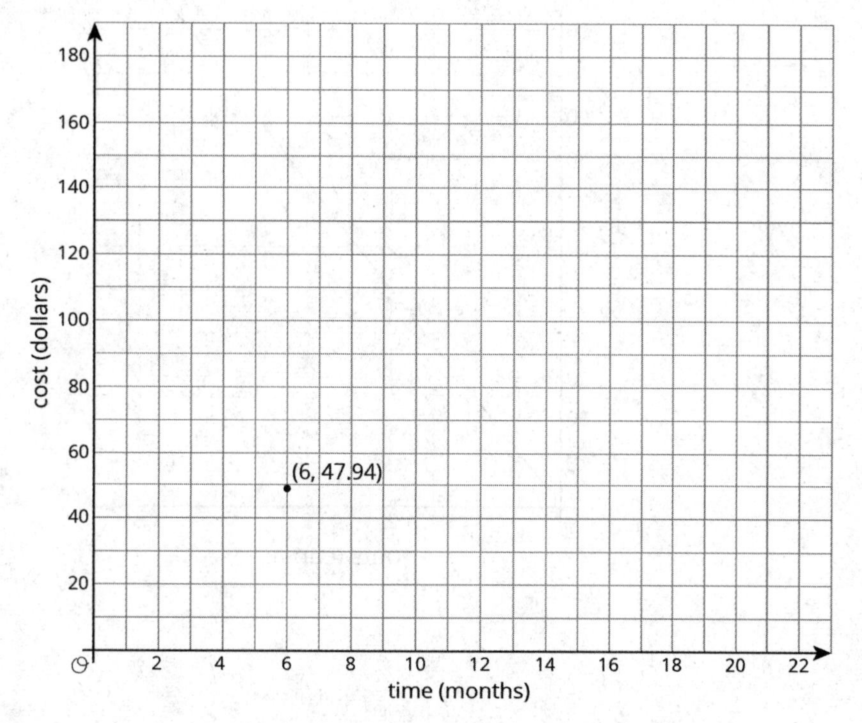


a. What is the constant of proportionality in this relationship?

b. What does the constant of proportionality tell us about the situation?

c. Add at least three more points to the graph and label them with their coordinates.

d. Write an equation that represents the relationship between $C$, the total cost of the subscription, and $m$, the number of months.

3. The graph shows the amounts of almonds, in grams, for different amounts of oats, in cups, in a granola mix. Label the point $(1, k)$ on the graph, find the value of $k$, and explain its meaning.

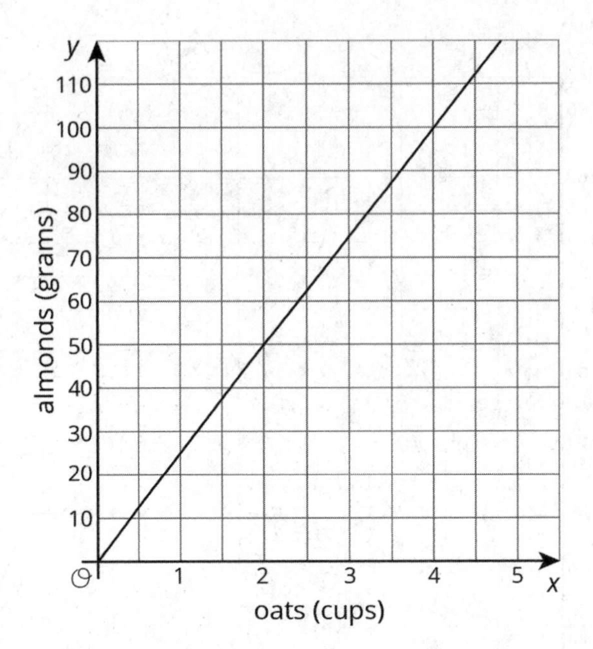


4. Select **all** the pieces of information that would tell you $x$ and $y$ have a proportional relationship. Let $y$ represent the distance in meters between a rock and a turtle's current position and $x$ represent the time in minutes the turtle has been moving.

   A. $y = 3x$

   B. After 4 minutes, the turtle has walked 12 feet away from the rock.

   C. The turtle walks for a bit, then stops for a minute before walking again.

   D. The turtle walks away from the rock at a constant rate.

   (From Unit 5, Lesson 6.)

5. What information do you need to know to write an equation relating two quantities that have a proportional relationship?

   (From Unit 5, Lesson 6.)

# Lesson 8: Using Graphs to Compare Relationships

Let's graph more than one relationship on the same grid.

## 8.1: Number Talk: Fraction Multiplication and Division

Find each product or quotient mentally.

$\frac{2}{3} \cdot \frac{1}{2}$

$\frac{4}{3} \cdot \frac{1}{4}$

$4 \div \frac{1}{5}$

$\frac{9}{6} \div \frac{1}{2}$

## 8.2: Race to the Bumper Cars

Diego, Lin, and Mai went from the ticket booth to the bumper cars.

1. Use each description to complete the table representing that person's journey.

    a. Diego left the ticket booth at the same time as Tyler. Diego jogged ahead at a steady pace and reached the bumper cars in 30 seconds.

    b. Lin left the ticket booth at the same time as Tyler. She ran at a steady pace and arrived at the bumper cars in 20 seconds.

    c. Mai left the booth 10 seconds later than Tyler. Her steady jog enabled her to catch up with Tyler just as he arrived at the bumper cars.

| Diego's time (seconds) | Diego's distance (meters) |
|---|---|
| 0 | |
| 15 | |
| 30 | 50 |
| 1 | |

| Lin's time (seconds) | Lin's distance (meters) |
|---|---|
| | 0 |
| | 25 |
| 20 | 50 |
| 1 | |

| Mai's time (seconds) | Mai's distance (meters) |
|---|---|
| | 0 |
| | 25 |
| 40 | 50 |
| 1 | |

2. Using a different color for each person, draw a graph of all four people's journeys (including Tyler's from the other day).

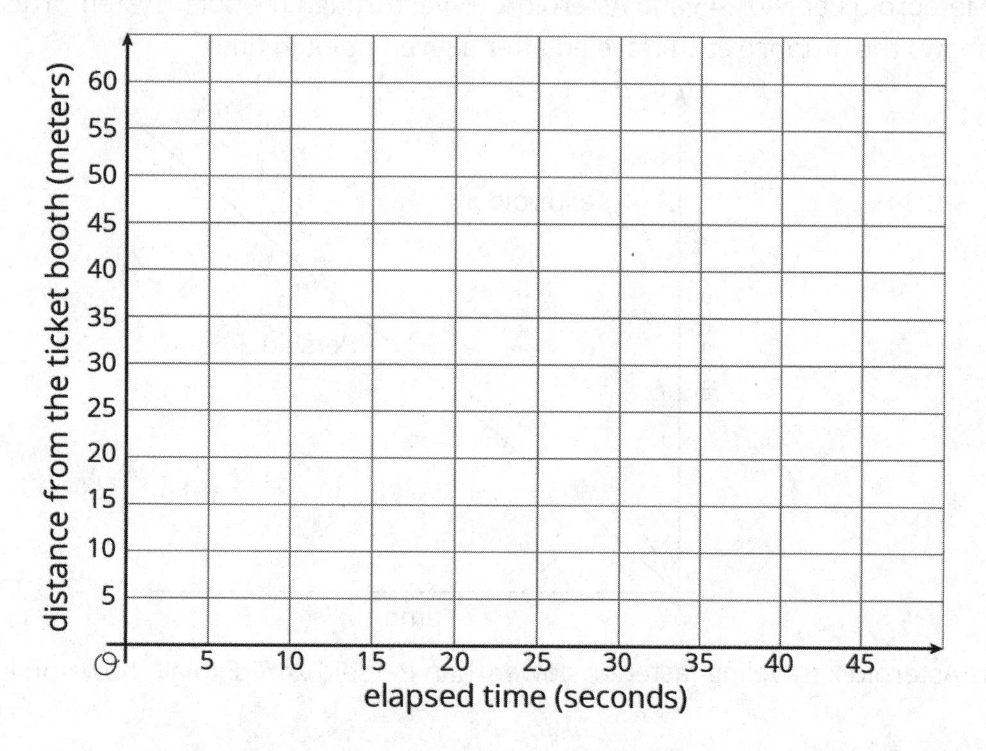


3. Which person is moving the most quickly? How is that reflected in the graph?

## Are you ready for more?

Write equations to represent each person's relationship between time and distance.

## 8.3: Space Rocks and the Price of Rope

1. Meteoroid Perseid 245 and Asteroid x travel through the solar system. The graph shows the distance each traveled after a given point in time.

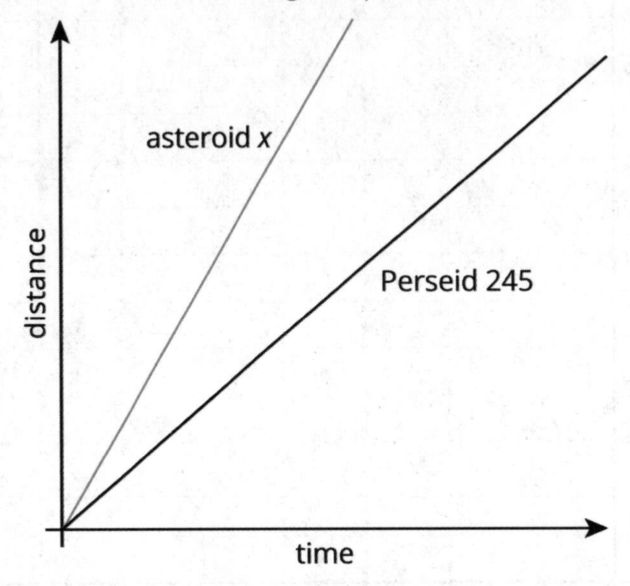


Is Asteroid x traveling faster or slower than Perseid 245? Explain how you know.

2. The graph shows the price of different lengths of two types of rope.

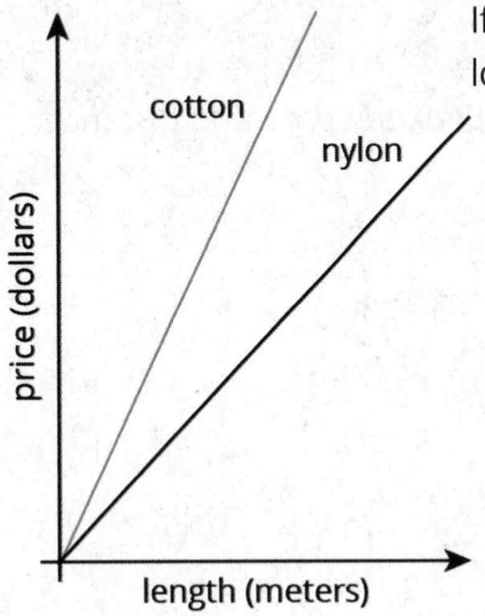


If you buy $1.00 of each kind of rope, which one will be longer? Explain how you know.

## Lesson 8 Summary

Here is a graph that shows the price of blueberries at two different stores. Which store has a better price?

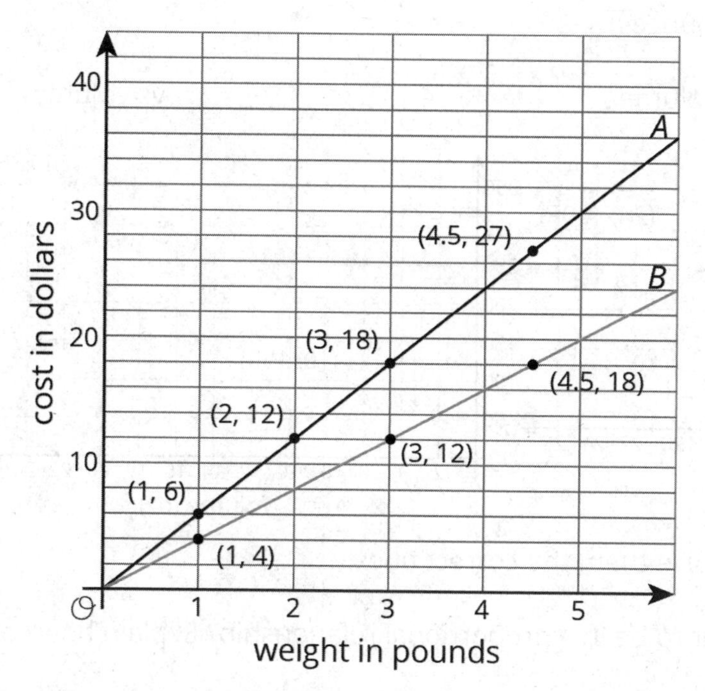


We can compare points that have the same $x$ value or the same $y$ value. For example, the points $(2, 12)$ and $(3, 12)$ tell us that at store B you can get more pounds of blueberries for the same price.

The points $(3, 12)$ and $(3, 18)$ tell us that at store A you have to pay more for the same quantity of blueberries. This means store B has the better price.

We can also use the graphs to compare the constants of proportionality. The line representing store B goes through the point $(1, 4)$, so the constant of proportionality is 4. This tells us that at store B the blueberries cost \$4 per pound. This is cheaper than the \$6 per pound unit price at store A.

# Lesson 8 Practice Problems

1. The graphs below show some data from a coffee shop menu. One of the graphs shows cost (in dollars) vs. drink volume (in ounces), and one of the graphs shows calories vs. drink volume (in ounces).

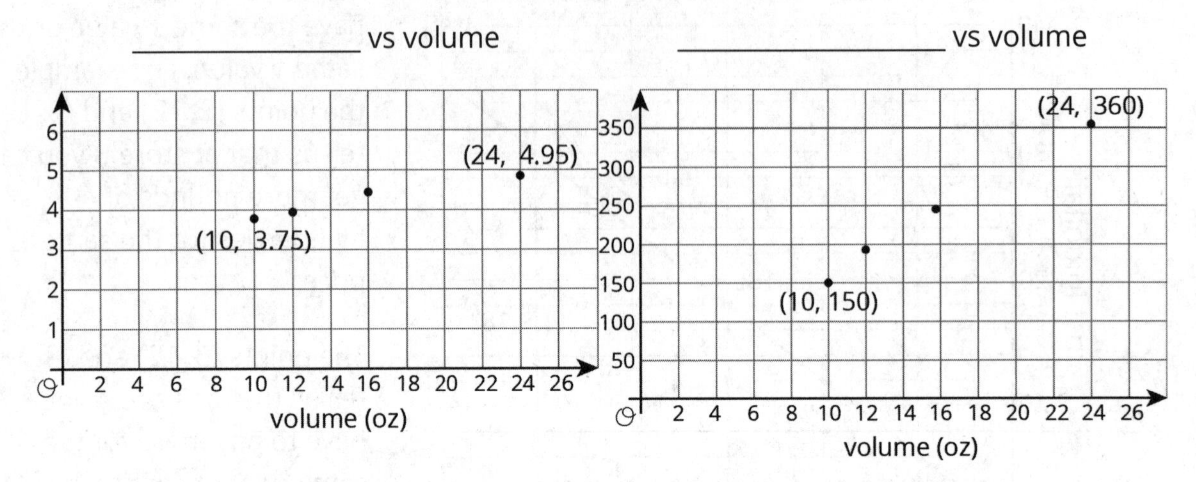


   a. Which graph is which? Give them the correct titles.

   b. Which quantities appear to be in a proportional relationship? Explain how you know.

   c. For the proportional relationship, find the constant of proportionality. What does that number mean?

2. Lin and Andre biked home from school at a steady pace. Lin biked 1.5 km and it took her 5 minutes. Andre biked 2 km and it took him 8 minutes.

   a. Draw a graph with two lines that represent the bike rides of Lin and Andre.

   b. For each line, highlight the point with coordinates $(1, k)$ and find $k$.

   c. Who was biking faster?

3. Match each equation to its graph.

a. $y = 2x$

b. $y = \frac{4}{5}x$

c. $y = \frac{1}{4}x$

d. $y = \frac{2}{3}x$

e. $y = \frac{4}{3}x$

f. $y = \frac{3}{2}x$

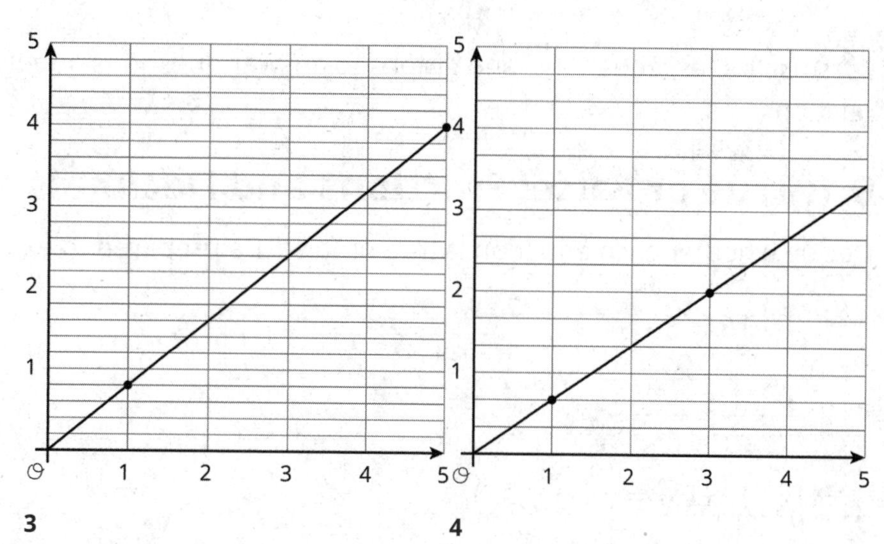

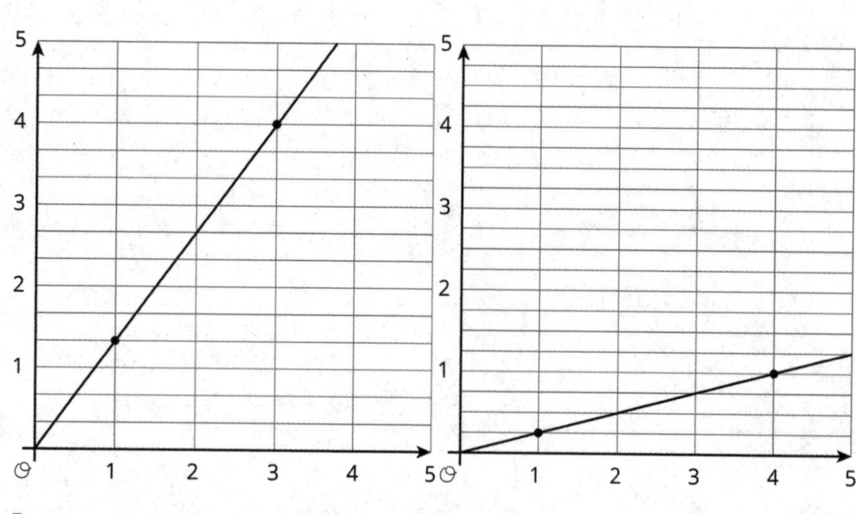


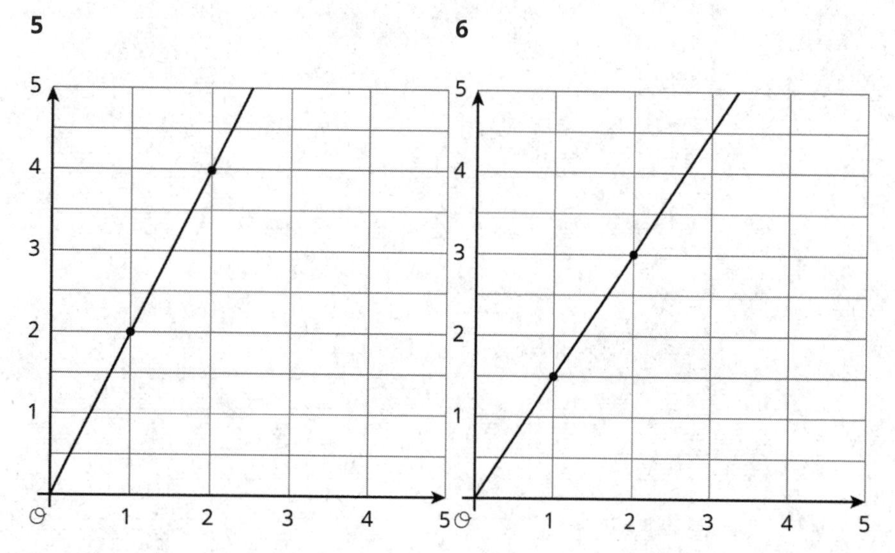

# Lesson 9: Two Graphs for Each Relationship

Let's use tables, equations, and graphs to answer questions about proportional relationships.

## 9.1: True or False: Fractions and Decimals

Decide whether each equation is true or false. Be prepared to explain your reasoning.

1. $\frac{3}{2} \cdot 16 = 3 \cdot 8$
2. $\frac{3}{4} \div \frac{1}{2} = \frac{6}{4} \div \frac{1}{4}$
3. $(2.8) \cdot (13) = (0.7) \cdot (52)$

## 9.2: Tables, Graphs, and Equations

Your teacher will assign you *one* of these three points:

$$A = (10, 4),\ B = (4, 5),\ C = (8, 5).$$

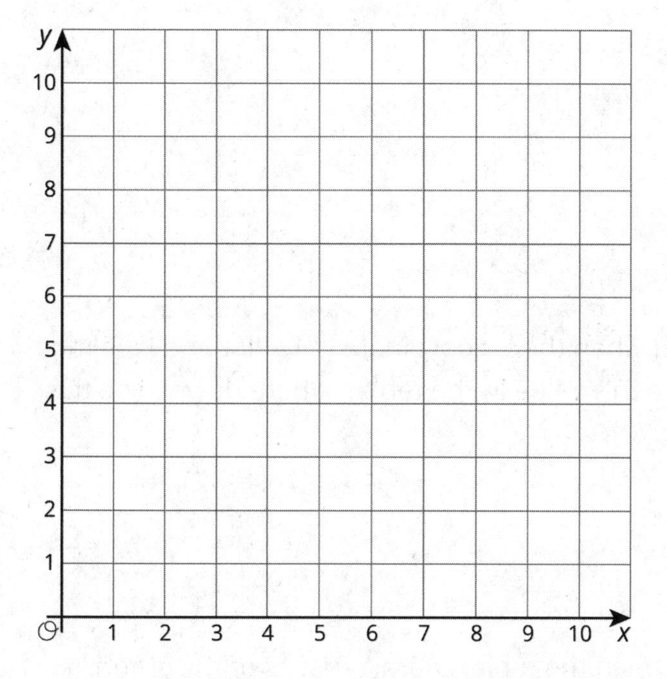


| $x$ | $y$ | $\frac{y}{x}$ |
|---|---|---|
| 0 | | NA |
| 1 | | |
| 2 | | |
| 3 | | |
| 4 | | |
| 5 | | |
| 6 | | |
| 7 | | |
| 8 | | |
| 9 | | |
| 10 | | |

1. On the graph, plot and label *only* your assigned point.
2. Use a ruler to line up your point with the origin, $(0, 0)$. Draw a line that starts at the origin, goes through your point, and continues to the edge of the graph.
3. Complete the table with the coordinates of points on your graph. Use a fraction to represent any value that is not a whole number.
4. Write an equation that represents the relationship between $x$ and $y$ defined by your point.

5. Compare your graph and table with the rest of your group. What is the same and what is different about:

   a. your tables?

   b. your equations?

   c. your graphs?

6. What is the $y$-coordinate of your graph when the $x$-coordinate is 1? Plot and label this point on your graph. Where do you see this value in the table? Where do you see this value in your equation?

7. Describe any connections you see between the table, characteristics of the graph, and the equation.

**Are you ready for more?**

The graph of an equation of the form $y = kx$, where $k$ is a positive number, is a line through $(0, 0)$ and the point $(1, k)$.

1. Name at least one line through $(0, 0)$ that cannot be represented by an equation like this.

2. If you could draw the graphs of *all* of the equations of this form in the same coordinate plane, what would it look like?

## 9.3: Hot Dog Eating Contest

Andre and Jada were in a hot dog eating contest. Andre ate 10 hot dogs in 3 minutes. Jada ate 12 hot dogs in 5 minutes.

Here are two different graphs that both represent this situation.

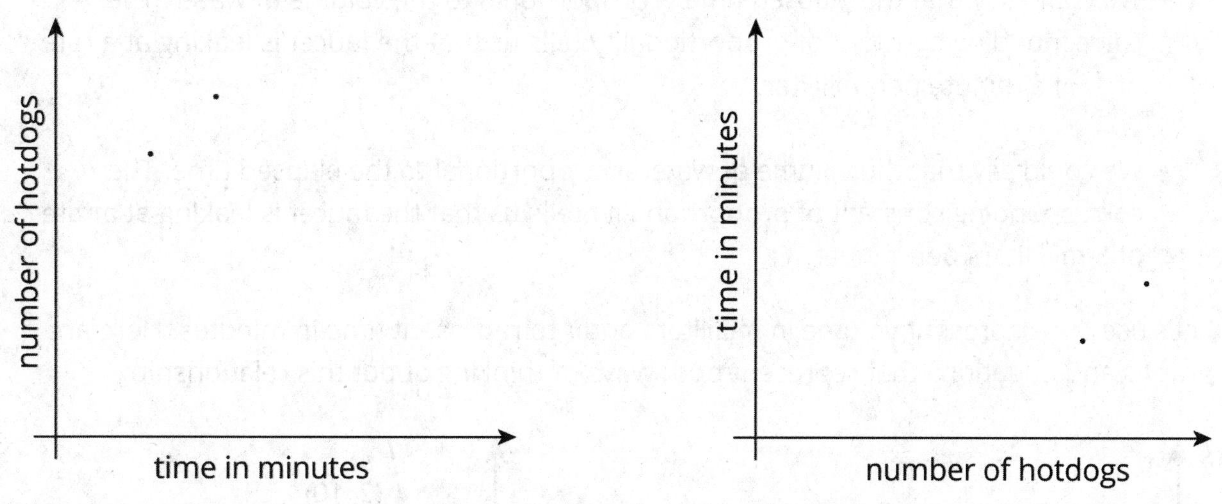


1. On the first graph, which point shows Andre's consumption and which shows Jada's consumption? Label them.

2. Draw two lines: one through the origin and Andre's point, and one through the origin and Jada's point.

3. Write an equation for each line. Use $t$ to represent time in minutes and $h$ to represent number of hot dogs.

   a. Andre:

   b. Jada:

4. For each equation, what does the constant of proportionality tell you?

5. Repeat the previous steps for the second graph.

   a. Andre:

   b. Jada:

## Lesson 9 Summary

Imagine that a faucet is leaking at a constant rate and that every 2 minutes, 10 milliliters of water leaks from the faucet. There is a proportional relationship between the volume of water and elapsed time.

- We could say that the elapsed time is proportional to the volume of water. The corresponding constant of proportionality tells us that the faucet is leaking at a rate of $\frac{1}{5}$ of a minute per milliliter.
- We could say that the volume of water is proportional to the elapsed time. The corresponding constant of proportionality tells us that the faucet is leaking at a rate of 5 milliliters per minute.

Let's use $v$ to represent volume in milliliters and $t$ to represent time in minutes. Here are graphs and equations that represent both ways of thinking about this relationship:

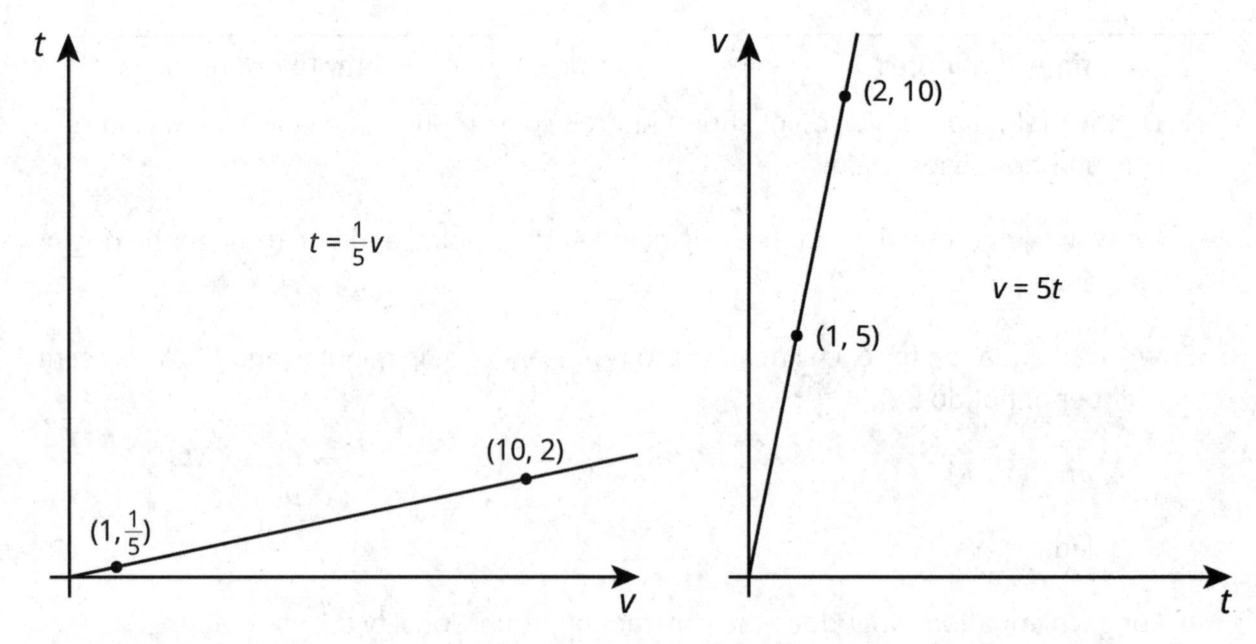


Even though the relationship between time and volume is the same, we are making a different choice in each case about which variable to view as the independent variable. The graph on the left has $v$ as the independent variable, and the graph on the right has $t$ as the independent variable.

# Lesson 9 Practice Problems

1. At the supermarket you can fill your own honey bear container. A customer buys 12 oz of honey for $5.40.

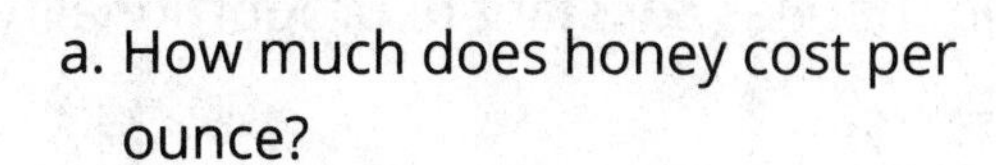

   a. How much does honey cost per ounce?

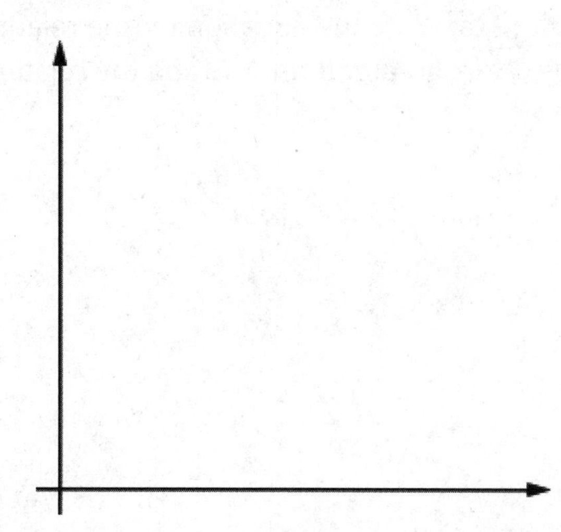

   b. How much honey can you buy per dollar?

   c. Write two different equations that represent this situation. Use $h$ for ounces of honey and $c$ for cost in dollars.

   - Choose one of your equations, and sketch its graph. Be sure to label the axes.

2. The point $(3, \frac{6}{5})$ lies on the graph representing a proportional relationship. Which of the following points also lie on the same graph? Select **all** that apply.

   A. $(1, 0.4)$

   B. $(1.5, \frac{6}{10})$

   C. $(\frac{6}{5}, 3)$

   D. $(4, \frac{11}{5})$

   E. $(15, 6)$

3. A trail mix recipe asks for 4 cups of raisins for every 6 cups of peanuts. There is proportional relationship between the amount of raisins, $r$ (cups), and the amount of peanuts, $p$ (cups), in this recipe.

   a. Write the equation for the relationship that has constant of proportionality greater than 1. Graph the relationship.

   b. Write the equation for the relationship that has constant of proportionality less than 1. Graph the relationship.

4. Here is a graph that represents a proportional relationship.

a. Come up with a situation that could be represented by this graph.

b. Label the axes with the quantities in your situation.

c. Give the graph a title.

d. Choose a point on the graph. What do the coordinates represent in your situation?

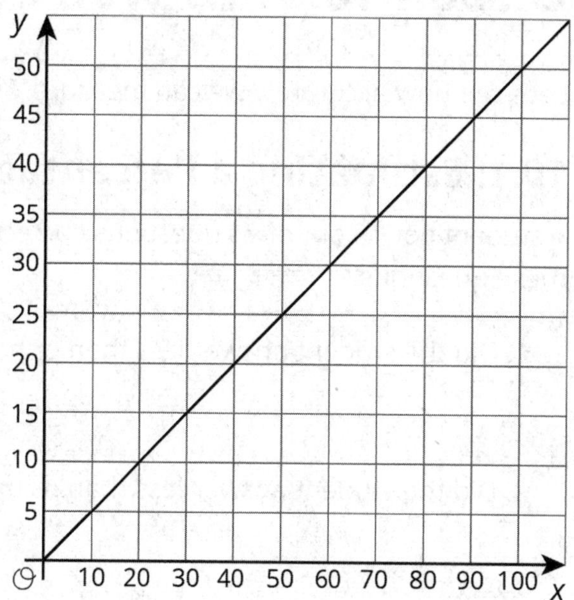


(From Unit 5, Lesson 7.)

# Lesson 10: How Well Can You Measure?

Let's see how accurately we can measure.

## 10.1: Estimating a Percentage

A student got 16 out of 21 questions correct on a quiz. Use mental estimation to answer these questions.

1. Did the student answer less than or more than 80% of the questions correctly?

2. Did the student answer less than or more than 75% of the questions correctly?

## 10.2: Perimeter of a Square

Your teacher will give you a picture of 9 different squares and will assign your group 3 of these squares to examine more closely.

1. For each of your assigned squares, measure the length of the diagonal and the perimeter of the square in centimeters.

   Check your measurements with your group. After you come to an agreement, record your measurements in the table.

| | diagonal (cm) | perimeter (cm) |
|---|---|---|
| square A | | |
| square B | | |
| square C | | |
| square D | | |
| square E | | |
| square F | | |
| square G | | |
| square H | | |
| square I | | |

2. Plot the diagonal and perimeter values from the table on the coordinate plane.

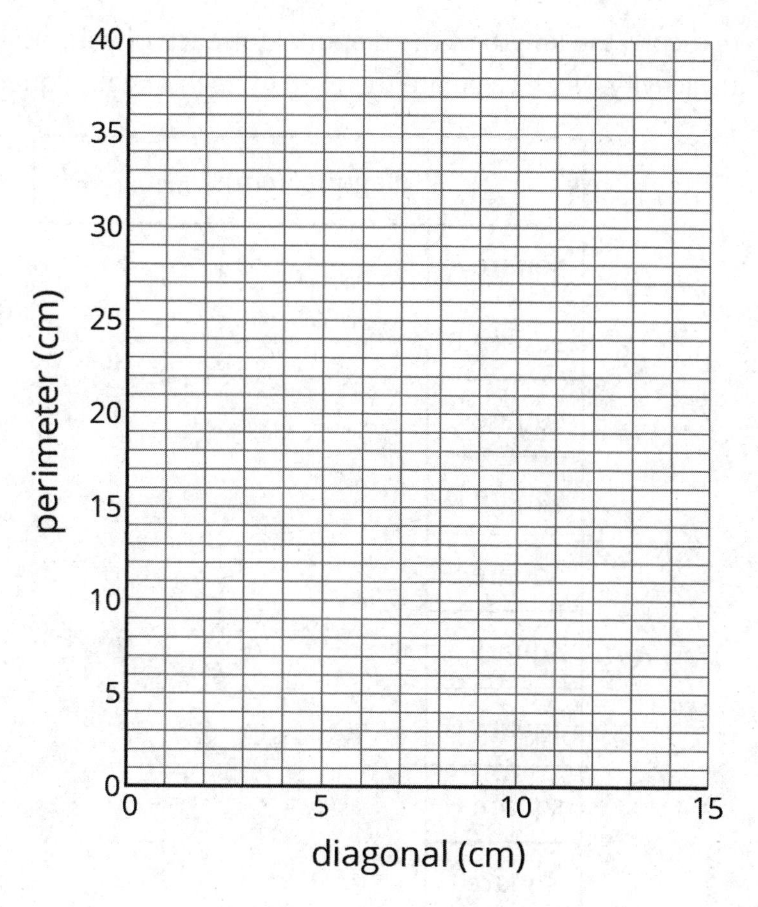


3. What do you notice about the points on the graph?

Pause here so your teacher can review your work.

4. Record measurements of the other squares to complete your table.

## 10.3: Area of a Square

1. In the table, record the length of the diagonal for each of your assigned squares from the previous activity. Next, calculate the area of each of your squares.

| | diagonal (cm) | area ($cm^2$) |
|---|---|---|
| square A | | |
| square B | | |
| square C | | |
| square D | | |
| square E | | |
| square F | | |
| square G | | |
| square H | | |
| square I | | |

Pause here so your teacher can review your work. Be prepared to share your values with the class.

2. Examine the class graph of these values. What do you notice?

3. How is the relationship between the diagonal and area of a square the same as the relationship between the diagonal and perimeter of a square from the previous activity? How is it different?

**Are you ready for more?**

Here is a rough map of a neighborhood.

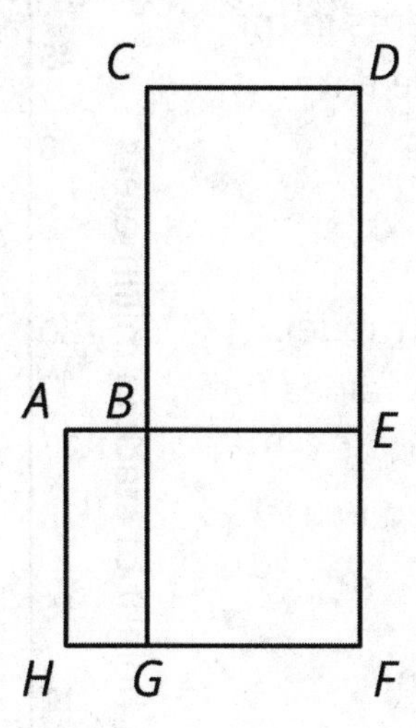


There are 4 mail routes during the week.

- On Monday, the mail truck follows the route A-B-E-F-G-H-A, which is 14 miles long.
- On Tuesday, the mail truck follows the route B-C-D-E-F-G-B, which is 22 miles long.
- On Wednesday, the truck follows the route A-B-C-D-E-F-G-H-A, which is 24 miles long.
- On Thursday, the mail truck follows the route B-E-F-G-B.

How long is the route on Thursdays?

## Lesson 10 Summary

When we measure the values for two related quantities, plotting the measurements in the coordinate plane can help us decide if it makes sense to model them with a proportional relationship. If the points are close to a line through $(0, 0)$, then a proportional relationship is a good model. For example, here is a graph of the values for the height, measured in millimeters, of different numbers of pennies placed in a stack.

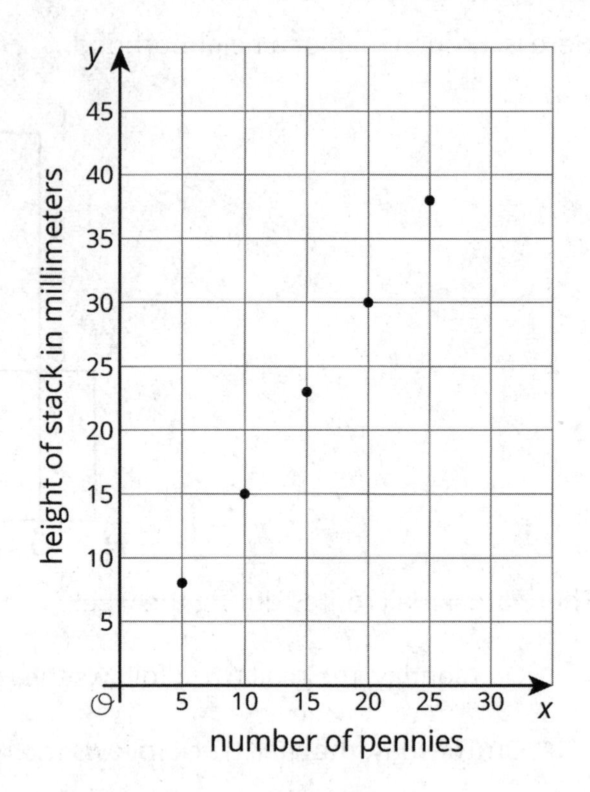


Because the points are close to a line through $(0, 0)$, the height of the stack of pennies appears to be proportional to the number of pennies in a stack. This makes sense because we can see that the heights of the pennies only vary a little bit.

An additional way to investigate whether or not a relationship is proportional is by making a table. Here is some data for the weight of different numbers of pennies in grams, along with the corresponding number of grams per penny.

| number of pennies | grams | grams per penny |
|---|---|---|
| 1 | 3.1 | 3.1 |
| 2 | 5.6 | 2.8 |
| 5 | 13.1 | 2.6 |
| 10 | 25.6 | 2.6 |

Though we might expect this relationship to be proportional, the quotients are not very close to one another. In fact, the metal in pennies changed in 1982, and older pennies are heavier. This explains why the weight per penny for different numbers of pennies are so different!

## Lesson 10 Practice Problems

1. Estimate the side length of a square that has a 9 cm long diagonal.

2. Select **all** quantities that are proportional to the diagonal length of a square.

   A. Area of the square

   B. Perimeter of the square

   C. Side length of the square

3. Diego made a graph of two quantities that he measured and said, "The points all lie on a line except one, which is a little bit above the line. This means that the quantities can't be proportional." Do you agree with Diego? Explain.

4. The graph shows that while it was being filled, the amount of water in gallons in a swimming pool was approximately proportional to the time that has passed in minutes.

   a. About how much water was in the pool after 25 minutes?

   b. Approximately when were there 500 gallons of water in the pool?

   c. Estimate the constant of proportionality for the gallons of water per minute going into the pool.

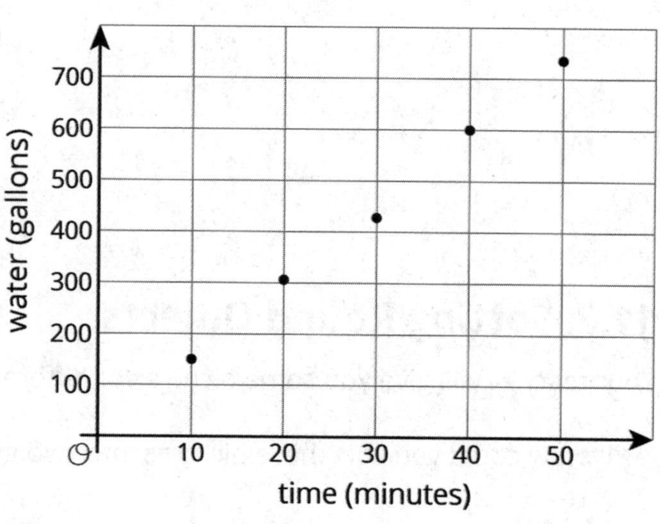

# Lesson 11: Exploring Circles

Let's explore circles.

## 11.1: How Do You Figure?

Here are two figures.

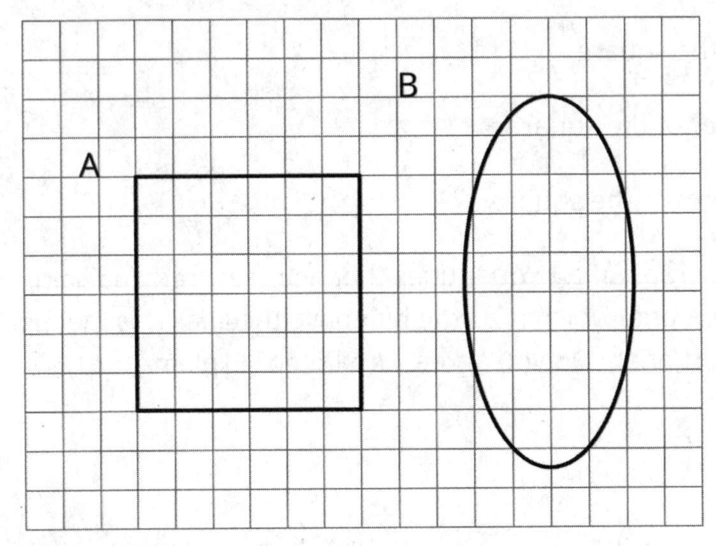


Figure C looks more like Figure A than like Figure B. Sketch what Figure C might look like. Explain your reasoning.

## 11.2: Sorting Round Objects

Your teacher will give you some pictures of different objects.

1. How could you sort these pictures into two groups? Be prepared to share your reasoning.

2. Work with your partner to sort the pictures into the categories that your class has agreed on. Pause here so your teacher can review your work.

3. What are some characteristics that all **circles** have in common?

4. Put the circular objects in order from smallest to largest.

5. Select one of the pictures of a circular object. What are some ways you could measure the actual size of your circle?

## Are you ready for more?

On January 3rd, Earth is 147,500,000 kilometers away from the Sun. On July 4th, Earth is 152,500,000 kilometers away from the Sun. The Sun has a radius of about 865,000 kilometers.

Could Earth's orbit be a circle with some point in the Sun as its center? Explain your reasoning.

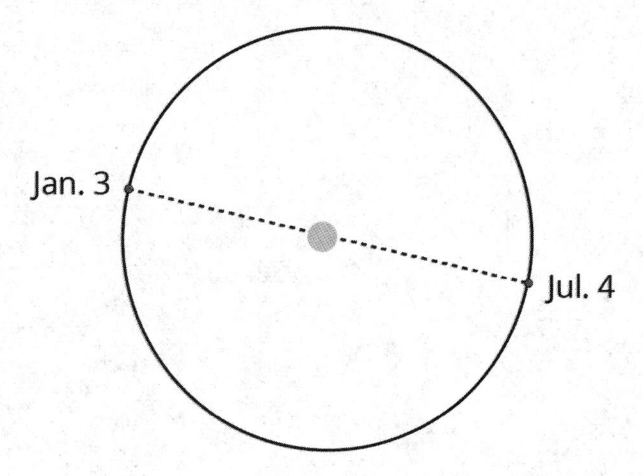

## 11.3: Measuring Circles

Priya, Han, and Mai each measured one of the circular objects from earlier.

- Priya says that the bike wheel is 24 inches.
- Han says that the yo-yo trick is 24 inches.
- Mai says that the glow necklace is 24 inches.

1. Do you think that all these circles are the same size?

2. What part of the circle did each person measure? Explain your reasoning.

## 11.4: Drawing Circles

Draw and label each circle.

1. Circle A, with a **diameter** of 6 cm.

2. Circle B, with a **radius** of 5 cm. Pause here so your teacher can review your work.

3. Circle C, with a radius that is equal to Circle A's diameter.

4. Circle D, with a diameter that is equal to Circle B's radius.

5. Use a compass to recreate one of these designs.

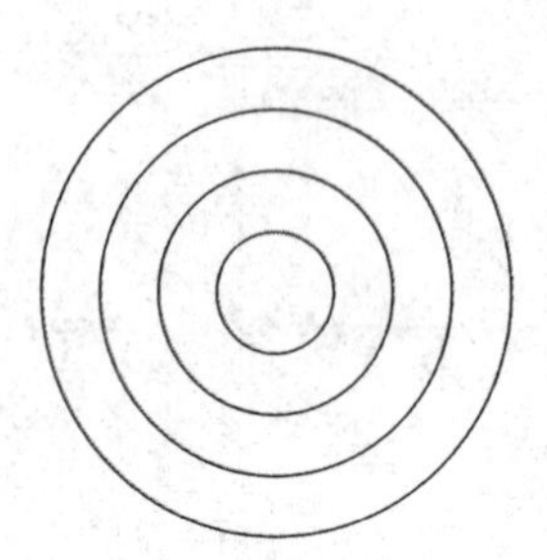

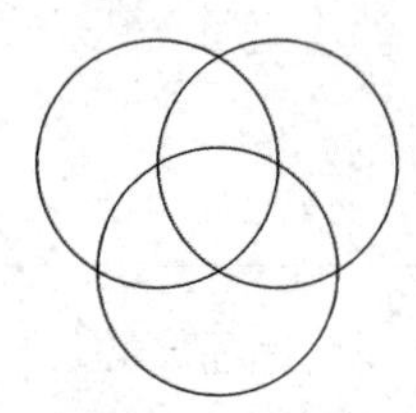

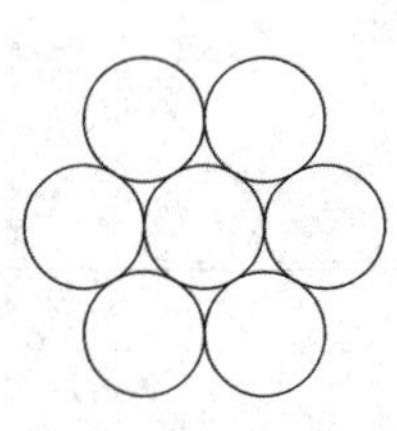

## Lesson 11 Summary

A **circle** consists of all of the points that are the same distance away from a particular point called the *center* of the circle.

A segment that connects the center with any point on the circle is called a **radius**. For example, segments $QG$, $QH$, $QI$, and $QJ$ are all radii of circle 2. (We say one radius and two radii.) The length of any radius is always the same for a given circle. For this reason, people also refer to this distance as the *radius* of the circle.

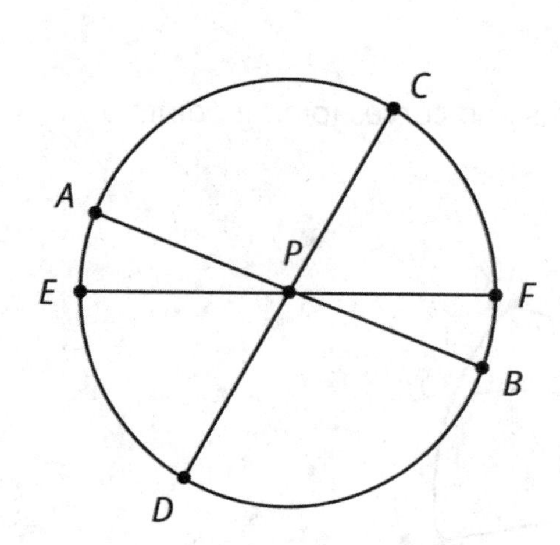


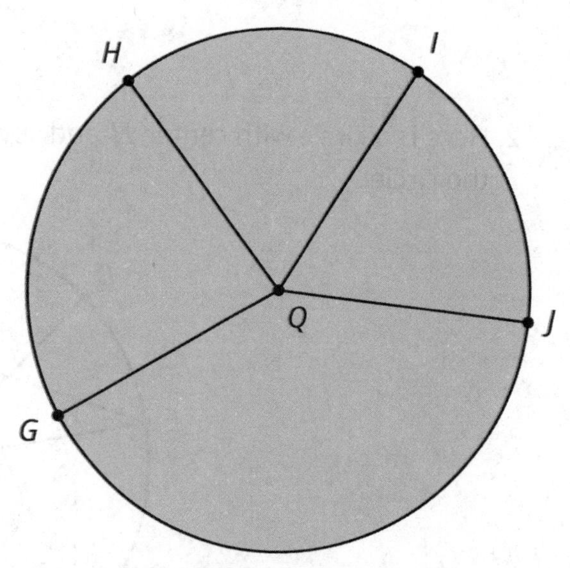


A segment that connects two opposite points on a circle (passing through the circle's center) is called a **diameter**. For example, segments $AB$, $CD$, and $EF$ are all diameters of circle 1. All diameters in a given circle have the same length because they are composed of two radii. For this reason, people also refer to the length of such a segment as the *diameter* of the circle.

The **circumference** of a circle is the distance around it. If a circle was made of a piece of string and we cut it and straightened it out, the circumference would be the length of that string. A circle always encloses a circular region. The region enclosed by circle 2 is shaded, but the region enclosed by circle 1 is not. When we refer to the area of a circle, we mean the area of the enclosed circular region.

## Glossary

- circle
- circumference
- diameter
- radius

## Lesson 11 Practice Problems

1. Use a geometric tool to draw a circle. Draw and measure a radius and a diameter of the circle.

2. Here is a circle with center $H$ and some line segments and curves joining points on the circle.

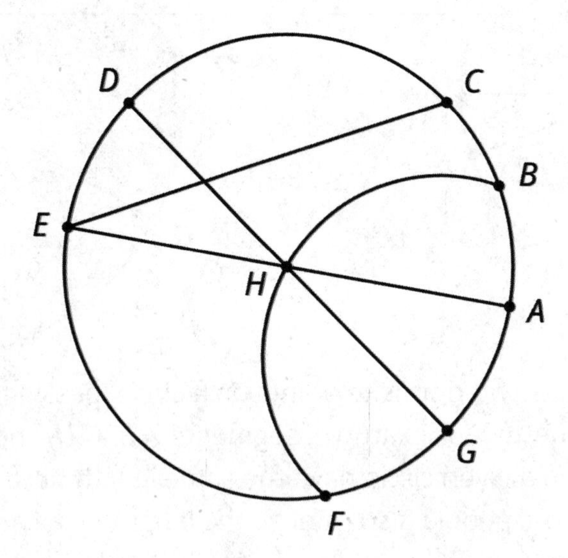


Identify examples of the following. Explain your reasoning.

a. Diameter

b. Radius

3. Lin measured the diameter of a circle in two different directions. Measuring vertically, she got 3.5 cm, and measuring horizontally, she got 3.6 cm. Explain some possible reasons why these measurements differ.

4. A small, test batch of lemonade used $\frac{1}{4}$ cup of sugar added to 1 cup of water and $\frac{1}{4}$ cup of lemon juice. After confirming it tasted good, a larger batch is going to be made with the same ratios using 10 cups of water. How much sugar should be added so that the large batch tastes the same as the test batch?

(From Unit 5, Lesson 1.)

5. The graph of a proportional relationship contains the point with coordinates $(3, 12)$. What is the constant of proportionality of the relationship?

(From Unit 5, Lesson 9.)

# Lesson 12: Exploring Circumference

Let's explore the circumference of circles.

## 12.1: Which Is Greater?

Clare wonders if the height of the toilet paper tube or the distance around the tube is greater. What information would she need in order to solve the problem? How could she find this out?

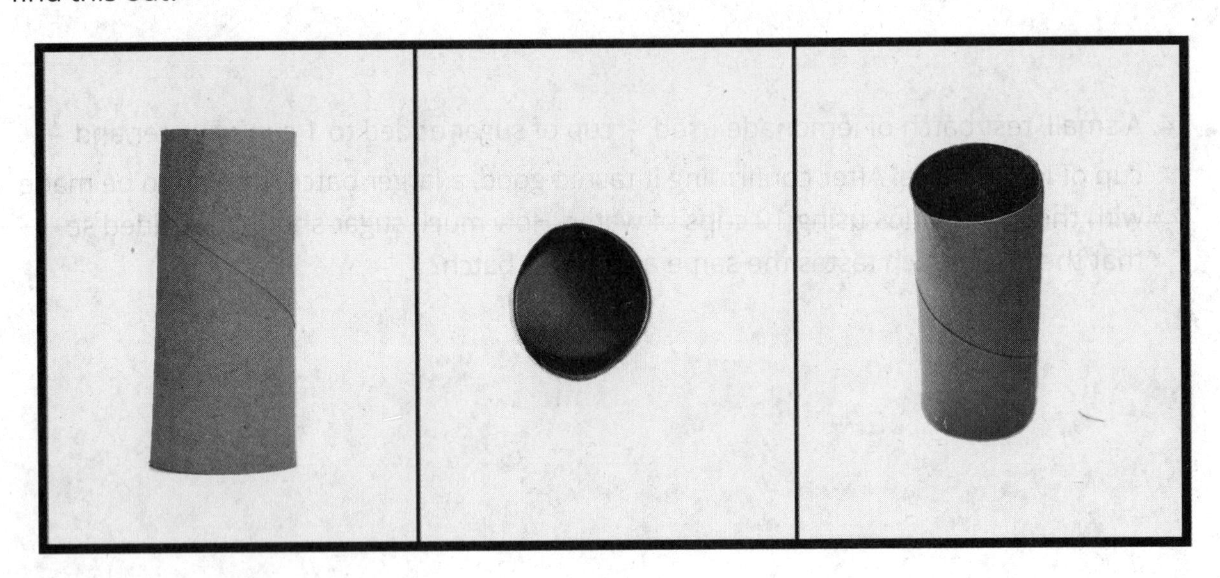

## 12.2: Measuring Circumference and Diameter

Your teacher will give you several circular objects.

1. Measure the diameter and the circumference of the circle in each object to the nearest tenth of a centimeter. Record your measurements in the table.

| object | diameter (cm) | circumference (cm) |
|---|---|---|
| | | |
| | | |
| | | |

2. Plot the diameter and circumference values from the table on the coordinate plane. What do you notice?

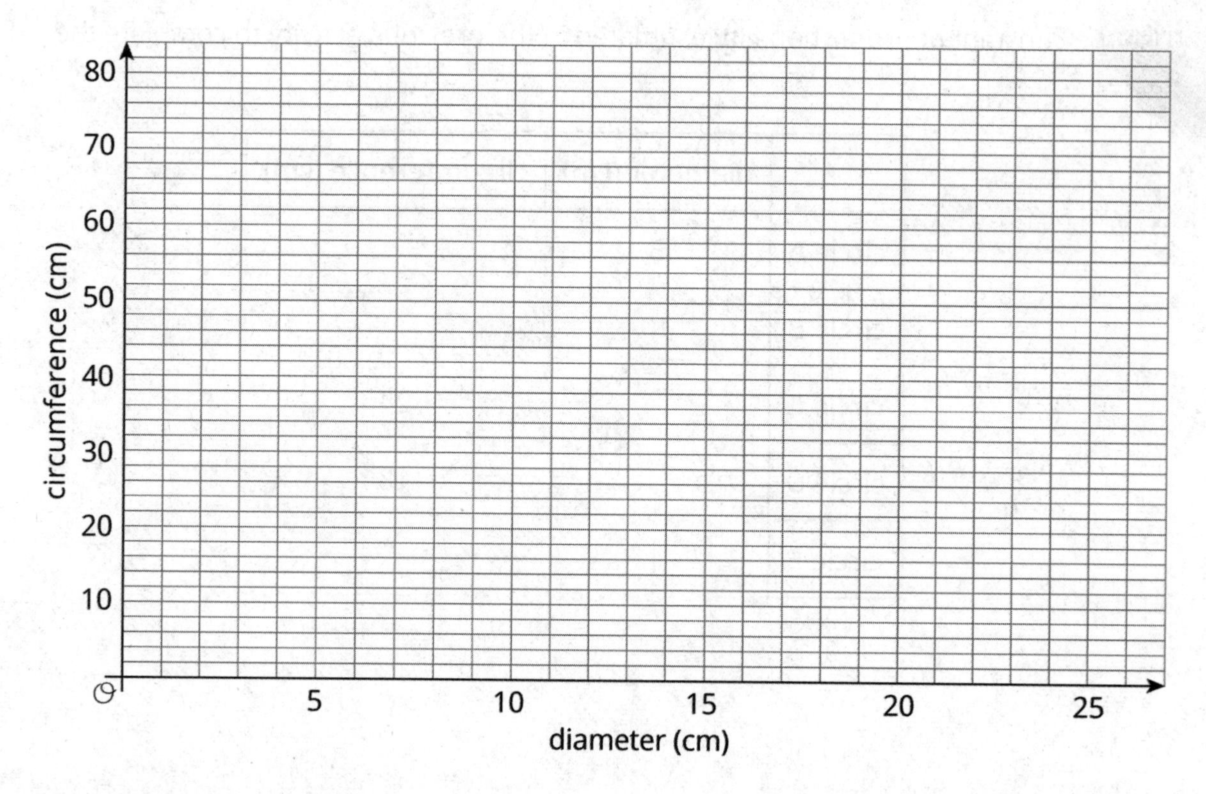


3. Plot the points from two other groups on the same coordinate plane. Do you see the same pattern that you noticed earlier?

## 12.3: Calculating Circumference and Diameter

Here are five circles. One measurement for each circle is given in the table.

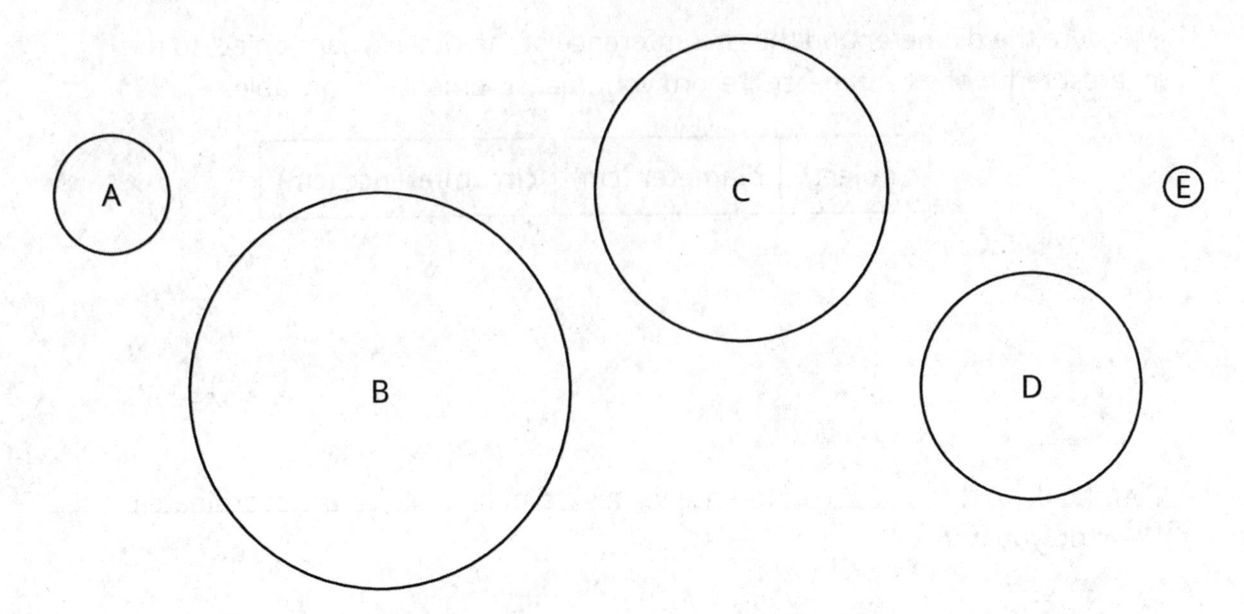


Use the constant of proportionality estimated in the previous activity to complete the table.

| | diameter (cm) | circumference (cm) |
|---|---|---|
| circle A | 3 | |
| circle B | 10 | |
| circle C | | 24 |
| circle D | | 18 |
| circle E | 1 | |

### Are you ready for more?

The circumference of Earth is approximately 40,000 km. If you made a circle of wire around the globe, that is only 10 meters (0.01 km) longer than the circumference of the globe, could a flea, a mouse, or even a person creep under it?

### Lesson 12 Summary

There is a proportional relationship between the diameter and circumference of any circle. That means that if we write $C$ for circumference and $d$ for diameter, we know that $C = kd$, where $k$ is the constant of proportionality.

The exact value for the constant of proportionality is called $\pi$. Some frequently used approximations for $\pi$ are $\frac{22}{7}$, 3.14, and 3.14159, but none of these is exactly $\pi$.

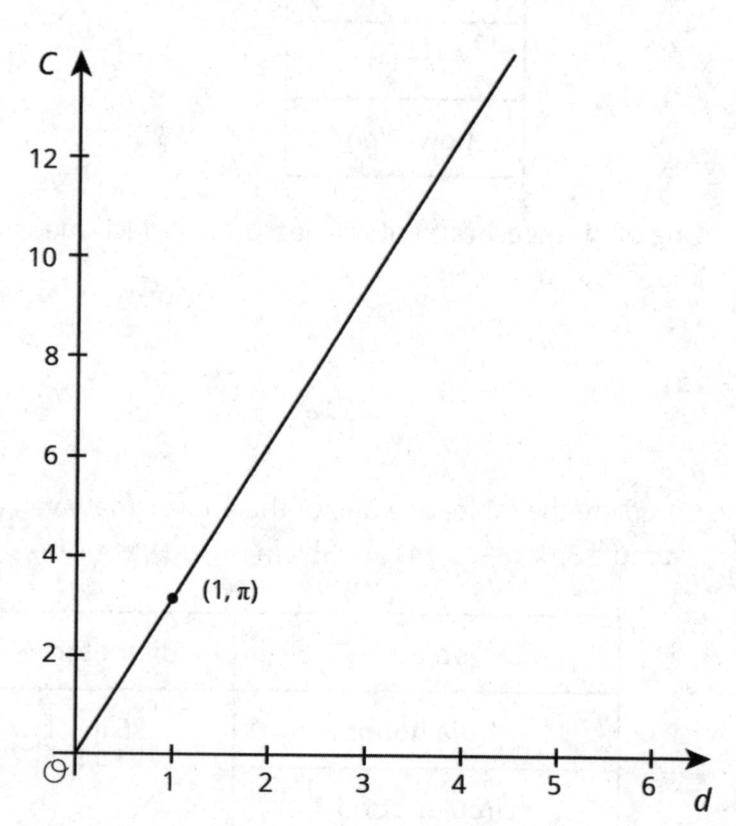


We can use this to estimate the circumference if we know the diameter, and vice versa. For example, using 3.1 as an approximation for $\pi$, if a circle has a diameter of 4 cm, then the circumference is about $(3.1) \cdot 4 = 12.4$ or 12.4 cm.

The relationship between the circumference and the diameter can be written as

$$C = \pi d$$

### Glossary

- pi ($\pi$)

## Lesson 12 Practice Problems

1. Diego measured the diameter and circumference of several circular objects and recorded his measurements in the table.

| object | diameter (cm) | circumference (cm) |
|---|---|---|
| half dollar coin | 3 | 10 |
| flying disc | 23 | 28 |
| jar lid | 8 | 25 |
| flower pot | 15 | 48 |

One of his measurements is inaccurate. Which measurement is it? Explain how you know.

2. Complete the table. Use one of the approximate values for $\pi$ discussed in class (for example 3.14, $\frac{22}{7}$, 3.1416). Explain or show your reasoning.

| object | diameter | circumference |
|---|---|---|
| hula hoop | 35 in | |
| circular pond | | 556 ft |
| magnifying glass | 5.2 cm | |
| car tire | | 71.6 in |



iM

3. $A$ is the center of the circle, and the length of $CD$ is 15 centimeters.

a. Name a segment that is a radius.
How long is it?

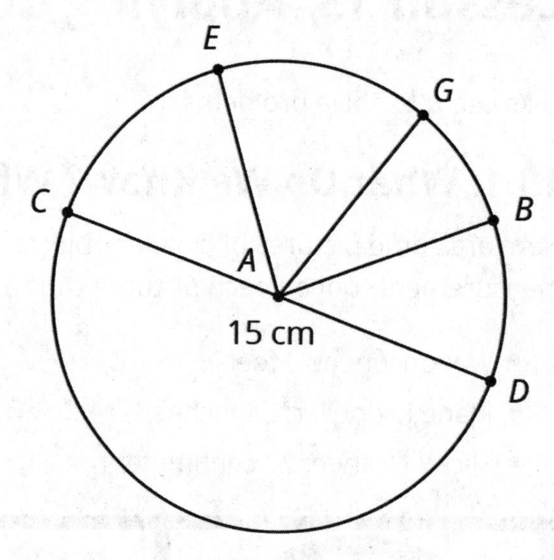


b. Name a segment that is a diameter.
How long is it?

(From Unit 5, Lesson 11.)

4. a. Consider the equation $y = 1.5x + 2$. Find four pairs of $x$ and $y$ values that make the equation true. Plot the points $(x, y)$ on the coordinate plane.

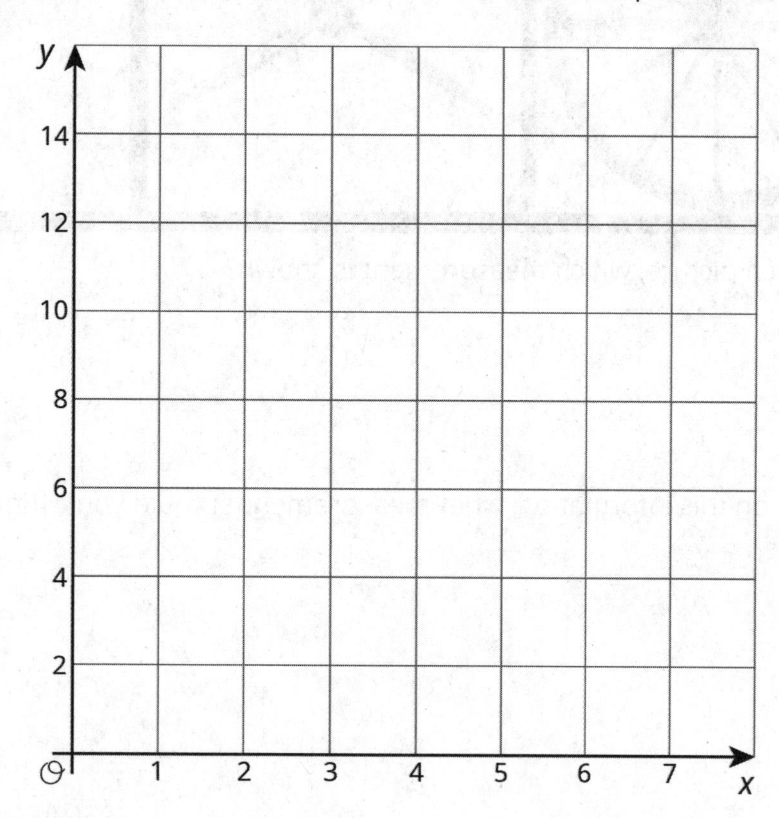


b. Based on the graph, can this be a proportional relationship? Why or why not?

(From Unit 5, Lesson 7.)

# Lesson 13: Applying Circumference

Let's use $\pi$ to solve problems.

## 13.1: What Do We Know? What Can We Estimate?

Here are some pictures of circular objects, with measurement tools shown. The measurement tool on each picture reads as follows:

- Wagon wheel: 3 feet
- Plane propeller: 24 inches
- Sliced Orange: 20 centimeters

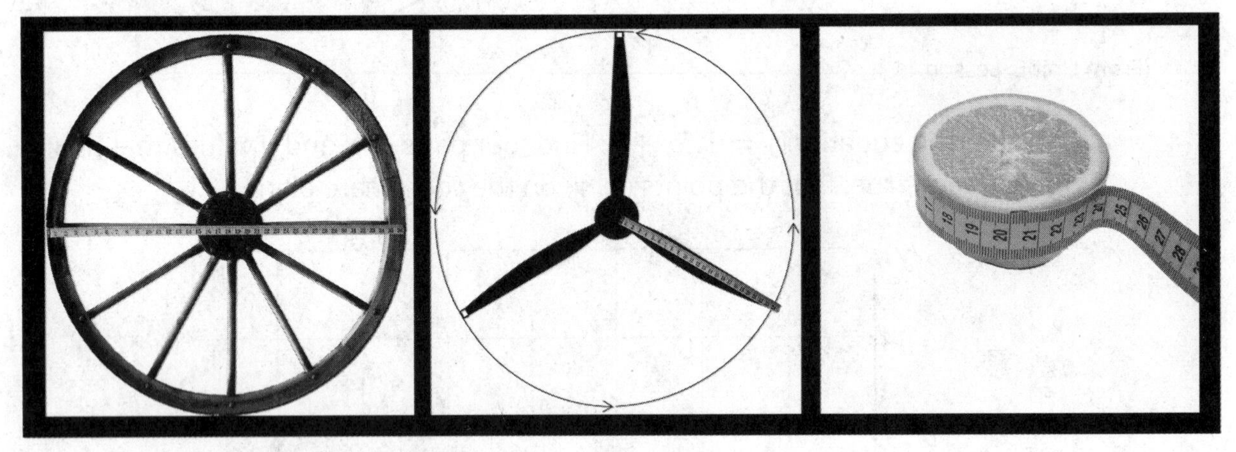

1. For each picture, which measurement is shown?

2. Based on this information, what measurement(s) could you estimate for each picture?

## 13.2: Using $\pi$

In the previous activity, we looked at pictures of circular objects. One measurement for each object is listed in the table.

Your teacher will assign you an approximation of $\pi$ to use for this activity.

1. Complete the table.

| object | radius | diameter | circumference |
|---|---|---|---|
| wagon wheel | | 3 ft | |
| airplane propeller | 24 in | | |
| orange slice | | | 20 cm |

2. A bug was sitting on the tip of the propeller blade when the propeller started to rotate. The bug held on for 5 rotations before flying away. How far did the bug travel before it flew off?

## 13.3: Around the Running Track

The field inside a running track is made up of a rectangle that is 84.39 m long and 73 m wide, together with a half-circle at each end.

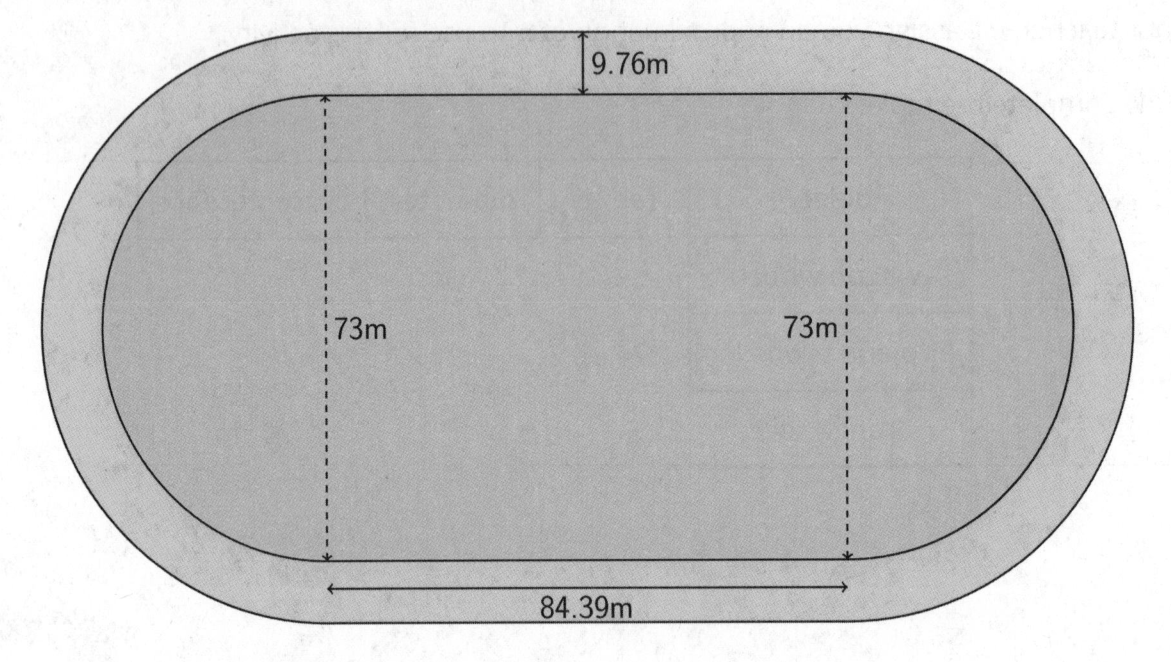


1. What is the distance around the inside of the track? Explain or show your reasoning.

2. The track is 9.76 m wide all the way around. What is the distance around the outside of the track? Explain or show your reasoning.

**Are you ready for more?**

This size running track is usually called a 400-meter track. However, if a person ran as close to the "inside" as possible on the track, they would run less than 400 meters in one lap. How far away from the inside border would someone have to run to make one lap equal exactly 400 meters?

## 13.4: Measuring a Picture Frame

Kiran bent some wire around a rectangle to make a picture frame. The rectangle is 8 inches by 10 inches.

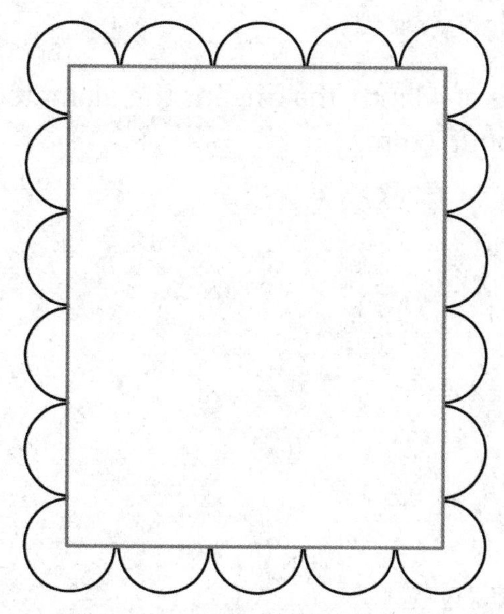

1. Find the perimeter of the wire picture frame. Explain or show your reasoning.

2. If the wire picture frame were stretched out to make one complete circle, what would its radius be?

## Lesson 13 Summary

The circumference of a circle, $C$, is $\pi$ times the diameter, $d$. The diameter is twice the radius, $r$. So if we know any one of these measurements for a particular circle, we can find the others. We can write the relationships between these different measures using equations:

$$d = 2r$$
$$C = \pi d$$
$$C = 2\pi r$$

If the diameter of a car tire is 60 cm, that means the radius is 30 cm and the circumference is $60 \cdot \pi$ or about 188 cm.

If the radius of a clock is 5 in, that means the diameter is 10 in, and the circumference is $10 \cdot \pi$ or about 31 in.

If a ring has a circumference of 44 mm, that means the diameter is $44 \div \pi$, which is about 14 mm, and the radius is about 7 mm.

# Lesson 13 Practice Problems

1. Here is a picture of a Ferris wheel. It has a diameter of 80 meters.

   a. On the picture, draw and label a diameter.

   b. How far does a rider travel in one complete rotation around the Ferris wheel?

2. Identify each measurement as the diameter, radius, or circumference of the circular object. Then, estimate the other two measurements for the circle.

   a. The length of the minute hand on a clock is 5 in.

   b. The distance across a sink drain is 3.8 cm.

   c. The tires on a mining truck are 14 ft tall.

   d. The fence around a circular pool is 75 ft long.

   e. The distance from the tip of a slice of pizza to the crust is 7 in.

   f. Breaking a cookie in half creates a straight side 10 cm long.

g. The length of the metal rim around a glass lens is 190 mm.

h. From the center to the edge of a DVD measures 60 mm.

3. A half circle is joined to an equilateral triangle with side lengths of 12 units. What is the perimeter of the resulting shape?

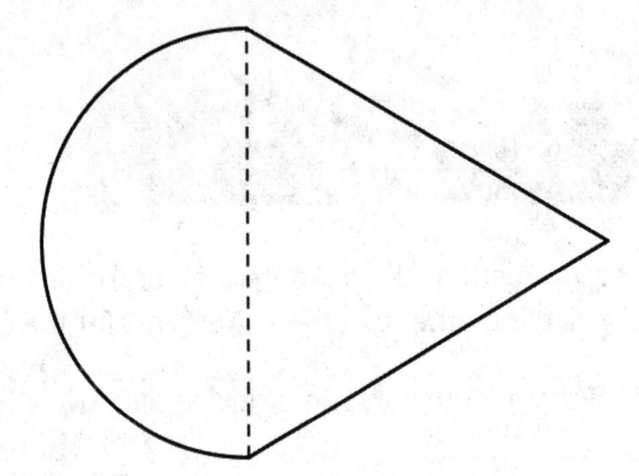

4. Circle A has a diameter of 1 foot. Circle B has a circumference of 1 meter. Which circle is bigger? Explain your reasoning. (1 inch = 2.54 centimeters)

5. The circumference of Tyler's bike tire is 72 inches. What is the diameter of the tire?

(From Unit 5, Lesson 12.)

# Lesson 14: Estimating Areas

Let's estimate the areas of weird shapes.

## 14.1: Mental Calculations

Find a strategy to make each calculation mentally:

$599 + 87$

$254 - 88$

$99 \cdot 75$

## 14.2: House Floorplan

Here is a floor plan of a house. Approximate lengths of the walls are given. What is the approximate area of the home, including the balcony? Explain or show your reasoning.

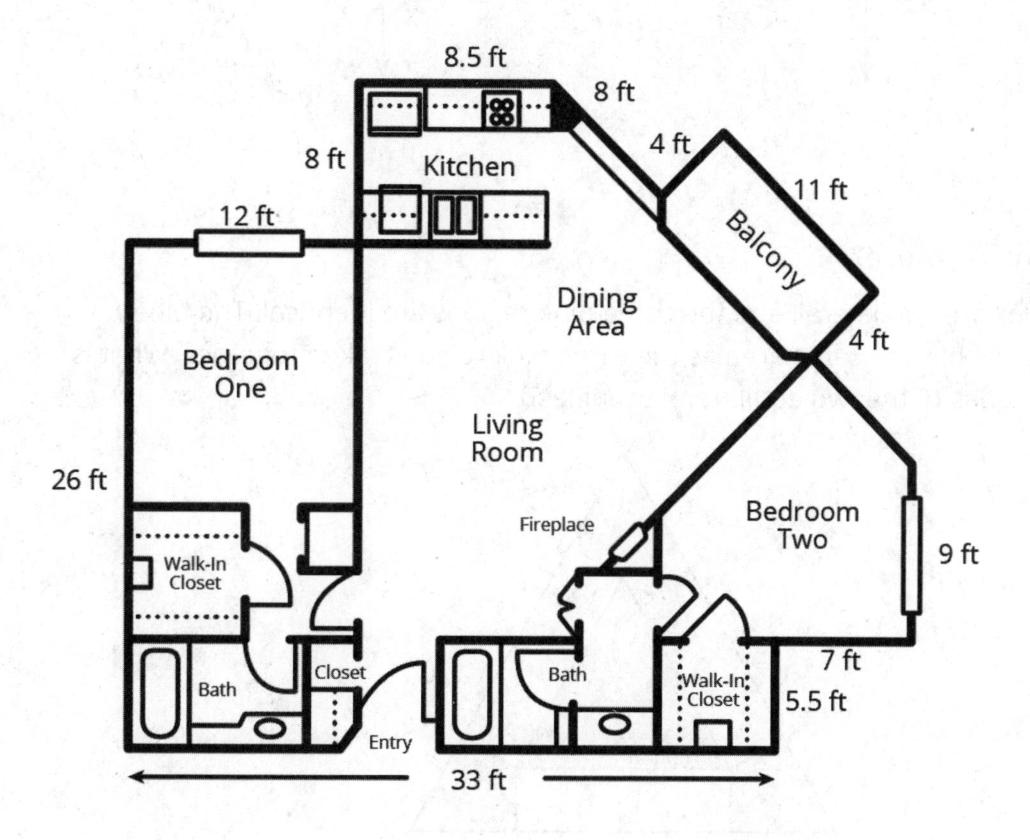

## 14.3: Area of Nevada

Estimate the area of Nevada in square miles. Explain or show your reasoning.

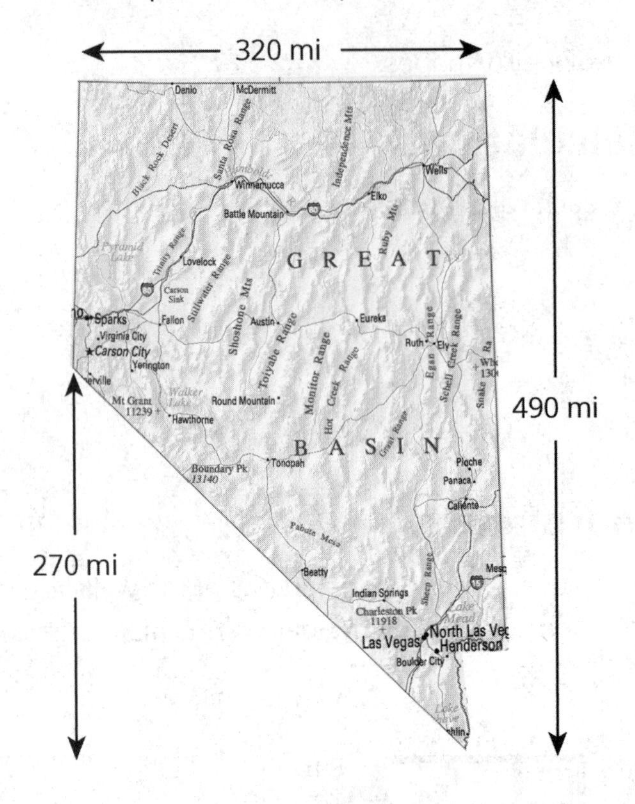


### Are you ready for more?

The two triangles are equilateral, and the three pink regions are identical. The blue equilateral triangle has the same area as the three pink regions taken together. What is the ratio of the sides of the two equilateral triangles?

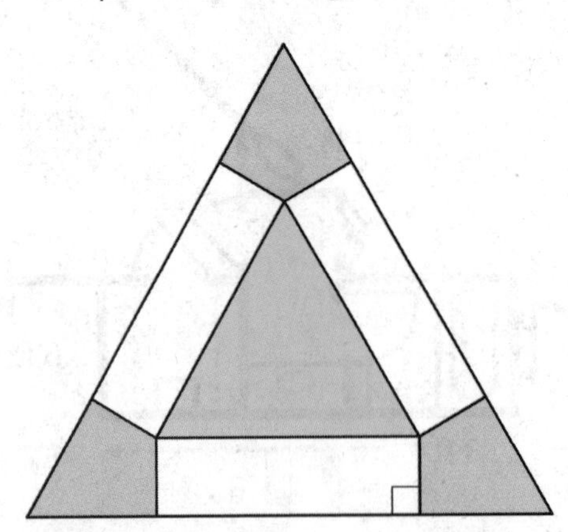

## Lesson 14 Summary

We can find the area of some complex polygons by surrounding them with a simple polygon like a rectangle. For example, this octagon is contained in a rectangle.

The rectangle is 20 units long and 16 units wide, so its area is 320 square units. To get the area of the octagon, we need to subtract the areas of the four right triangles in the corners. These triangles are each 8 units long and 5 units wide, so they each have an area of 20 square units. The area of the octagon is

$$320 - (4 \cdot 20)$$

or 240 square units.

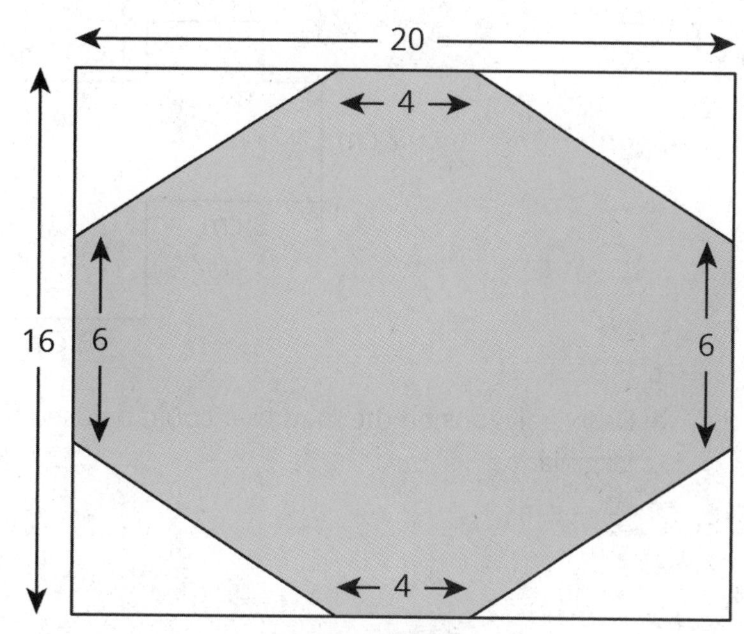


We can estimate the area of irregular shapes by approximating them with a polygon and finding the area of the polygon. For example, here is a satellite picture of Lake Tahoe with some one-dimensional measurements around the lake.

The area of the rectangle is 160 square miles, and the area of the triangle is 17.5 square miles for a total of 177.5 square miles. We recognize that this is an approximation, and not likely the exact area of the lake.

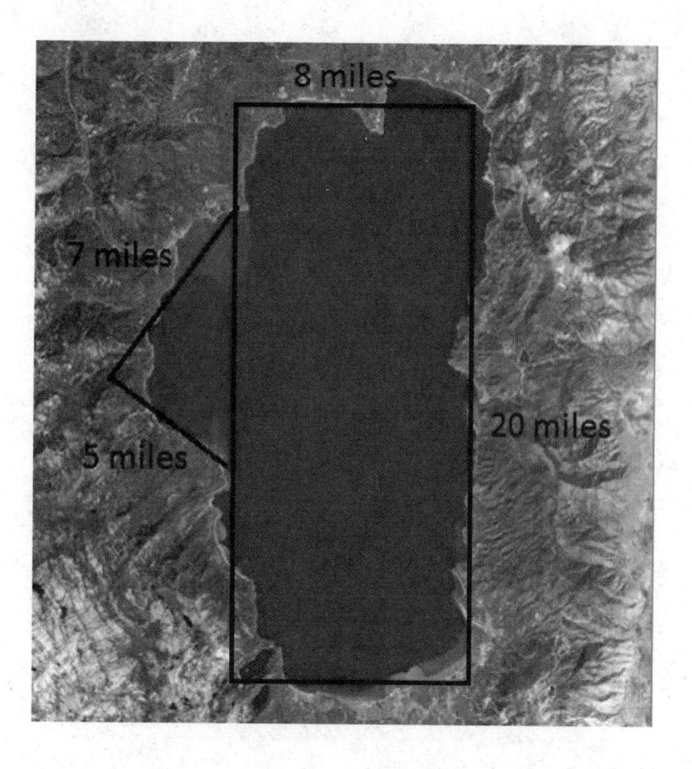

## Lesson 14 Practice Problems

1. Find the area of the polygon.

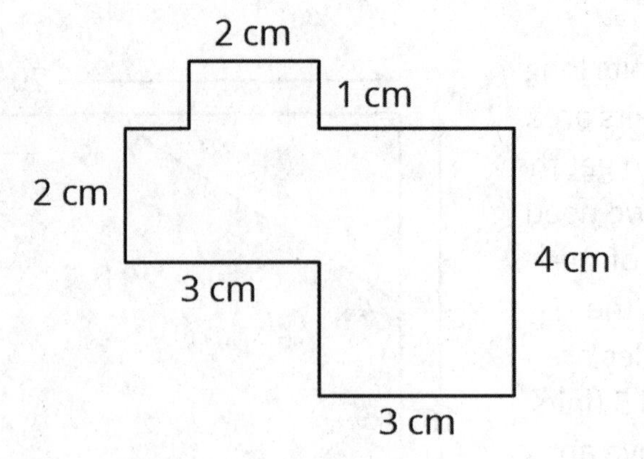


2. a. Draw polygons on the map that could be used to approximate the area of Virginia.

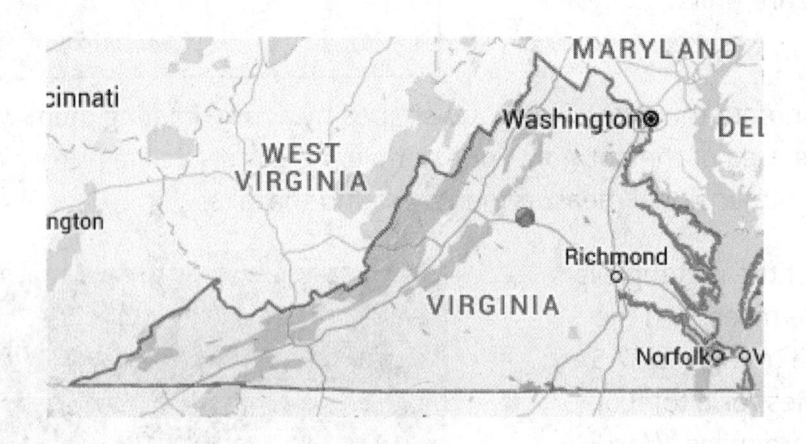


b. Which measurements would you need to know in order to calculate an approximation of the area of Virginia? Label the sides of the polygons whose measurements you would need. (Note: You aren't being asked to calculate anything.)

3. Jada's bike wheels have a diameter of 20 inches. How far does she travel if the wheels rotate 37 times?

This problem is from an earlier lesson

4. The radius of Earth is approximately 6,400 km. The equator is the circle around Earth dividing it into the northern and southern hemispheres. (The center of the earth is also the center of the equator.) What is the length of the equator?

(From Unit 5, Lesson 13.)

5. Here are several recipes for sparkling lemonade. For each recipe, describe how many tablespoons of lemonade mix it takes per cup of sparkling water.

    a. 4 tablespoons of lemonade mix and 12 cups of sparkling water

    b. 4 tablespoons of lemonade mix and 6 cups of sparkling water

    c. 3 tablespoons of lemonade mix and 5 cups of sparkling water

    d. $\frac{1}{2}$ tablespoon of lemonade mix and $\frac{3}{4}$ cups of sparkling water

(From Unit 5, Lesson 1.)

# Lesson 15: Area of a Circle

Let's investigate the areas of circles.

## 15.1: Irrigating a Field

A circular field is set into a square with an 800 m side length. Estimate the field's area.

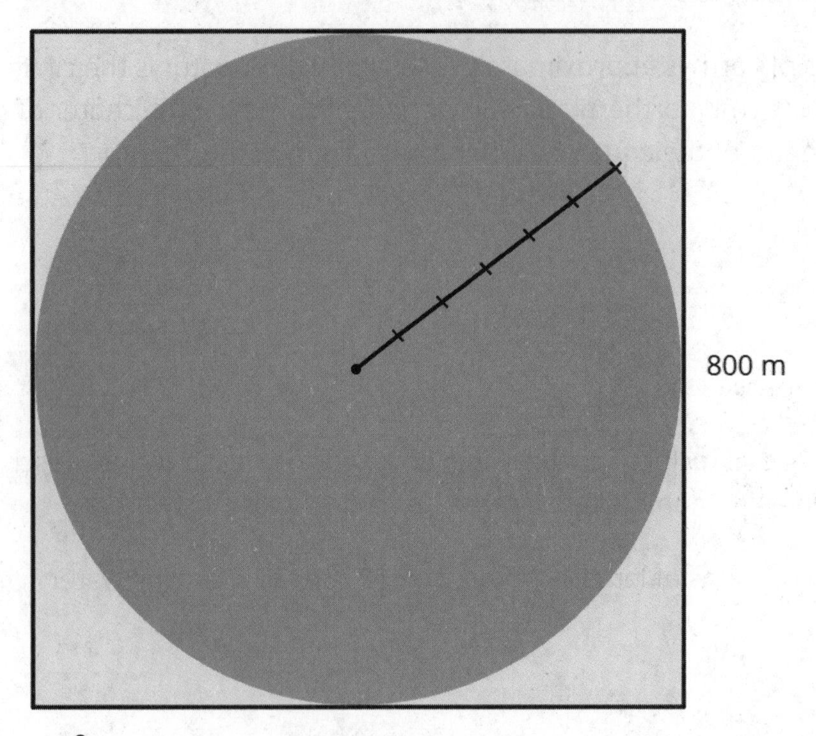


- About 5,000 $m^2$
- About 50,000 $m^2$
- About 500,000 $m^2$
- About 5,000,000 $m^2$
- About 50,000,000 $m^2$

## 15.2: Estimating Areas of Circles

Your teacher will give your group two circles of different sizes.

1. For each circle, use the squares on the graph paper to measure the diameter and estimate the **area of the circle.** Record your measurements in the table.

| diameter (cm) | estimated area ($cm^2$) |
|---|---|
| | |
| | |

2. Plot the values from the table on the class coordinate plane. Then plot the class's data points on your coordinate plane.

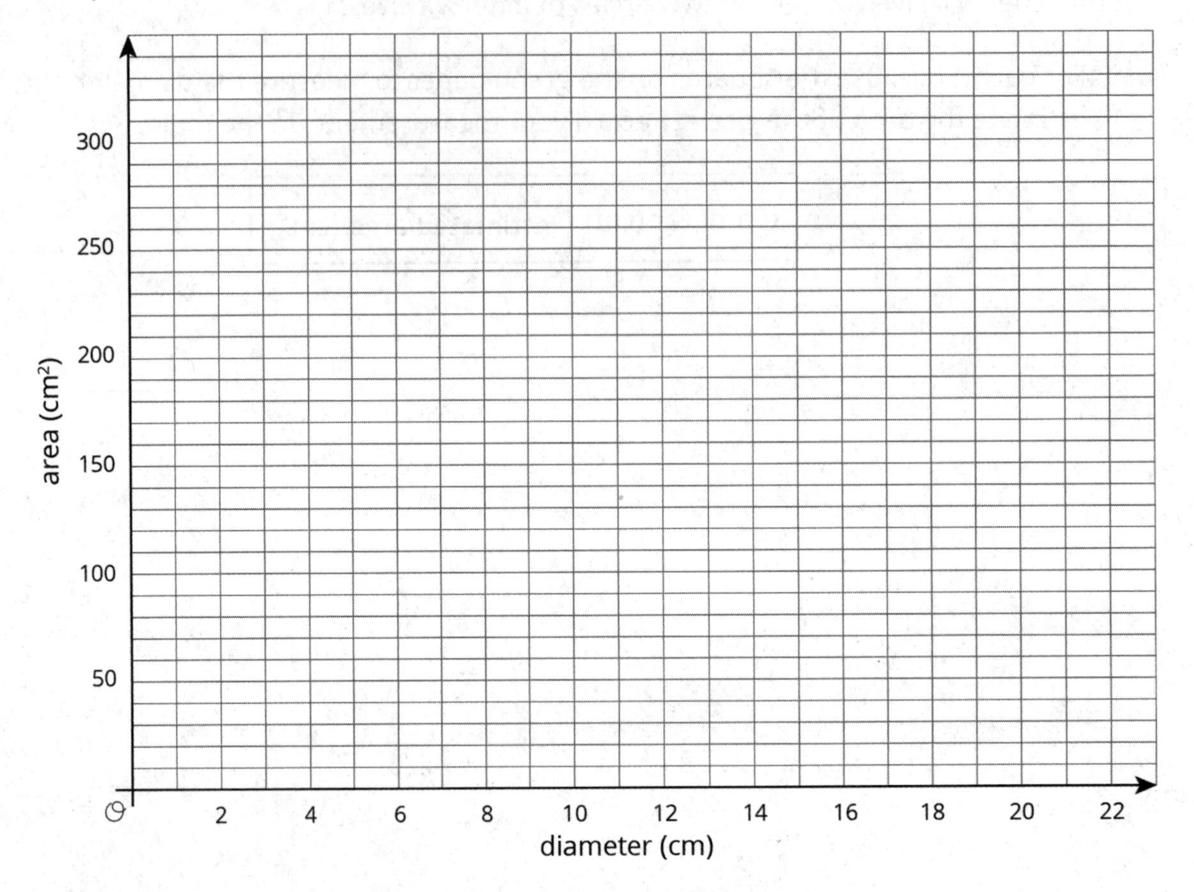


3. In a previous lesson, you graphed the relationship between the diameter and circumference of a circle. How is this graph the same? How is it different?

**Are you ready for more?**

How many circles of radius 1 unit can you fit inside each of the following so that they do not overlap?

1. a circle of radius 2 units?
2. a circle of radius 3 units?
3. a circle of radius 4 units?

If you get stuck, consider using coins or other circular objects.

## 15.3: Making a Polygon out of a Circle

Your teacher will give you a circular object, a marker, and two pieces of paper of different colors.

Follow these instructions to create a visual display:

1. Using a thick marker, trace your circle in two separate places on the same piece of paper.
2. Cut out both circles, cutting around the marker line.
3. Fold and cut one of the circles into fourths.
4. Arrange the fourths so that straight sides are next to each other, but the curved edges are alternately on top and on bottom. Pause here so your teacher can review your work.
5. Fold and cut the fourths in half to make eighths. Arrange the eighths next to each other, like you did with the fourths.
6. If your pieces are still large enough, repeat the previous step to make sixteenths.
7. Glue the remaining circle and the new shape onto a piece of paper that is a different color.

After you finish gluing your shapes, answer the following questions.

1. How do the areas of the two shapes compare?
2. What polygon does the shape made of the circle pieces most resemble?
3. How could you find the area of this polygon?

## 15.4: Making Another Polygon out of a Circle

Imagine a circle made of rings that can bend, but not stretch.

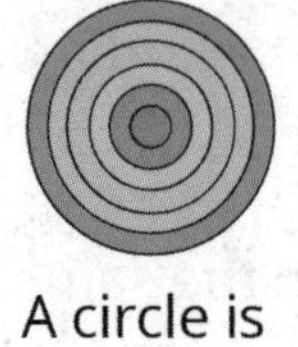

A circle is made of rings.

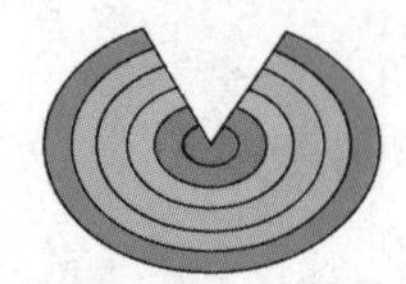

The rings are unrolled.

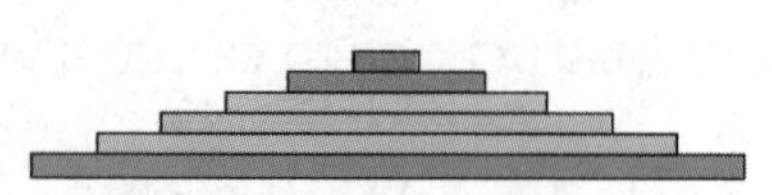

The circle has been made into a new shape.

1. What polygon does the new shape resemble?

2. How does the area of the polygon compare to the area of the circle?

3. How can you find the area of the polygon?

4. Show, in detailed steps, how you could find the polygon's area in terms of the circle's measurements. Show your thinking. Organize it so it can be followed by others.

5. After you finish, trade papers with a partner and check each other's work. If you disagree, work to reach an agreement. Discuss:

   - Do you agree or disagree with each step?
   - Is there a way to make the explanation clearer?

6. Return your partner's work, and revise your explanation based on the feedback you received.

## Lesson 15 Summary

The circumference $C$ of a circle is proportional to the diameter $d$, and we can write this relationship as $C = \pi d$. The circumference is also proportional to the radius of the circle, and the constant of proportionality is $2 \cdot \pi$ because the diameter is twice as long as the radius, so $C = 2\pi r$. However, the **area of a circle** is *not* proportional to the diameter (or the radius).

The area of a circle can be found by taking the product of half the circumference and the radius. If $A$ is the area of the circle, this gives the equation:

$$A = \frac{1}{2}(2\pi r) \cdot r$$

This equation can be rewritten as:

$$A = \pi r^2$$

(Remember that when we have $r \cdot r$ we can write $r^2$ and we can say "$r$ **squared**.")

This means that if we know the radius, we can find the area. For example, if a circle has radius 10 cm, then the area is about $(3.14) \cdot 100$ which is 314 cm$^2$.

If we know the diameter, we can figure out the radius, and then we can find the area. For example, if a circle has a diameter of 30 ft, then the radius is 15 ft, and the area is about $(3.14) \cdot 225$ which is approximately 707 ft$^2$.

# Lesson 15 Practice Problems

1. The $x$-axis of each graph has the diameter of a circle in meters. Label the $y$-axis on each graph with the appropriate measurement of a circle: radius (m), circumference (m), or area ($m^2$).

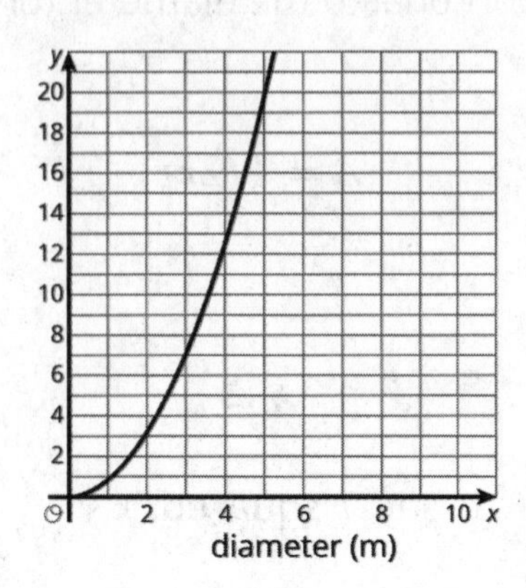


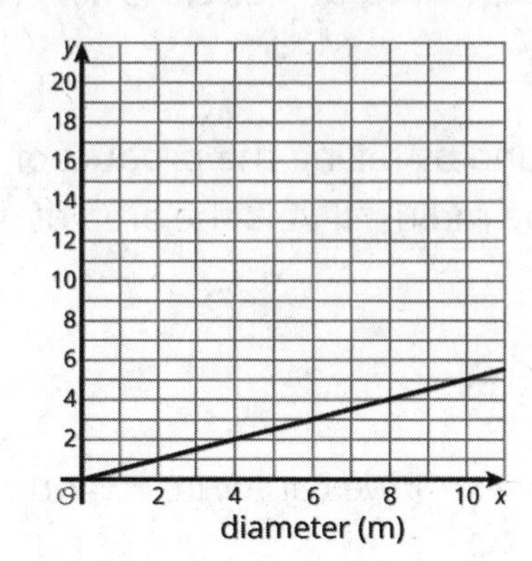


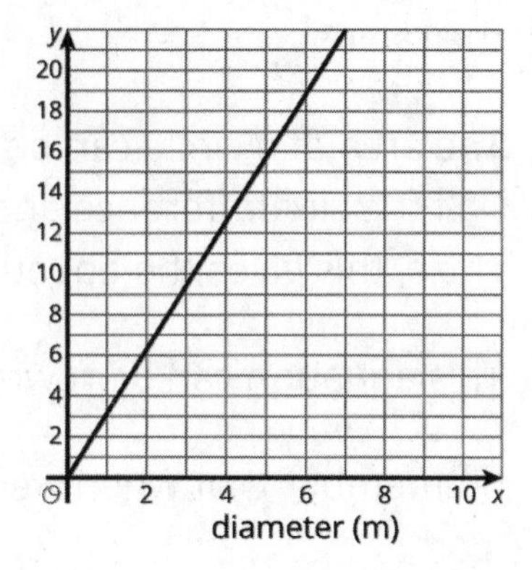


2. A circle's circumference is approximately 76 cm. Estimate the radius, diameter, and area of the circle.

3. Jada paints a circular table that has a diameter of 37 inches. What is the area of the table?

4. Point $A$ is the center of the circle, and the length of $CD$ is 15 centimeters. Find the circumference of this circle.

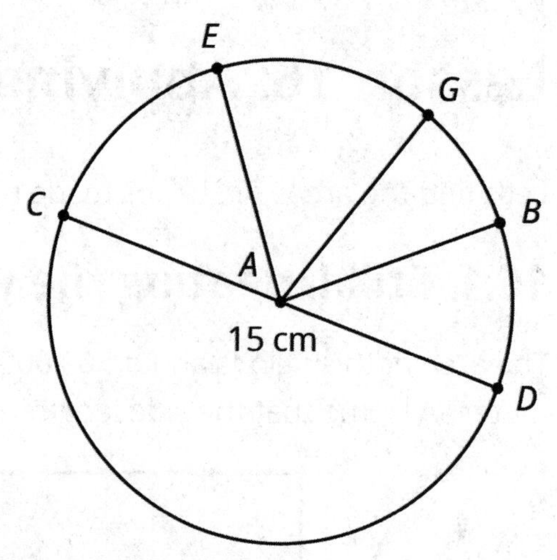


(From Unit 5, Lesson 12.)

5. The Carousel on the National Mall has 4 rings of horses. Kiran is riding on the inner ring, which has a radius of 9 feet. Mai is riding on the outer ring, which is 8 feet farther out from the center than the inner ring is.

   a. In one rotation of the carousel, how much farther does Mai travel than Kiran?

   b. One rotation of the carousel takes 12 seconds. How much faster does Mai travel than Kiran?

(From Unit 5, Lesson 13.)

# Lesson 16: Applying Area of Circles

Let's find the areas of shapes made up of circles.

## 16.1: Still Irrigating the Field

The area of this field is about 500,000 $m^2$. What is the field's area to the nearest square meter? Assume that the side lengths of the square are exactly 800 m.

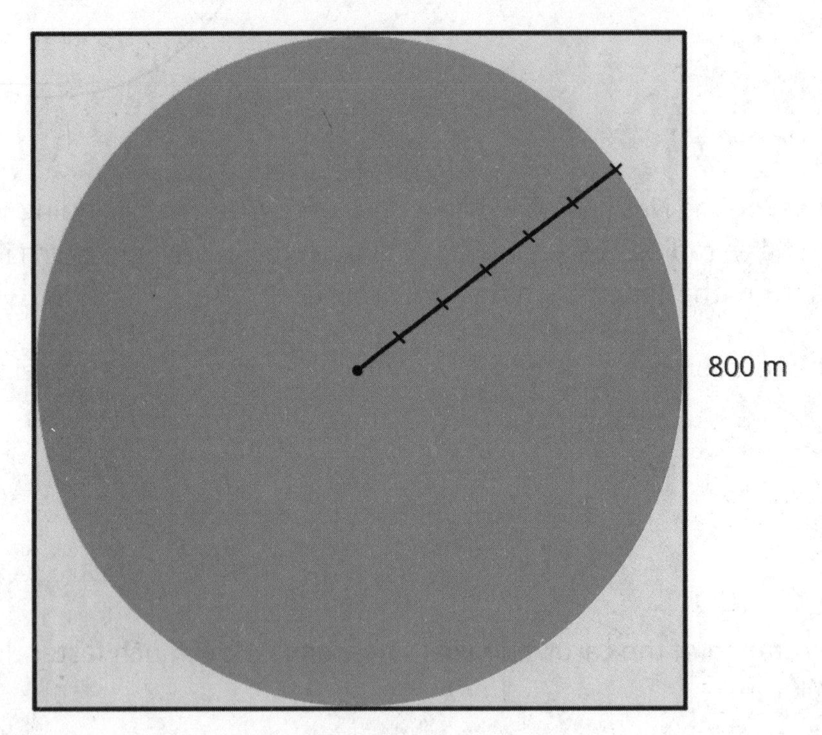


- 502,400 $m^2$
- 502,640 $m^2$
- 502,655 $m^2$
- 502,656 $m^2$
- 502,857 $m^2$

## 16.2: Comparing Areas Made of Circles

1. Each square has a side length of 12 units. Compare the areas of the shaded regions in the 3 figures. Which figure has the largest shaded region? Explain or show your reasoning.

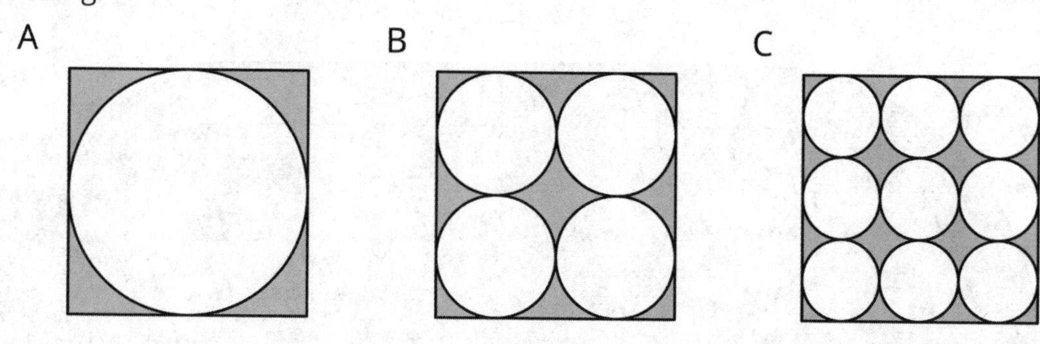


2. Each square in Figures D and E has a side length of 1 unit. Compare the area of the two figures. Which figure has more area? How much more? Explain or show your reasoning.

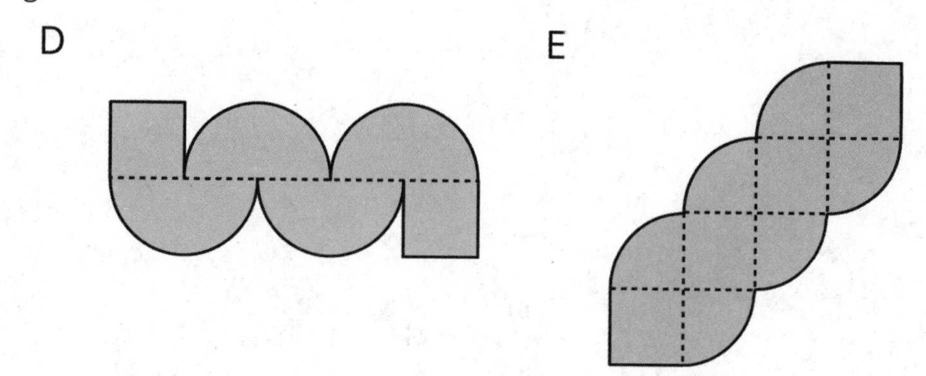


**Are you ready for more?**

Which figure has a longer perimeter, Figure D or Figure E? How much longer?

## 16.3: The Running Track Revisited

The field inside a running track is made up of a rectangle 84.39 m long and 73 m wide, together with a half-circle at each end. The running lanes are 9.76 m wide all the way around.

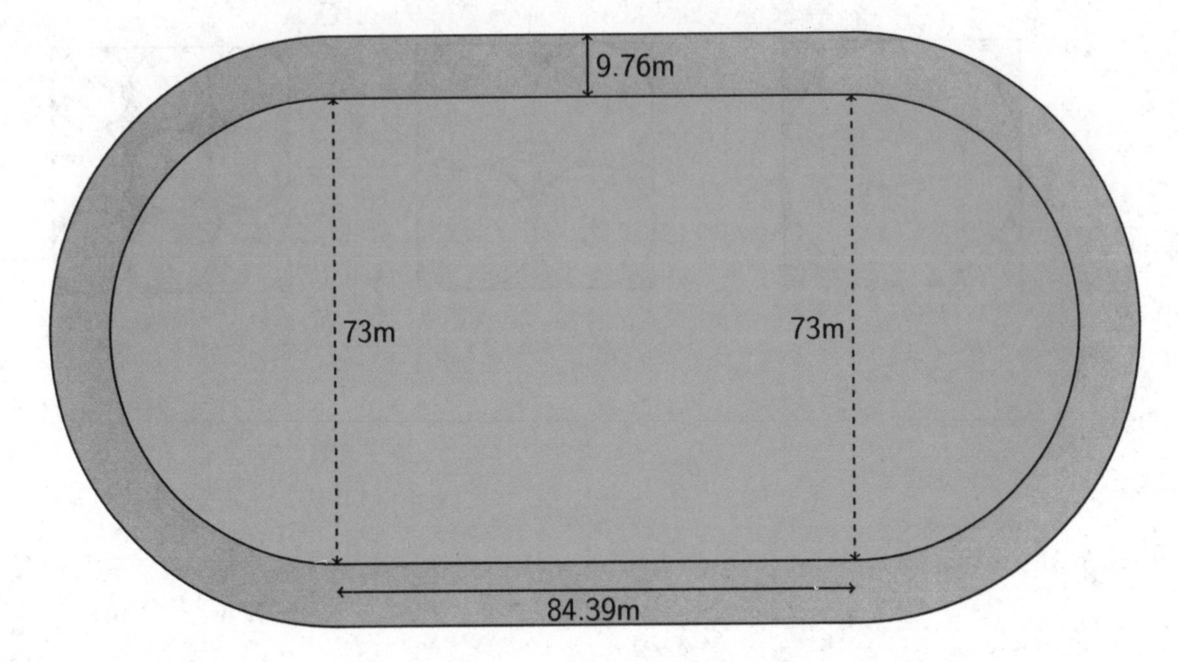


What is the area of the running track that goes around the field? Explain or show your reasoning.

## Lesson 16 Summary

The relationship between $A$, the area of a circle, and $r$, its radius, is $A = \pi r^2$. We can use this to find the area of a circle if we know the radius. For example, if a circle has a radius of 10 cm, then the area is $\pi \cdot 10^2$ or $100\pi$ cm$^2$. We can also use the formula to find the radius of a circle if we know the area. For example, if a circle has an area of $49\pi$ m$^2$ then its radius is 7 m and its diameter is 14 m.

Sometimes instead of leaving $\pi$ in expressions for the area, a numerical approximation can be helpful. For the examples above, a circle of radius 10 cm has area about 314 cm$^2$. In a similar way, a circle with area 154 m$^2$ has radius about 7 m.

We can also figure out the area of a fraction of a circle. For example, the figure shows a circle divided into 3 pieces of equal area. The shaded part has an area of $\frac{1}{3}\pi r^2$.

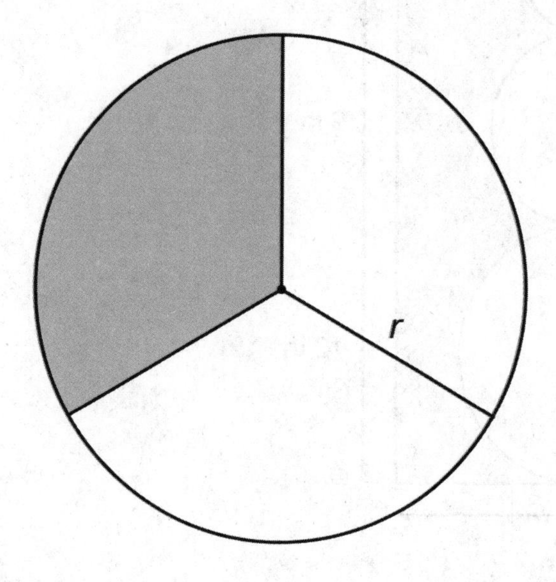

## Lesson 16 Practice Problems

1. A circle with a 12-inch diameter is folded in half and then folded in half again. What is the area of the resulting shape?

2. Find the area of the shaded region. Express your answer in terms of $\pi$.

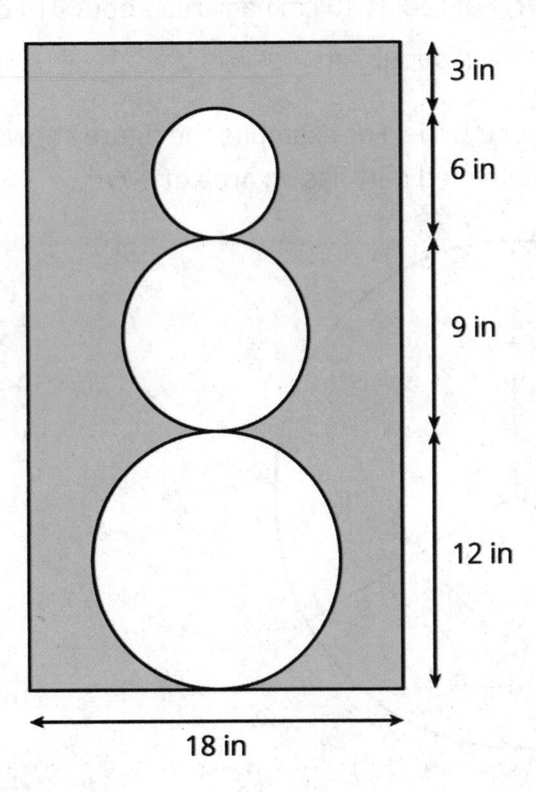


3. The face of a clock has a circumference of 63 in. What is the area of the face of the clock?

(From Unit 5, Lesson 15.)

4. Which of these pairs of quantities are proportional to each other? For the quantities that are proportional, what is the constant of proportionality?

   a. Radius and diameter of a circle

   b. Radius and circumference of a circle

   c. Radius and area of a circle

   d. Diameter and circumference of a circle

   e. Diameter and area of a circle

(From Unit 5, Lesson 15.)

5. Find the area of this shape in two different ways.

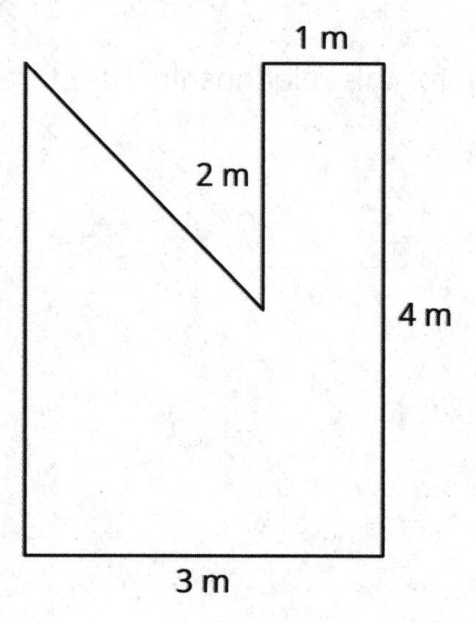


(From Unit 5, Lesson 14.)

6. Elena and Jada both read at a constant rate, but Elena reads more slowly. For every 4 pages that Elena can read, Jada can read 5.

   a. Complete the table.

| pages read by Elena | pages read by Jada |
|---|---|
| 4 | 5 |
| 1 | |
| 9 | |
| $e$ | |
| | 15 |
| | $j$ |

   b. Here is an equation for the table: $j = 1.25e$. What does the 1.25 mean?

   c. Write an equation for this relationship that starts $e = ...$

(From Unit 5, Lesson 2.)

# Lesson 17: Four Representations

Let's contrast relationships that are and are not proportional in four different ways.

## 17.1: Which is the Bluest?

1. Which group of blocks is the bluest?

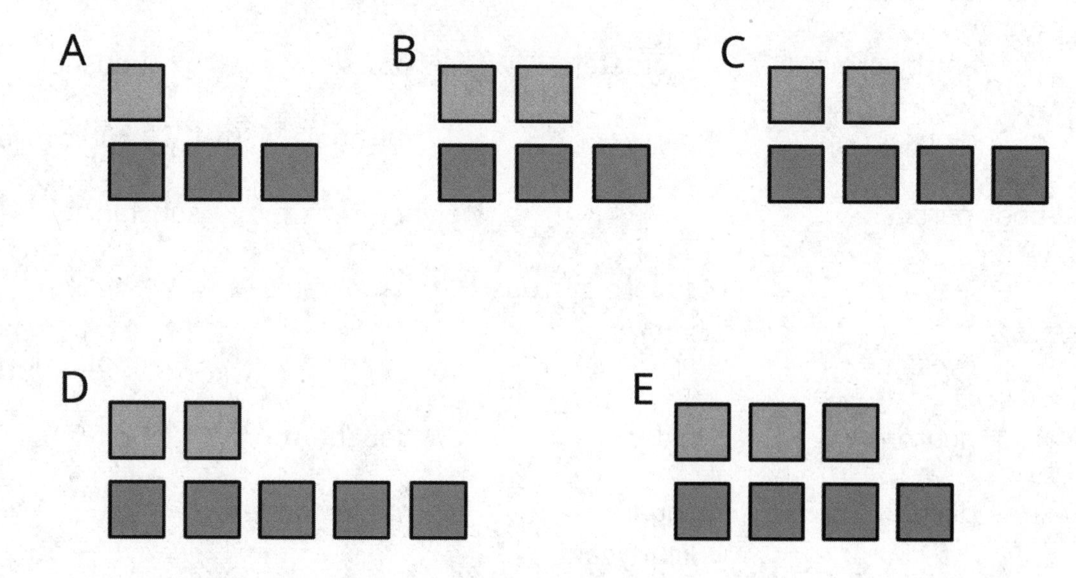


2. Order the groups of blocks from least blue to bluest.

## 17.2: One Scenario, Four Representations

1. Select two things from different lists. Make up a situation where there is a *proportional relationship* between quantities that involve these things.

creatures
- starfish
- centipedes
- earthworms
- dinosaurs

length
- centimeters
- cubits
- kilometers
- parsecs

time
- nanoseconds
- minutes
- years
- millennia

volume
- milliliters
- gallons
- bushels
- cubic miles

body parts
- legs
- eyes
- neurons
- digits

area
- square microns
- acres
- hides
- square light-years

weight
- nanograms
- ounces
- deben
- metric tonnes

substance
- helium
- oobleck
- pitch
- glue

2. Select two other things from the lists, and make up a situation where there is a relationship between quantities that involve these things, but the relationship is *not* proportional.

3. Your teacher will give you two copies of the "One Scenario, Four Representations" sheet. For each of your situations, describe the relationships in detail. If you get stuck, consider asking your teacher for a copy of the sample response.

   a. Write one or more sentences describing the relationship between the things you chose.

   b. Make a table with titles in each column and at least 6 pairs of numbers relating the two things.

   c. Graph the situation and label the axes.

   d. Write an equation showing the relationship and explain in your own words what each number and letter in your equation means.

   e. Explain how you know whether each relationship is proportional or not proportional. Give as many reasons as you can.

## 17.3: Make a Poster

Create a visual display of your two situations that includes all the information from the previous activity.

## Lesson 17 Summary

The constant of proportionality for a proportional relationship can often be easily identified in a graph, a table, and an equation that represents it. Here is an example of all three representations for the same relationship. The constant of proportionality is circled:

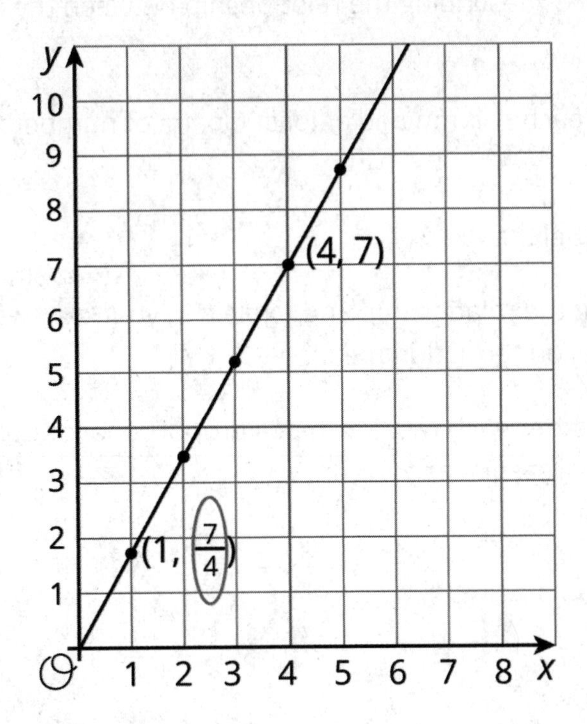


$y = \frac{7}{4}x$

| $x$ | $y$ |
|---|---|
| 0 | 0 |
| 1 | $\frac{7}{4}$ |
| 2 | $\frac{7}{2}$ |
| 3 | $\frac{21}{4}$ |
| 4 | 7 |

On the other hand, some relationships are not proportional. If the graph of a relationship is not a straight line through the origin, if the equation cannot be expressed in the form $y = kx$, or if the table does not have a constant of proportionality that you can multiply by any number in the first column to get the associated number in the second column, then the relationship between the quantities is not a proportional relationship.

## Lesson 17 Practice Problems

1. The equation $c = 2.95g$ shows how much it costs to buy gas at a gas station on a certain day. In the equation, $c$ represents the cost in dollars, and $g$ represents how many gallons of gas were purchased.

   a. Write down at least four (gallons of gas, cost) pairs that fit this relationship.

   b. Create a graph of the relationship.

   c. What does 2.95 represent in this situation?

   d. Jada's mom remarks, "You can get about a third of a gallon of gas for a dollar." Is she correct? How did she come up with that?

2. There is a proportional relationship between a volume measured in cups and the same volume measured in tablespoons. 3 cups is equivalent to 48 tablespoons, as shown in the graph.

   a. Plot and label at least two more points that represent the relationship.

   b. Use a straightedge to draw a line that represents this proportional relationship.

   c. For which value y is $(1, y)$ on the line you just drew?

   d. What is the constant of proportionality for this relationship?

   e. Write an equation representing this relationship. Use $c$ for cups and $t$ for tablespoons.

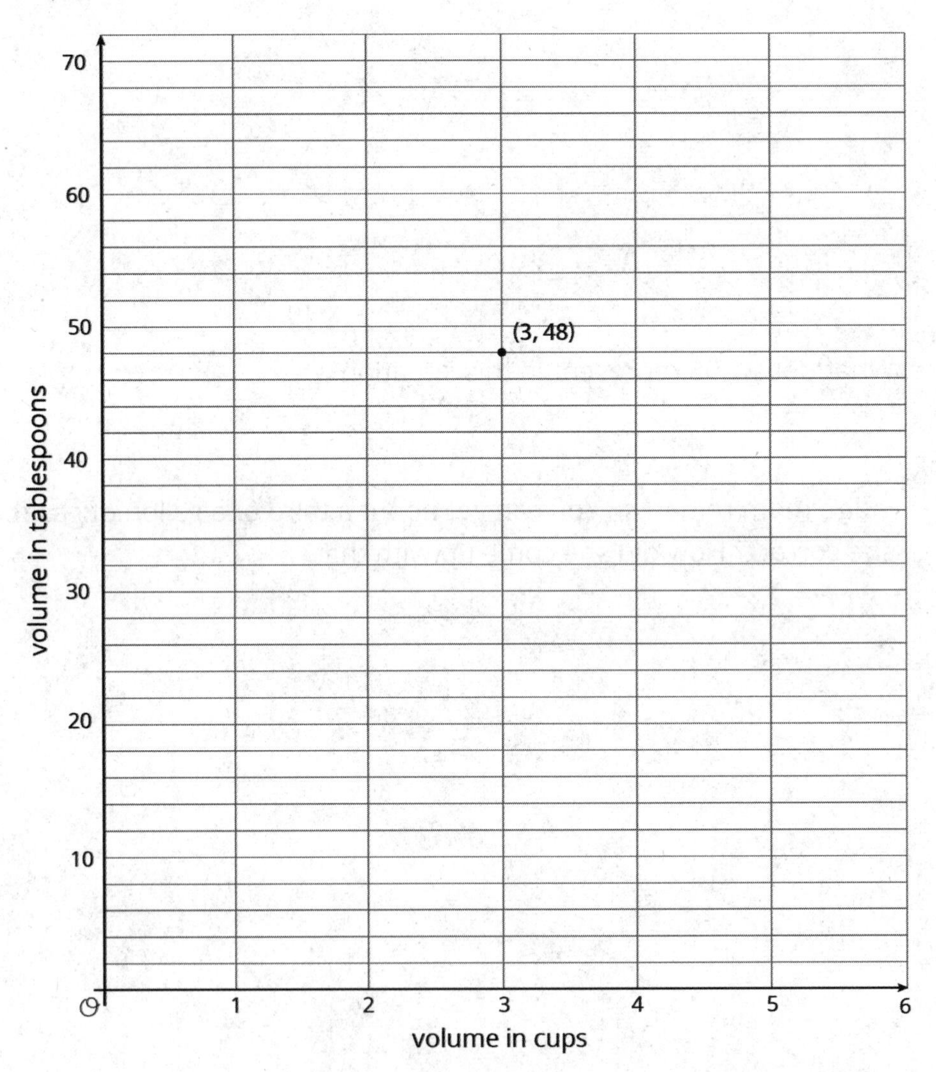

# Lesson 18: Using Water Efficiently

Let's investigate saving water.

## 18.1: Comparing Baths and Showers

Some people say that it uses more water to take a bath than a shower. Others disagree.

1. What information would you collect in order to answer the question?

2. Estimate some reasonable values for the things you suggest.

## 18.2: Saving Water: Bath or Shower?

1. Describe a method for comparing the water usage for a bath and a shower.

2. Find out values for the measurements needed to use the method you described. You may ask your teacher or research them yourself.

3. Under what conditions does a bath use more water? Under what conditions does a shower use more water?

## 18.3: Representing Water Usage

1. Continue considering the problem from the previous activity. Name two quantities that are in a proportional relationship. Explain how you know they are in a proportional relationship.

2. What are two constants of proportionality for the proportional relationship? What do they tell us about the situation?

3. On graph paper, create a graph that shows how the two quantities are related. Make sure to label the axes.

4. Write two equations that relate the quantities in your graph. Make sure to record what each variable represents.

# Lesson 19: Distinguishing Circumference and Area

Let's contrast circumference and area.

## 19.1: Filling the Plate

About how many cheese puffs can fit on the plate in a single layer? Be prepared to explain your reasoning.

## 19.2: Card Sort: Circle Problems

Your teacher will give you cards with questions about circles.

1. Sort the cards into two groups based on whether you would use the circumference or the area of the circle to answer the question. Pause here so your teacher can review your work.

2. Your teacher will assign you a card to examine more closely. What additional information would you need in order to answer the question on your card?

3. Estimate measurements for the circle on your card.

4. Use your estimates to calculate the answer to the question.

## 19.3: Visual Display of Circle Problem

In the previous activity you estimated the answer to a question about circles.

Create a visual display that includes:

- The question you were answering
- A diagram of a circle labeled with your estimated measurements
- Your thinking, organized so that others can follow it
- Your answer, expressed in terms of $\pi$ and also expressed as a decimal approximation

## 19.4: Analyzing Circle Claims

Here are two students' answers for each question. Do you agree with either of them? Explain or show your reasoning.

1. How many feet are traveled by a person riding once around the merry-go-round?

- Clare says, "The radius of the merry-go-round is about 4 feet, so the distance around the edge is about $8\pi$ feet."
- Andre says, "The diameter of the merry-go-round is about 4 feet, so the distance around the edge is about $4\pi$ feet."

2. How much room is there to spread frosting on the cookie?

- Clare says "The radius of the cookie is about 3 centimeters, so the space for frosting is about $6\pi$ cm$^2$."
- Andre says "The diameter of the cookie is about 3 inches, so the space for frosting is about $2.25\pi$ in$^2$."

3. How far does the unicycle move when the wheel makes 5 full rotations?

- Clare says, "The diameter of the unicycle wheel is about 0.5 meters. In 5 complete rotations it will go about $\frac{5}{2}\pi$ m$^2$."
- Andre says, "I agree with Clare's estimate of the diameter, but that means the unicycle will go about $\frac{5}{4}\pi$ m."

## Are you ready for more?

A goat (point $G$) is tied with a 6-foot rope to the corner of a shed. The floor of the shed is a square whose sides are each 3 feet long. The shed is closed and the goat can't go inside. The space all around the shed is flat, grassy, and the goat can't reach any other structures or objects. What is the area over which the goat can roam?

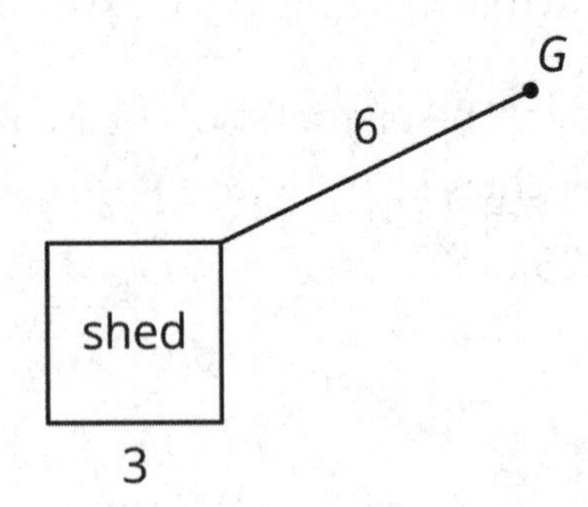

## Lesson 19 Summary

Sometimes we need to find the circumference of a circle, and sometimes we need to find the area. Here are some examples of quantities related to the circumference of a circle:

- The length of a circular path.
- The distance a wheel will travel after one complete rotation.
- The length of a piece of rope coiled in a circle.

Here are some examples of quantities related to the area of a circle:

- The amount of land that is cultivated on a circular field.
- The amount of frosting needed to cover the top of a round cake.
- The number of tiles needed to cover a round table.

In both cases, the radius (or diameter) of the circle is all that is needed to make the calculation. The circumference of a circle with radius $r$ is $2\pi r$ while its area is $\pi r^2$. The circumference is measured in linear units (such as cm, in, km) while the area is measured in square units (such as $cm^2$, $in^2$, $km^2$).

## Lesson 19 Practice Problems

1. For each problem, decide whether the circumference of the circle or the area of the circle is most useful for finding a solution. Explain your reasoning.

    a. A car's wheels spin at 1000 revolutions per minute. The diameter of the wheels is 23 inches. You want to know how fast the car is travelling.

    b. A circular kitchen table has a diameter of 60 inches. You want to know how much fabric is needed to cover the table top.

    c. A circular puzzle is 20 inches in diameter. All of the pieces are about the same size. You want to know about how many pieces there are in the puzzle.

    d. You want to know about how long it takes to walk around a circular pond.

2. The city of Paris, France is completely contained within an almost circular road that goes around the edge. Use the map with its scale to:

    a. Estimate the circumference of Paris.

    b. Estimate the area of Paris.

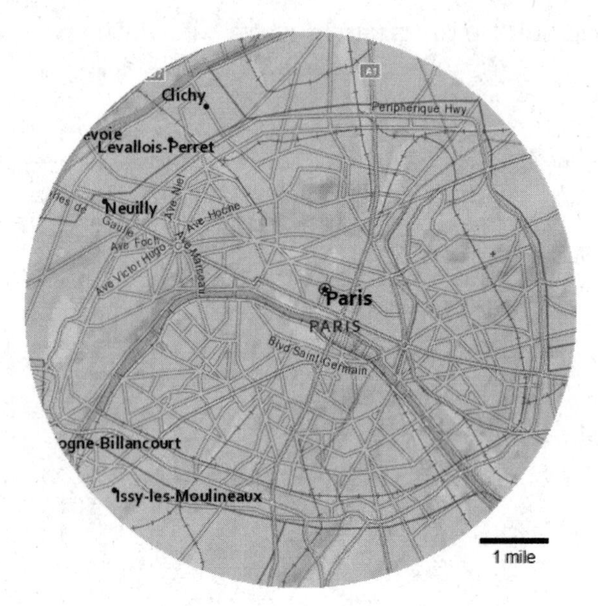

3. Here is a diagram of a softball field:

   a. About how long is the fence around the field?

   b. About how big is the outfield?

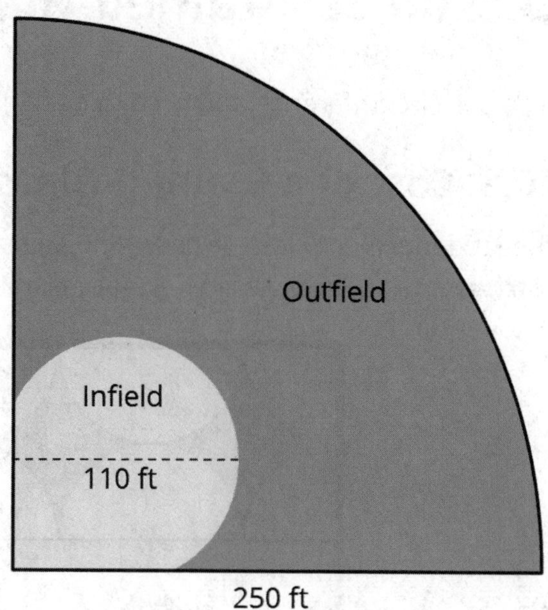


4. While in math class, Priya and Kiran come up with two ways of thinking about the proportional relationship shown in the table.

| $x$ | $y$ |
|---|---|
| 2 | ? |
| 5 | 1750 |

Both students agree that they can solve the equation $5k = 1750$ to find the constant of proportionality.

- Priya says, "I can solve this equation by dividing 1750 by 5."
- Kiran says, "I can solve this equation by multiplying 1750 by $\frac{1}{5}$."

   a. What value of $k$ would each student get using their own method?

   b. How are Priya and Kiran's approaches related?

   c. Explain how each student might approach solving the equation $\frac{2}{3}k = 50$.

(From Unit 5, Lesson 2.)

# Lesson 20: Stained-Glass Windows

Let's use circumference and area to design stained-glass windows.

## 20.1: Cost of a Stained-Glass Window

The students in art class are designing a stained-glass window to hang in the school entryway. The window will be 3 feet tall and 4 feet wide. Here is their design.

They have raised $100 for the project. The colored glass costs $5 per square foot and the clear glass costs $2 per square foot. The material they need to join the pieces of glass together costs 10 cents per foot and the frame around the window costs $4 per foot.

Do they have enough money to cover the cost of making the window?

## 20.2: A Bigger Window

A local community member sees the school's stained-glass window and really likes the design. They ask the students to create a larger copy of the window using a scale factor of 3. Would $450 be enough to buy the materials for the larger window? Explain or show your reasoning.

## 20.3: Invent Your Own Design

Draw a stained-glass window design that could be made for less than $450. Show your thinking. Organize your work so it can be followed by others.

# Learning Targets

### Lesson 1: Proportional Relationships and Equations

- I can write an equation of the form $y = kx$ to represent a proportional relationship described by a table or a story.
- I can write the constant of proportionality as an entry in a table.

### Lesson 2: Two Equations for Each Relationship

- I can find two constants of proportionality for a proportional relationship.
- I can write two equations representing a proportional relationship described by a table or story.

### Lesson 3: Using Equations to Solve Problems

- I can find missing information in a proportional relationship using the constant of proportionality.
- I can relate all parts of an equation like $y = kx$ to the situation it represents.

### Lesson 4: Comparing Relationships with Tables

- I can decide if a relationship represented by a table could be proportional and when it is definitely not proportional.

### Lesson 5: Comparing Relationships with Equations

- I can decide if a relationship represented by an equation is proportional or not.

### Lesson 6: Solving Problems about Proportional Relationships

- I can ask questions about a situation to determine whether two quantities are in a proportional relationship.
- I can solve all kinds of problems involving proportional relationships.

### Lesson 7: Graphs of Proportional Relationships

- I can find the constant of proportionality from a graph.
- I know that the graph of a proportional relationship lies on a line through $(0, 0)$.

### Lesson 8: Using Graphs to Compare Relationships

- I can compare two, related proportional relationships based on their graphs.
- I know that the steeper graph of two proportional relationships has a larger constant of proportionality.

### Lesson 9: Two Graphs for Each Relationship

- I can interpret a graph of a proportional relationship using the situation.
- I can write an equation representing a proportional relationship from a graph.

### Lesson 10: How Well Can You Measure?

- I can examine quotients and use a graph to decide whether two associated quantities are in a proportional relationship.
- I understand that it can be difficult to measure the quantities in a proportional relationship accurately.

### Lesson 11: Exploring Circles

- I can describe the characteristics that make a shape a circle.
- I can identify the diameter, center, radius, and circumference of a circle.

### Lesson 12: Exploring Circumference

- I can describe the relationship between circumference and diameter of any circle.
- I can explain what $\pi$ means.

### Lesson 13: Applying Circumference

- I can choose an approximation for $\pi$ based on the situation or problem.
- If I know the radius, diameter, or circumference of a circle, I can find the other two.

### Lesson 14: Estimating Areas

- I can calculate the area of a complicated shape by breaking it into shapes whose area I know how to calculate.

**Lesson 15: Area of a Circle**

- I know the formula for area of a circle.
- I know whether or not the relationship between the diameter and area of a circle is proportional and can explain how I know.

**Lesson 16: Applying Area of Circles**

- I can calculate the area of more complicated shapes that include fractions of circles.
- I can write exact answers in terms of $\pi$.

**Lesson 17: Four Representations**

- I can make connections between the graphs, tables, and equations of a proportional relationship.
- I can use units to help me understand information about proportional relationships.

**Lesson 18: Using Water Efficiently**

- I can answer a question by representing a situation using proportional relationships.

**Lesson 19: Distinguishing Circumference and Area**

- I can decide whether a situation about a circle has to do with area or circumference.
- I can use formulas for circumference and area of a circle to solve problems.

**Lesson 20: Stained-Glass Windows**

- I can apply my understanding of area and circumference of circles to solve more complicated problems.

# Illustrative Mathematics

# Unit 6

STUDENT EDITION

Book 2

# Lesson 1: Half as Much Again

Let's use fractions to describe increases and decreases.

## 1.1: Notice and Wonder: Tape Diagrams

What do you notice? What do you wonder?

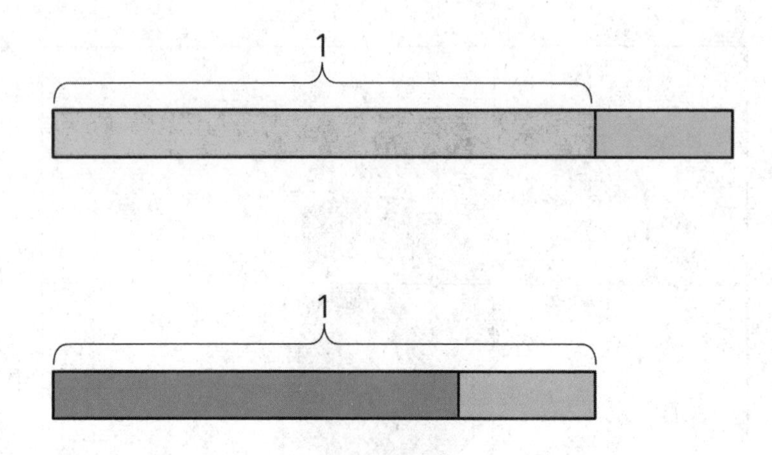


## 1.2: Walking Half as Much Again

1. Complete the table to show the total distance walked in each case.

    a. Jada's pet turtle walked 10 feet, and then half that length again.

    b. Jada's baby brother walked 3 feet, and then half that length again.

    c. Jada's hamster walked 4.5 feet, and then half that length again.

    d. Jada's robot walked 1 foot, and then half that length again.

    e. A person walked $x$ feet and then half that length again.

| initial distance | total distance |
|---|---|
| 10 | 15 |
| 3 | 4.5 |
| 4.5 | 6.75 |
| 1 | 1.5 |
| $x$ | $\frac{x}{2}+x$ |



iM

2. Explain how you computed the total distance in each case.

3. Two students each wrote an equation to represent the relationship between the initial distance walked ($x$) and the total distance walked ($y$).

   - Mai wrote $y = x + \frac{1}{2}x$.
   - Kiran wrote $y = \frac{3}{2}x$.

   Do you agree with either of them? Explain your reasoning.

**Are you ready for more?**

Zeno jumped 8 meters. Then he jumped half as far again (4 meters). Then he jumped half as far again (2 meters). So after 3 jumps, he was $8 + 4 + 2 = 14$ meters from his starting place.

1. Zeno kept jumping half as far again. How far would he be after 4 jumps? 5 jumps? 6 jumps?

2. Before he started jumping, Zeno put a mark on the floor that was exactly 16 meters from his starting place. How close can Zeno get to the mark if he keeps jumping half as far again?

3. If you enjoyed thinking about this problem, consider researching Zeno's Paradox.

## 1.3: More and Less

1. Match each situation with a diagram. A diagram may not have a match.

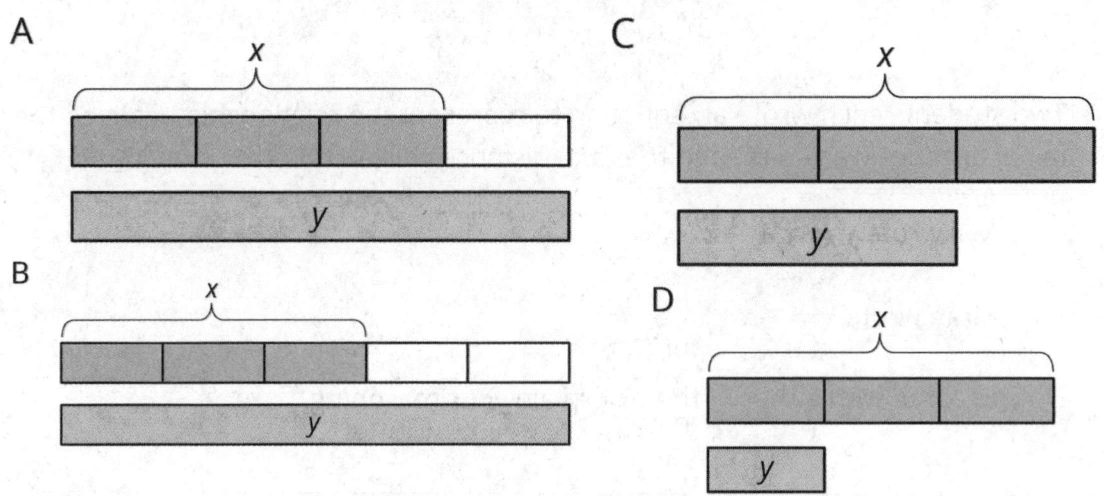


C ○ Han ate $x$ ounces of blueberries. Mai ate $\frac{1}{3}$ less than that.

B ○ Mai biked $x$ miles. Han biked $\frac{2}{3}$ more than that.

C ○ Han bought $x$ pounds of apples. Mai bought $\frac{2}{3}$ of that.

2. For each diagram, write an equation that represents the relationship between $x$ and $y$.

    a. Diagram A: $Y = \frac{4}{3}x$

    b. Diagram B: $Y = \frac{5}{3}x$

    c. Diagram C: $y = \frac{2}{3}x$

    d. Diagram D: $y = \frac{1}{3}x$

3. Write a story for one of the diagrams that doesn't have a match.

## 1.4: Card Sort: Representations of Proportional Relationships

Your teacher will give you a set of cards that have proportional relationships represented three different ways: as descriptions, equations, and tables. Mix up the cards and place them all face-up.

1. Take turns with a partner to match a description with an equation and a table.
   a. For each match you find, explain to your partner how you know it's a match.
   b. For each match your partner finds, listen carefully to their explanation, and if you disagree, explain your thinking.
2. When you agree on all of the matches, check your answers with the answer key. If there are any errors, discuss why and revise your matches.

### Lesson 1 Summary

Using the distributive property provides a shortcut for calculating the final amount in situations that involve adding or subtracting a fraction of the original amount.

For example, one day Clare runs 4 miles. The next day, she plans to run that same distance plus half as much again. How far does she plan to run the next day?

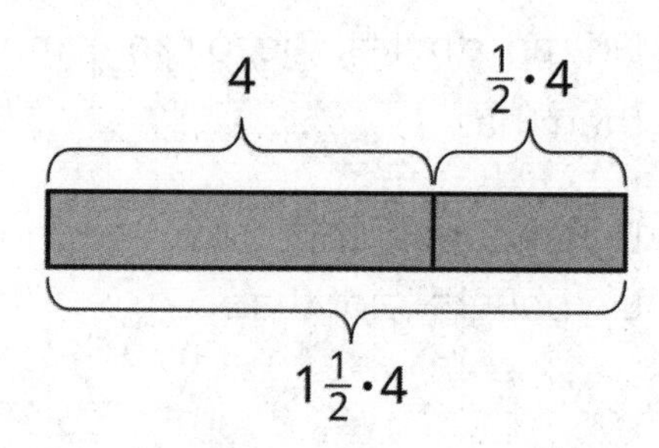


Tomorrow she will run 4 miles plus $\frac{1}{2}$ of 4 miles. We can use the distributive property to find this in one step: $1 \cdot 4 + \frac{1}{2} \cdot 4 = \left(1 + \frac{1}{2}\right) \cdot 4$

Clare plans to run $1\frac{1}{2} \cdot 4$, or 6 miles.

This works when we decrease by a fraction, too. If Tyler spent $x$ dollars on a new shirt, and Noah spent $\frac{1}{3}$ less than Tyler, then Noah spent $\frac{2}{3}x$ dollars since $x - \frac{1}{3}x = \frac{2}{3}x$.

### Glossary

- tape diagram

## Lesson 1 Practice Problems

1. Match each situation with a diagram.

A

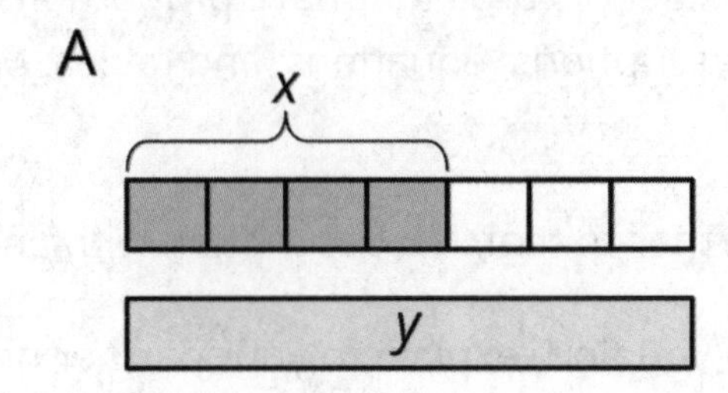


B

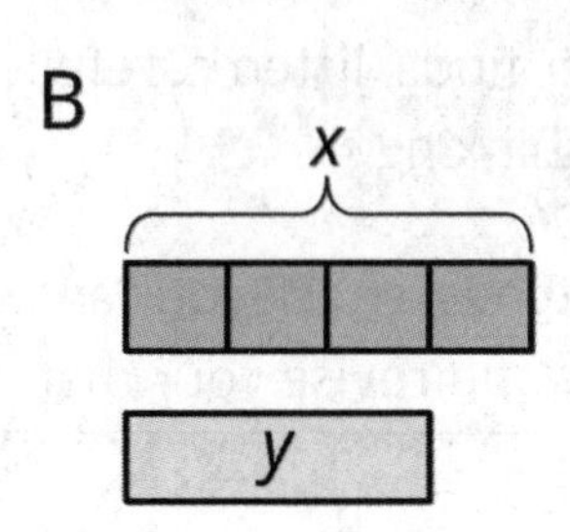


C

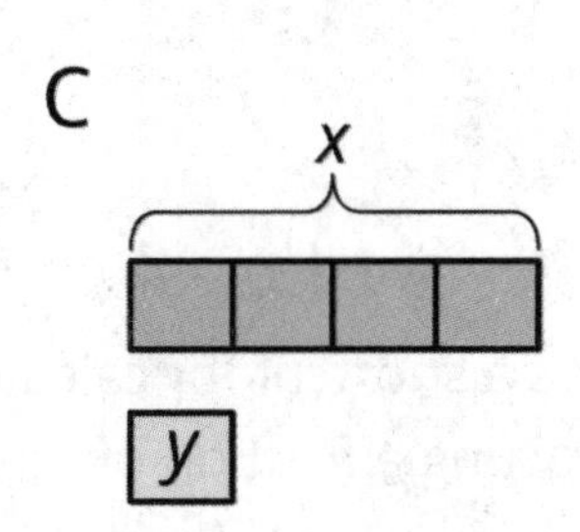


A. Diagram A

B. Diagram B

C. Diagram C

1. Diego drank $x$ ounces of juice. Lin drank $\frac{1}{4}$ less than that.
2. Lin ran $x$ miles. Diego ran $\frac{3}{4}$ more than that.
3. Diego bought $x$ pounds of almonds. Lin bought $\frac{1}{4}$ of that.

2. Elena walked 12 miles. Then she walked $\frac{1}{4}$ that distance. How far did she walk all together? Select **all** that apply.

A. $12 + \frac{1}{4}$

B. $12 \cdot \frac{1}{4}$

C. $12 + \frac{1}{4} \cdot 12$

D. $12\left(1 + \frac{1}{4}\right)$

E. $12 \cdot \frac{3}{4}$

F. $12 \cdot \frac{5}{4}$

3. Write a story that can be represented by the equation $y = x + \frac{1}{4}x$.

4. Select **all** ratios that are equivalent to $4 : 5$.

A. $2 : 2.5$

B. $2 : 3$

C. $3 : 3.75$

D. $7 : 8$

E. $8 : 10$

F. $14 : 27.5$

5. Jada is making circular birthday invitations for her friends. The diameter of the circle is 12 cm. She bought 180 cm of ribbon to glue around the edge of each invitation. How many invitations can she make?

(From Unit 5, Lesson 19.)

# Lesson 2: Say It with Decimals

Let's use decimals to describe increases and decreases.

## 2.1: Notice and Wonder: Fractions to Decimals

A calculator gives the following decimal representations for some unit fractions:

$\frac{1}{2} = 0.5$

$\frac{1}{3} = 0.3333333$

$\frac{1}{4} = 0.25$

$\frac{1}{5} = 0.2$

$\frac{1}{6} = 0.1666667$

$\frac{1}{7} = 0.142857143$

$\frac{1}{8} = 0.125$

$\frac{1}{9} = 0.1111111$

$\frac{1}{10} = 0.1$

$\frac{1}{11} = 0.0909091$

What do you notice? What do you wonder?

## 2.2: Repeating Decimals

1. Use **long division** to express each fraction as a decimal.

$\frac{9}{25}$ $\qquad$ $\frac{11}{30}$ $\qquad$ $\frac{4}{11}$

2. What is similar about your answers to the previous question? What is different?

3. Use the decimal representations to decide which of these fractions has the greatest value. Explain your reasoning.

### Are you ready for more?

One common approximation for $\pi$ is $\frac{22}{7}$. Express this fraction as a decimal. How does this approximation compare to 3.14?

## 2.3: More and Less with Decimals

1. Match each diagram with a description and an equation.

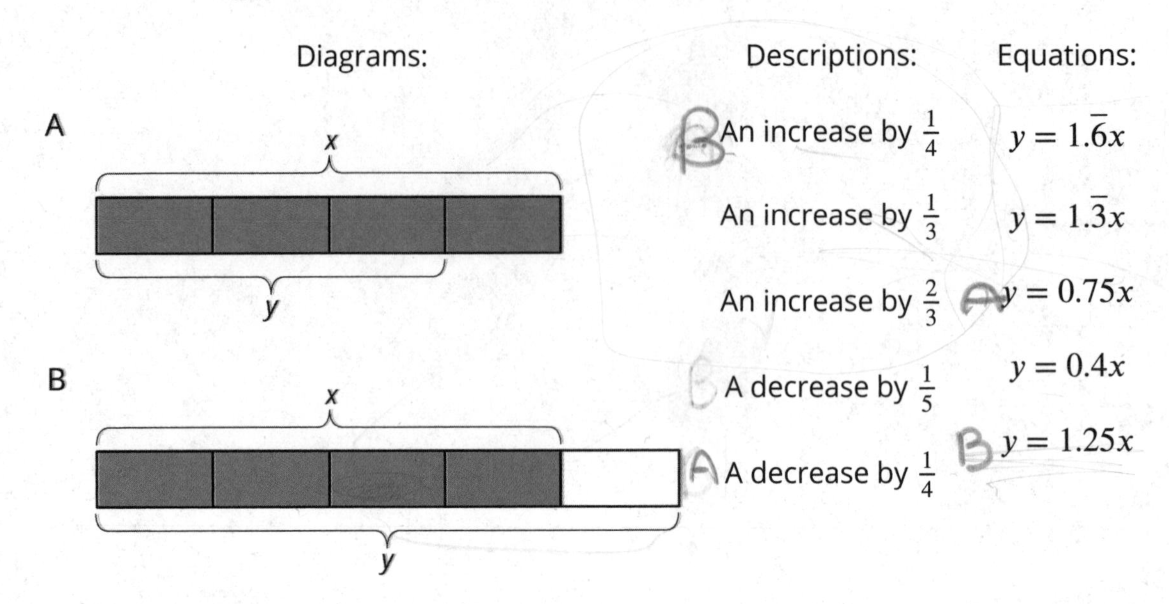


2. Draw a diagram for one of the unmatched equations.

## 2.4: Card Sort: More Representations

Your teacher will give you a set of cards that have proportional relationships represented 2 different ways: as descriptions and equations. Mix up the cards and place them all face-up.

Take turns with a partner to match a description with an equation.

1. For each match you find, explain to your partner how you know it's a match.
2. For each match your partner finds, listen carefully to their explanation, and if you disagree, explain your thinking.
3. When you have agreed on all of the matches, check your answers with the answer key. If there are any errors, discuss why and revise your matches.

## Lesson 2 Summary

**Long division** gives us a way of finding decimal representations for fractions.

For example, to find a decimal representation for $\frac{9}{8}$, we can divide 9 by 8.

$$\begin{array}{r} 1.125 \\ 8\overline{)9.000} \\ \underline{8\phantom{.000}} \\ 1\,0\phantom{00} \\ \underline{8\phantom{00}} \\ 20\phantom{0} \\ \underline{16\phantom{0}} \\ 40 \\ \underline{40} \\ 0 \end{array}$$

So $\frac{9}{8} = 1.125$.

Sometimes it is easier to work with the decimal representation of a number, and sometimes it is easier to work with its fraction representation. It is important to be able to work with both. For example, consider the following pair of problems:

- Priya earned $x$ dollars doing chores, and Kiran earned $\frac{6}{5}$ as much as Priya. How much did Kiran earn?
- Priya earned $x$ dollars doing chores, and Kiran earned 1.2 times as much as Priya. How much did Kiran earn?

Since $\frac{6}{5} = 1.2$, these are both exactly the same problem, and the answer is $\frac{6}{5}x$ or $1.2x$. When we work with percentages in later lessons, the decimal representation will come in especially handy.

## Glossary

- long division
- repeating decimal

## Lesson 2 Practice Problems

1. a. Match each diagram with a description and an equation.

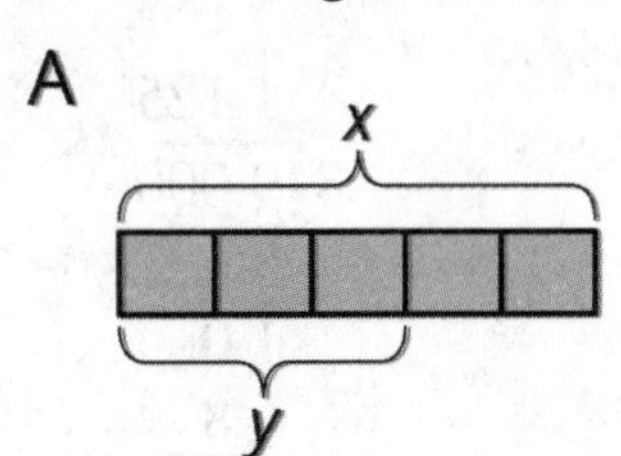


B

x

y

Descriptions:

An increase by $\frac{2}{3}$

An increase by $\frac{5}{6}$

A decrease by $\frac{2}{5}$

A decrease by $\frac{5}{11}$

Equations:

$y = 1.8\overline{3}x$

$y = 1.\overline{6}x$

$y = 0.6x$

$y = 0.4x$

b. Draw a diagram for one of the unmatched equations.

2. At the beginning of the month, there were 80 ounces of peanut butter in the pantry. Since then, the family ate 0.3 of the peanut butter. How many ounces of peanut butter are in the pantry now?

A. $0.7 \cdot 80$

B. $0.3 \cdot 80$

C. $80 - 0.3$

D. $(1 + 0.3) \cdot 80$

3. a. On a hot day, a football team drank an entire 50-gallon cooler of water and half as much again. How much water did they drink?

   b. Jada has 12 library books checked out and Han has $\frac{1}{3}$ less than that. How many books does Han have checked out?

(From Unit 6, Lesson 1.)

4. If $x$ represents a positive number, select **all** expressions whose value is greater than $x$.

   A. $\left(1 - \frac{1}{4}\right) x$

   B. $\left(1 + \frac{1}{4}\right) x$

   C. $\frac{7}{8}x$

   D. $\frac{9}{8}x$

(From Unit 6, Lesson 1.)

5. A person's resting heart rate is typically between 60 and 100 beats per minute. Noah looks at his watch, and counts 8 heartbeats in 10 seconds.

   a. Is his heart rate typical? Explain how you know.

   b. Write an equation for $h$, the number of times Noah's heart beats (at this rate) in $m$ minutes.

(From Unit 5, Lesson 3.)

# Lesson 3: Increasing and Decreasing

Let's use percentages to describe increases and decreases.

## 3.1: Improving Their Game

Here are the scores from 3 different sports teams from their last 2 games.

| sports team | total points in game 1 | total points in game 2 |
|---|---|---|
| football team | 22 | 30 |
| basketball team | 100 | 108 |
| baseball team | 4 | 12 |

1. What do you notice about the teams' scores? What do you wonder?

2. Which team improved the most? Explain your reasoning.

## 3.2: More Cereal and a Discounted Shirt

1. A cereal box says that now it contains 20% more. Originally, it came with 18.5 ounces of cereal. How much cereal does the box come with now?

18.5 = 100 $\frac{18.5}{\text{whole}}$

18.5 + .2(18.5)

22.2oz

whole + 20% whole

20% MORE

2. The price of a shirt is $18.50, but you have a coupon that lowers the price by 20%. What is the price of the shirt after using the coupon?

18.50 = 100%
− 20%
80%

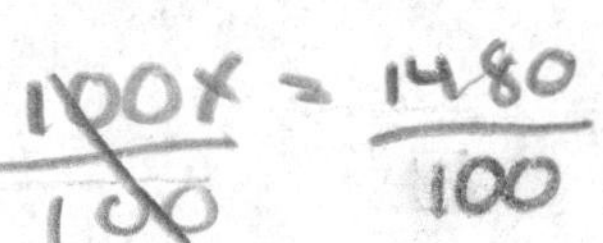


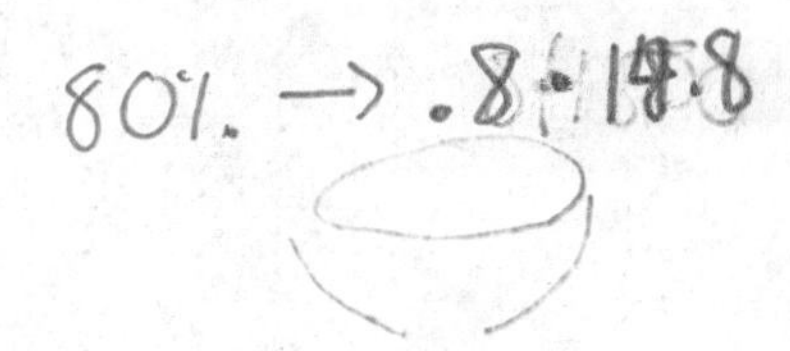


x = $14.80

## 3.3: Using Tape Diagrams

1. Match each situation to a diagram. Be prepared to explain your reasoning.

   a. Compared with last year's strawberry harvest, this year's strawberry harvest is a 25% increase.

   b. This year's blueberry harvest is 75% of last year's.

   c. Compared with last year, this year's peach harvest decreased 25%.

   d. This year's plum harvest is 125% of last year's plum harvest.

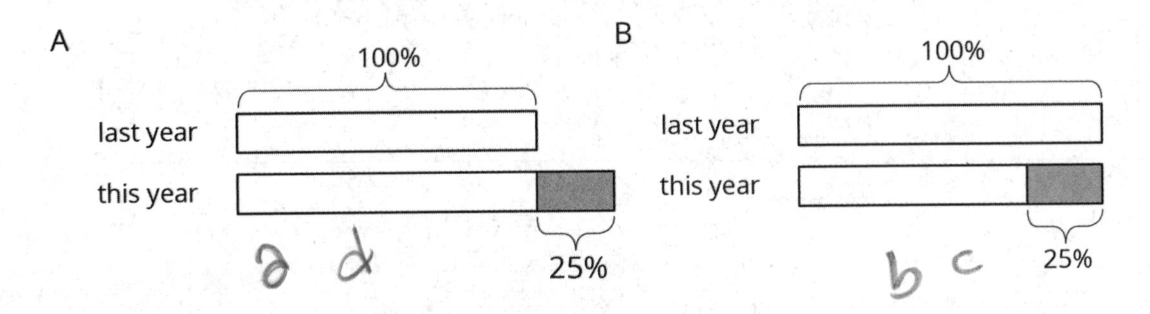


2. Draw a diagram to represent these situations.

   a. The number of ducks living at the pond increased by 40%.

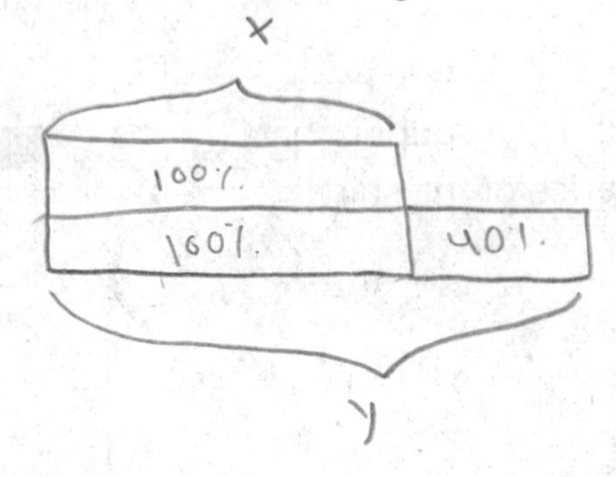


   b. The number of mosquitoes decreased by 80%.

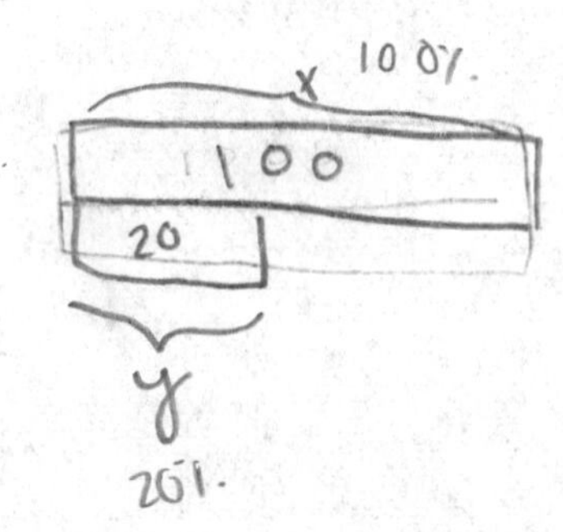

## Are you ready for more?

What could it mean to say there is a 100% decrease in a quantity? Give an example of a quantity where this makes sense.

# 3.4: Agree or Disagree: Percentages

Do you agree or disagree with each statement? Explain your reasoning.

1. Employee A gets a pay raise of 50%. Employee B gets a pay raise of 45%. So Employee A gets the bigger pay raise.

I agree if Employees A and B make the same amount of money.

$50 > 45$

2. Shirts are on sale for 20% off. You buy two of them. As you pay, the cashier says, "20% off of each shirt means 40% off of the total price."

I disagree because we don't add

## Lesson 3 Summary

Imagine that it takes Andre $\frac{3}{4}$ more than the time it takes Jada to get to school. Then we know that Andre's time is $1\frac{3}{4}$ or 1.75 times Jada's time. We can also describe this in terms of percentages:

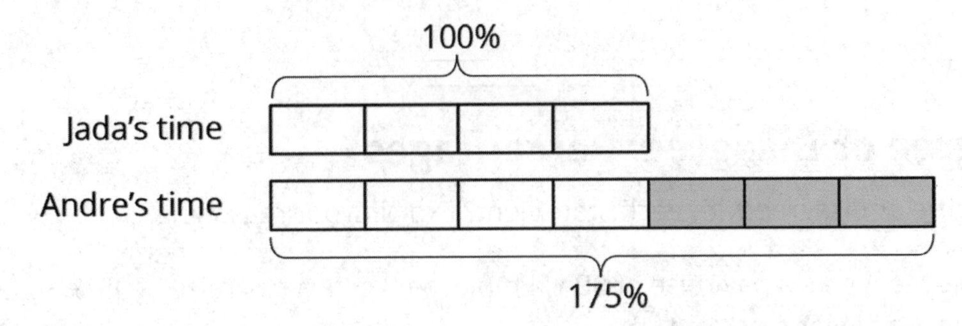


We say that Andre's time is 75% more than Jada's time. We can also see that Andre's time is 175% of Jada's time. In general, the terms **percent increase** and **percent decrease** describe an increase or decrease in a quantity as a percentage of the starting amount.

For example, if there were 500 grams of cereal in the original package, then "20% more" means that 20% of 500 grams has been added to the initial amount, $500 + (0.2) \cdot 500 = 600$, so there are 600 grams of cereal in the new package.

We can see that the new amount is 120% of the initial amount because

$$500 + (0.2) \cdot 500 = (1 + 0.2)500$$

100% 20%

120%

### Glossary

- percentage decrease
- percentage increase

## Lesson 3 Practice Problems

1. For each diagram, decide if $y$ is an increase or a decrease relative to $x$. Then determine the percent increase or decrease.

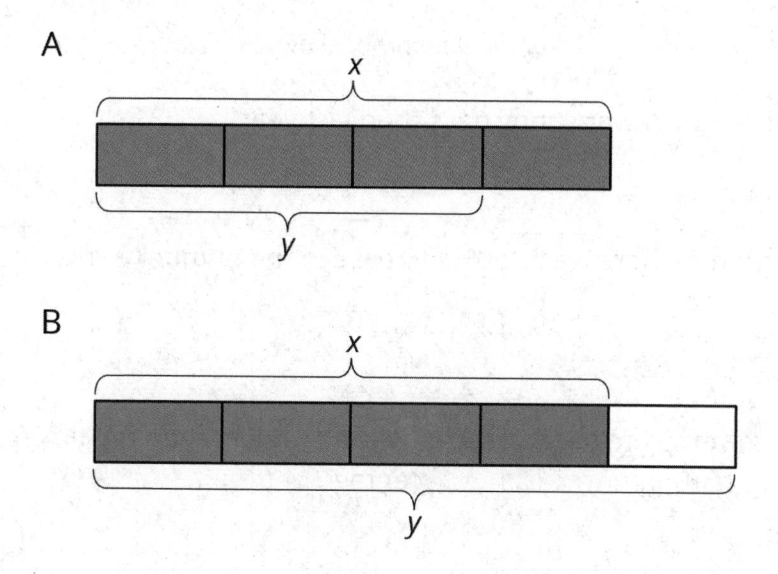


2. Draw diagrams to represent the following situations.

   a. The amount of flour that the bakery used this month was a 50% increase relative to last month.

   b. The amount of milk that the bakery used this month was a 75% decrease relative to last month.

3. Write each percent increase or decrease as a percentage of the initial amount. The first one is done for you.

    a. This year, there was 40% more snow than last year.

    *The amount of snow this year is 140% of the amount of snow last year.*

    b. This year, there were 25% fewer sunny days than last year.

    c. Compared to last month, there was a 50% increase in the number of houses sold this month.

    d. The runner's time to complete the marathon was a 10% less than the time to complete the last marathon.

4. The graph shows the relationship between the diameter and the circumference of a circle with the point $(1, \pi)$ shown. Find 3 more points that are on the line.

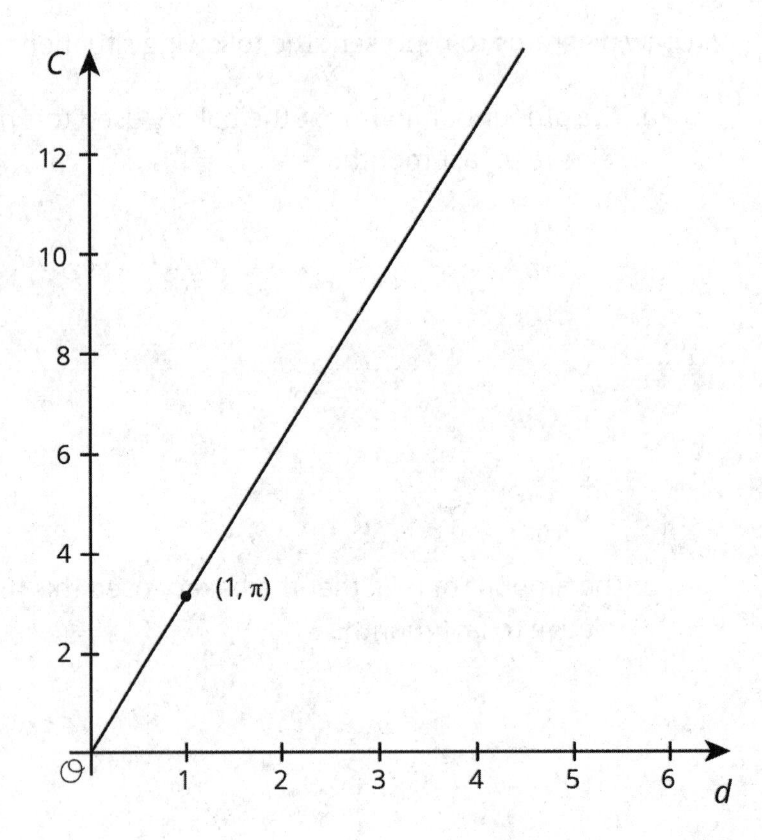


(From Unit 5, Lesson 12.)

5. Priya bought $x$ grams of flour. Clare bought $\frac{3}{8}$ more than that. Select **all** equations that represent the relationship between the amount of flour that Priya bought, $x$, and the amount of flour that Clare bought, $y$.

A. $y = \frac{3}{8}x$

B. $y = \frac{5}{8}x$

C. $y = x + \frac{3}{8}x$

D. $y = x - \frac{3}{8}x$

E. $y = \frac{11}{8}x$

(From Unit 6, Lesson 1.)

# Lesson 4: One Hundred Percent

Let's solve more problems about percent increase and percent decrease.

## 4.1: Notice and Wonder: Double Number Line

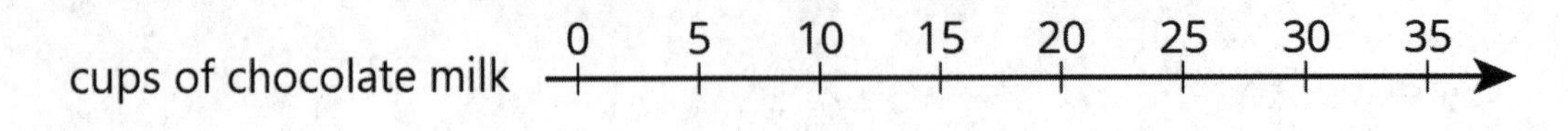


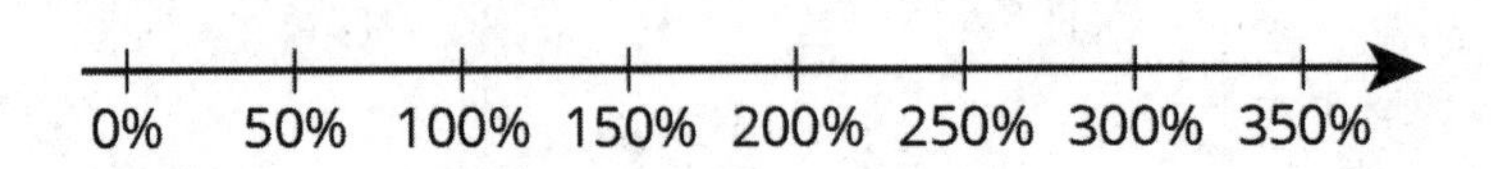


What do you notice? What do you wonder?

## 4.2: Double Number Lines

For each problem, complete the double number line diagram to show the percentages that correspond to the original amount and to the new amount.

1. The gas tank in dad's car holds 12 gallons. The gas tank in mom's truck holds 50% more than that. How much gas does the truck's tank hold?

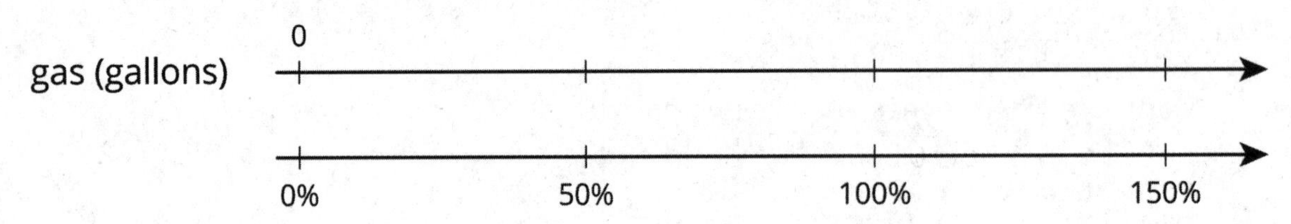


2. At a movie theater, the size of popcorn bags decreased 20%. If the old bags held 15 cups of popcorn, how much do the new bags hold?

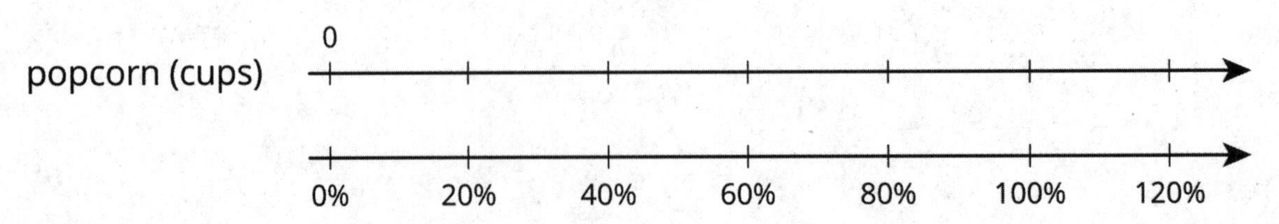

3. A school had 1,200 students last year and only 1,080 students this year. What was the percentage decrease in the number of students?

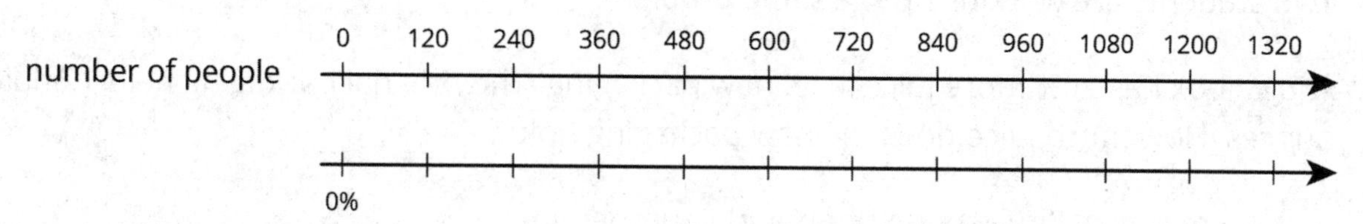


4. One week gas was $1.25 per gallon. The next week gas was $1.50 per gallon. By what percentage did the price increase?

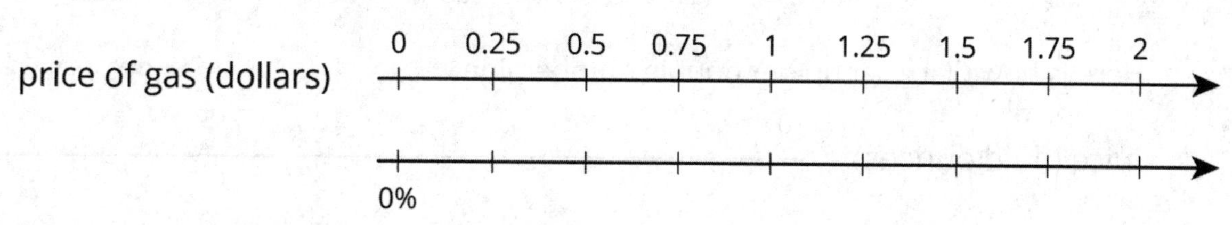


5. After a 25% discount, the price of a T-shirt was $12. What was the price before the discount?

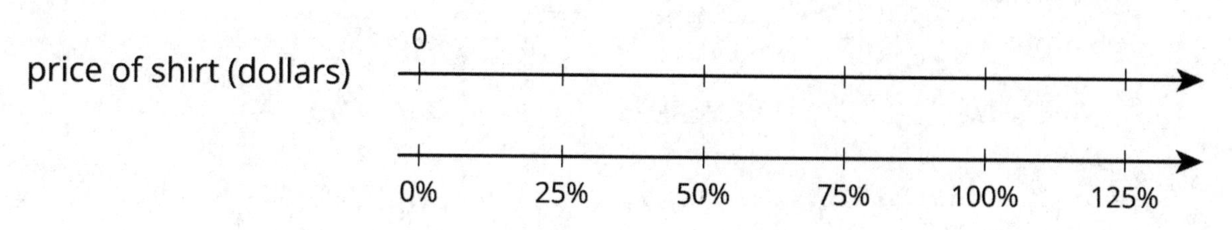


6. Compared to last year, the population of Boom Town has increased 25%.The population is now 6,600. What was the population last year?

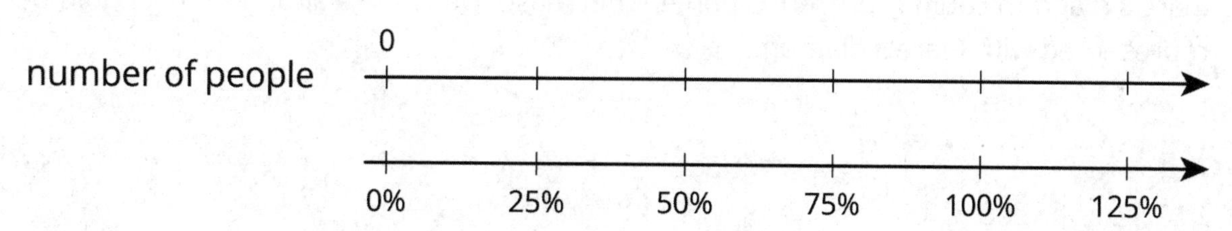

## 4.3: Representing More Juice

Two students are working on the same problem:

A juice box has 20% more juice in its new packaging. The original packaging held 12 fluid ounces. How much juice does the new packaging hold?

- Here is how Priya set up her double number line.

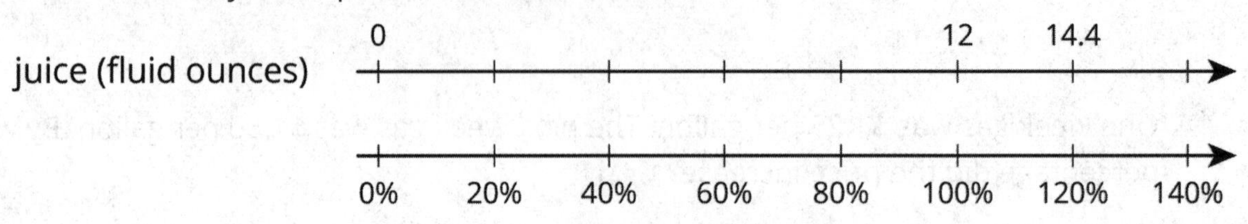


- Here is how Clare set up her double number line.

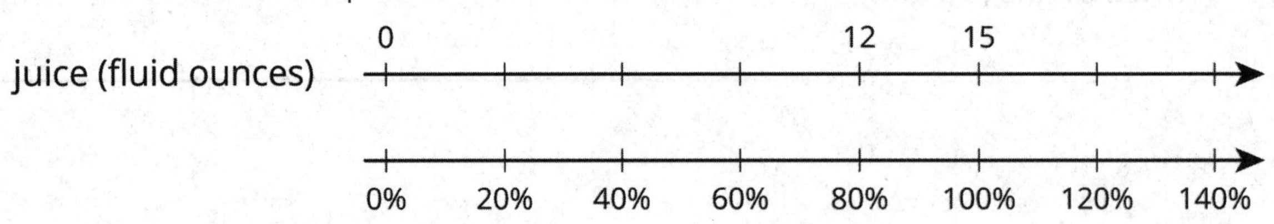


Do you agree with either of them? Explain or show your reasoning.

### Are you ready for more?

Clare's diagram could represent a percent decrease. Describe a situation that could be represented with Clare's diagram.

## 4.4: Protecting the Green Sea Turtle

Green sea turtles live most of their lives in the ocean, but come ashore to lay their eggs. Some beaches where turtles often come ashore have been made into protected sanctuaries so the eggs will not be disturbed.

1. One sanctuary had 180 green sea turtles come ashore to lay eggs last year. This year, the number of turtles increased by 10%. How many turtles came ashore to lay eggs in the sanctuary this year?

2. At another sanctuary, the number of nesting turtles decreased by 10%. This year there were 234 nesting turtles. How many nesting turtles were at this sanctuary last year?

## Lesson 4 Summary

We can use a double number line diagram to show information about percent increase and percent decrease:

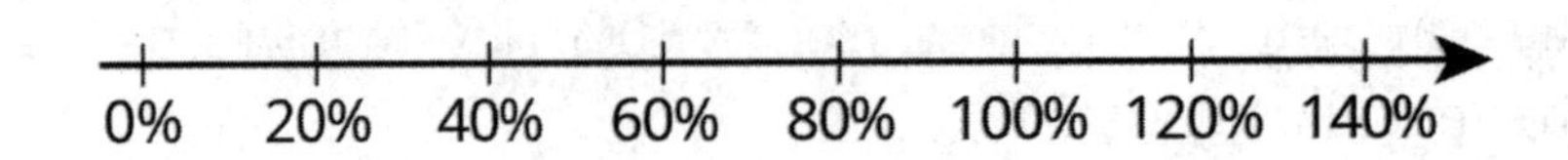


The initial amount of cereal is 500 grams, which is lined up with 100% in the diagram. We can find a 20% *increase* to 500 by adding 20% of 500:

$$500 + (0.2) \cdot 500 = (1.20) \cdot 500$$
$$= 600$$

In the diagram, we can see that 600 corresponds to 120%.

If the initial amount of 500 grams is *decreased* by 40%, we can find how much cereal there is by subtracting 40% of the 500 grams:

$$500 - (0.4) \cdot 500 = (0.6) \cdot 500$$
$$= 300$$

So a 40% decrease is the same as 60% of the initial amount. In the diagram, we can see that 300 is lined up with 60%.

To solve percentage problems, we need to be clear about what corresponds to 100%. For example, suppose there are 20 students in a class, and we know this is an increase of 25% from last year. In this case, the number of students in the class *last* year corresponds to 100%. So the initial amount (100%) is unknown and the final amount (125%) is 20 students.

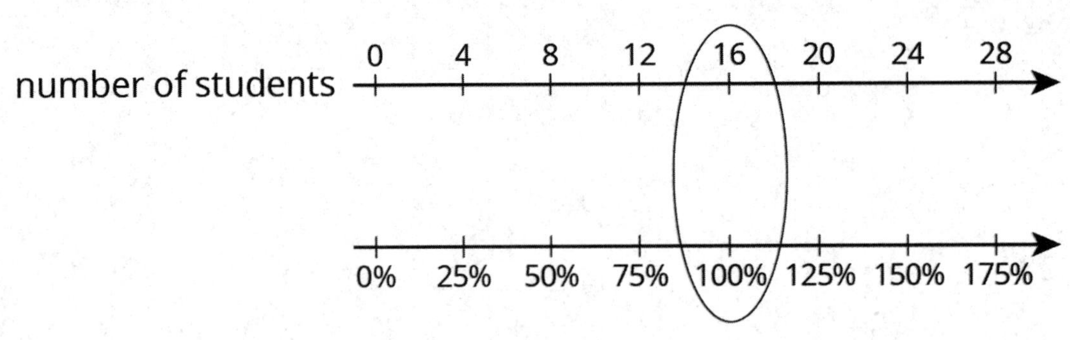


Looking at the double number line, if 20 students is a 25% increase from the previous year, then there were 16 students in the class last year.

## Lesson 4 Practice Problems

1. A bakery used 25% more butter this month than last month. If the bakery used 240 kilograms of butter last month, how much did it use this month?

2. Last week, the price of oranges at the farmer's market was $1.75 per pound. This week, the price has decreased by 20%. What is the price of oranges this week?

3. Noah thinks the answers to these two questions will be the same. Do you agree with him? Explain your reasoning.

    - This year, a herd of bison had a 10% increase in population. If there were 550 bison in the herd last year, how many are in the herd this year?

    - This year, another herd of bison had a 10% decrease in population. If there are 550 bison in the herd this year, how many bison were there last year?

4. Elena walked 12 miles. Then she walked 0.25 that distance. How far did she walk all together? Select **all** that apply.

   A. $12 + 0.25 \cdot 12$

   B. $12(1 + 0.25)$

   C. $12 \cdot 1.25$

   D. $12 \cdot 0.25$

   E. $12 + 0.25$

   (From Unit 6, Lesson 2.)

5. A circle's circumference is 600 m. What is a good approximation of the circle's area?

   A. 300 m$^2$

   B. 3,000 m$^2$

   C. 30,000 m$^2$

   D. 300,000 m$^2$

   (From Unit 5, Lesson 15.)

6. The equation $d = 3t$ represents the relationship between the distance ($d$) in inches that a snail is from a certain rock and the time ($t$) in minutes.

   a. What does the number 3 represent?

   b. How many minutes does it take the snail to get 9 inches from the rock?

   c. How far will the snail be from the rock after 9 minutes?

   (From Unit 5, Lesson 3.)

# Lesson 5: Percent Increase and Decrease with Equations

Let's use equations to represent increases and decreases.

## 5.1: Number Talk: From 100 to 106

How do you get from one number to the next using multiplication or division?

From 100 to 106

From 100 to 90

From 90 to 100

From 106 to 100

## 5.2: Interest and Depreciation

1. Money in a particular savings account increases by about 6% after a year. How much money will be in the account after one year if the initial amount is \$100? \$50? \$200? \$125? $x$ dollars? If you get stuck, consider using diagrams or a table to organize your work.

2. The value of a new car decreases by about 15% in the first year. How much will a car be worth after one year if its initial value was \$1,000? \$5,000? \$5,020? $x$ dollars? If you get stuck, consider using diagrams or a table to organize your work.

## 5.3: Matching Equations

Match an equation to each of these situations. Be prepared to share your reasoning.

1. The water level in a reservoir is now 52 meters. If this was a 23% increase, what was the initial depth?
2. The snow is now 52 inches deep. If this was a 77% decrease, what was the initial depth?

$0.23x = 52$

$0.77x = 52$

$1.23x = 52$

$1.77x = 52$

**Are you ready for more?**

An astronaut was exploring the moon of a distant planet, and found some glowing goo at the bottom of a very deep crater. She brought a 10-gram sample of the goo to her laboratory. She found that when the goo was exposed to light, the total amount of goo increased by 100% every hour.

1. How much goo will she have after 1 hour? After 2 hours? After 3 hours? After $n$ hours?

2. When she put the goo in the dark, it shrank by 75% every hour. How many hours will it take for the goo that was exposed to light for $n$ hours to return to the original size?

## 5.4: Representing Percent Increase and Decrease: Equations

1. The gas tank in dad's car holds 12 gallons. The gas tank in mom's truck holds 50% more than that. How much gas does the truck's tank hold? Explain why this situation can be represented by the equation $(1.5) \cdot 12 = t$. Make sure that you explain what $t$ represents.

2. Write an equation to represent each of the following situations.

   a. A movie theater decreased the size of its popcorn bags by 20%. If the old bags held 15 cups of popcorn, how much do the new bags hold?

   b. After a 25% discount, the price of a T-shirt was $12. What was the price before the discount?

   c. Compared to last year, the population of Boom Town has increased by 25%.The population is now 6,600. What was the population last year?

## Lesson 5 Summary

We can use equations to express percent increase and percent decrease. For example, if $y$ is 15% more than $x$,

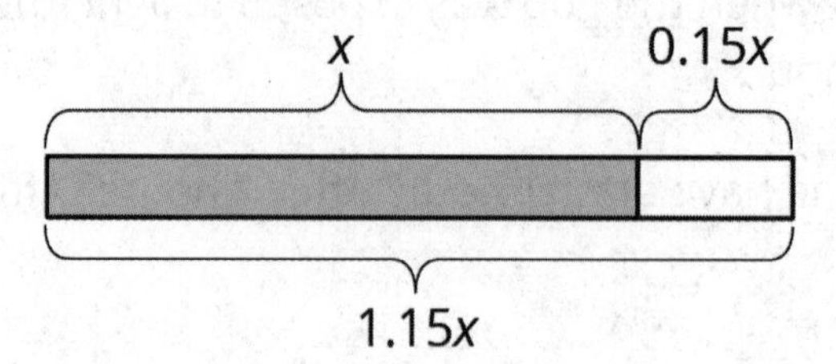


we can represent this using any of these equations:

$y = x + 0.15x$ $\qquad$ $y = (1 + 0.15)x$ $\qquad$ $y = 1.15x$

So if someone makes an investment of $x$ dollars, and its value increases by 15% to \$1250, then we can write and solve the equation $1.15x = 1250$ to find the value of the initial investment.

Here is another example: if $a$ is 7% less than $b$,

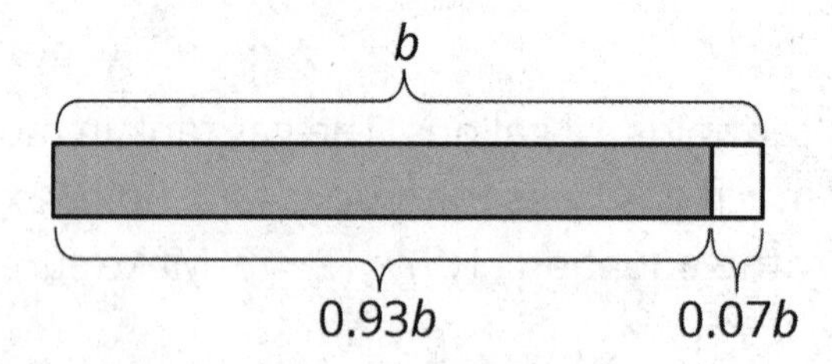


we can represent this using any of these equations:

$a = b - 0.07b$ $\qquad$ $a = (1 - 0.07)b$ $\qquad$ $a = 0.93b$

So if the amount of water in a tank decreased 7% from its starting value of $b$ to its ending value of 348 gallons, then you can write $0.93b = 348$.

Often, an equation is the most efficient way to solve a problem involving percent increase or percent decrease.

# Lesson 5 Practice Problems

1. A pair of designer sneakers was purchased for $120. Since they were purchased, their price has increased by 15%. What is the new price?

2. Elena's aunt bought her a $150 savings bond when she was born. When Elena is 20 years old, the bond will have earned 105% in interest. How much will the bond be worth when Elena is 20 years old?

3. In a video game, Clare scored 50% more points than Tyler. If $c$ is the number of points that Clare scored and $t$ is the number of points that Tyler scored, which equations are correct? Select **all** that apply.

   A. $c = 1.5t$

   B. $c = t + 0.5$

   C. $c = t + 0.5t$

   D. $c = t + 50$

   E. $c = (1 + 0.5)t$

4. Draw a diagram to represent each situation:

   a. The number of miles driven this month was a 30% decrease of the number of miles driven last month.

   b. The amount of paper that the copy shop used this month was a 25% increase of the amount of paper they used last month.

(From Unit 6, Lesson 3.)

5. Which decimal is the best estimate of the fraction $\frac{29}{40}$?

   A. 0.5

   B. 0.6

   C. 0.7

   D. 0.8

(From Unit 6, Lesson 2.)

6. Could 7.2 inches and 28 inches be the diameter and circumference of the same circle? Explain why or why not.

(From Unit 5, Lesson 12.)

# Lesson 6: More and Less than 1%

Let's explore percentages smaller than 1%.

## 6.1: Number Talk: What Percentage?

Determine the percentage mentally.

10 is what percentage of 50?

5 is what percentage of 50?

1 is what percentage of 50?

17 is what percentage of 50?

## 6.2: Waiting Tables

During one waiter's shift, he delivered 13 appetizers, 17 entrées, and 10 desserts.

1. What percentage of the dishes he delivered were:

    a. desserts?

    b. appetizers?

    c. entrées?

2. What do your percentages add up to?

## 6.3: Fractions of a Percent

1. Find each percentage of 60. What do you notice about your answers?

   30% of 60   3% of 60   0.3% of 60   0.03% of 60

2. 20% of 5,000 is 1,000 and 21% of 5,000 is 1,050. Find each percentage of 5,000 and be prepared to explain your reasoning. If you get stuck, consider using the double number line diagram.

   a. 1% of 5,000

   b. 0.1% of 5,000

   c. 20.1% of 5,000

   d. 20.4% of 5,000

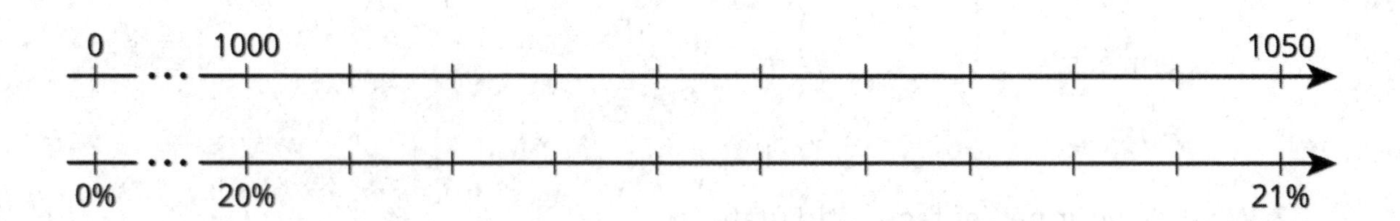

3. 15% of 80 is 12 and 16% of 80 is 12.8. Find each percentage of 80 and be prepared to explain your reasoning.

    a. 15.1% of 80

    b. 15.7% of 80

## Are you ready for more?

To make Sierpinski's triangle,

- Start with an equilateral triangle. This is step 1.
- Connect the midpoints of every side, and remove the middle triangle, leaving three smaller triangles. This is step 2.
- Do the same to each of the remaining triangles. This is step 3.
- Keep repeating this process.

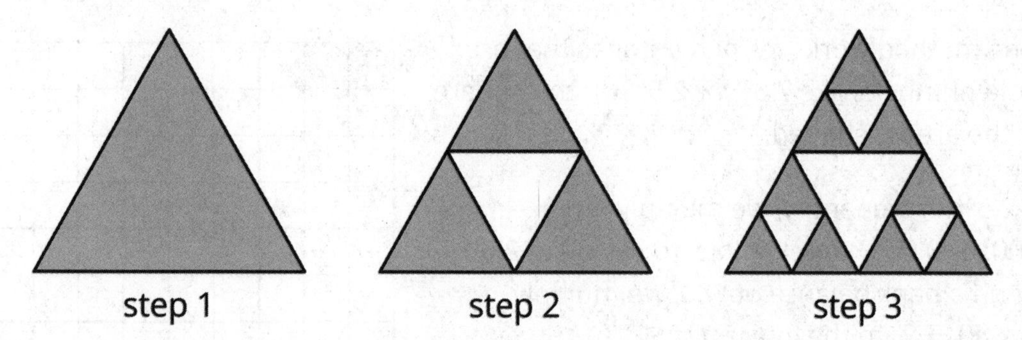


1. What percentage of the area of the original triangle is left after step 2? Step 3? Step 10?

2. At which step does the percentage first fall below 1%?

## 6.4: Population Growth

1. The population of City A was approximately 243,000 people, and it increased by 8% in one year. What was the new population?

2. The population of city B was approximately 7,150,000, and it increased by 0.8% in one year. What was the new population?

### Lesson 6 Summary

A percentage, such as 30%, is a rate per 100. To find 30% of a quantity, we multiply it by $30 \div 100$, or 0.3.

The same method works for percentages that are not whole numbers, like 7.8% or 2.5%. In the square, 2.5% of the area is shaded.

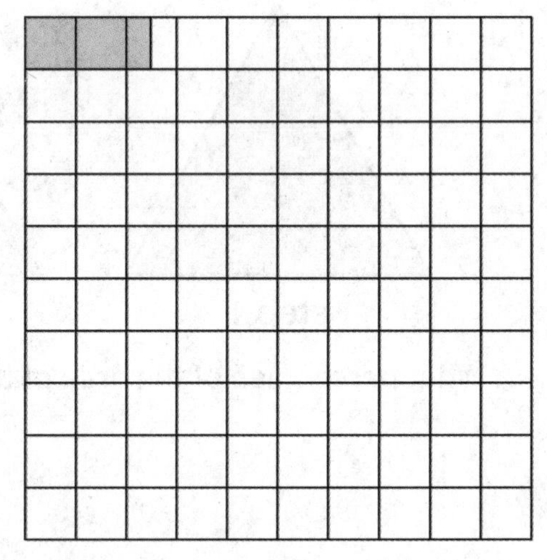

To find 2.5% of a quantity, we multiply it by $2.5 \div 100$, or 0.025. For example, to calculate 2.5% interest on a bank balance of \$80, we multiply $(0.025) \cdot 80 = 2$, so the interest is \$2.

We can sometimes find percentages like 2.5% mentally by using convenient whole number percents. For example, 25% of 80 is one fourth of 80, which is 20. Since 2.5 is one tenth of 25, we know that 2.5% of 80 is one tenth of 20, which is 2.

## Lesson 6 Practice Problems

1. The student government snack shop sold 32 items this week. For each snack type, what percentage of all snacks sold were of that type?

| snack type | number of items sold |
|---|---|
| fruit cup | 8 |
| veggie sticks | 6 |
| chips | 14 |
| water | 4 |

2. Select **all** the options that have the same value as $3\frac{1}{2}\%$ of 20.

   A. 3.5% of 20

   B. $3\frac{1}{2} \cdot 20$

   C. $(0.35) \cdot 20$

   D. $(0.035) \cdot 20$

   E. 7% of 10

3. 22% of 65 is 14.3. What is 22.6% of 65? Explain your reasoning.

4. A bakery used 30% more sugar this month than last month. If the bakery used 560 kilograms of sugar last month, how much did it use this month?

   (From Unit 6, Lesson 4.)

5. Match each situation to a diagram. The diagrams can be used more than once.

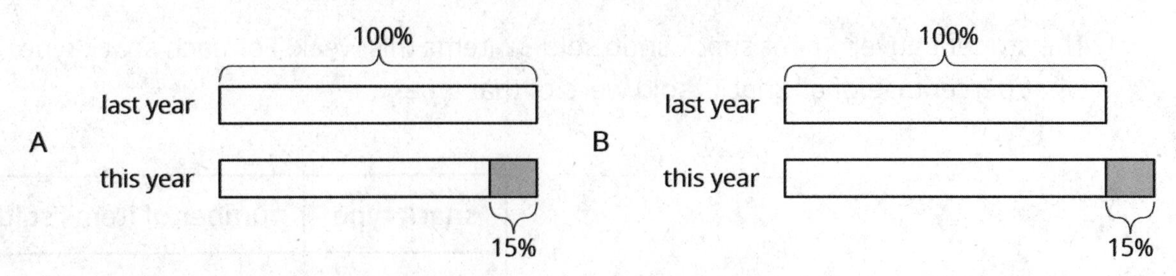


A. The amount of apples this year decreased by 15% compared with last year's amount.

B. The amount of pears this year is 85% of last year's amount.

C. The amount of cherries this year increased by 15% compared with last year's amount.

D. The amount of oranges this year is 115% of last year's amount.

1. Diagram A
2. Diagram B

(From Unit 6, Lesson 3.)

6. A certain type of car has room for 4 passengers.

   a. Write an equation relating the number of cars ($n$) to the number of passengers ($p$).

   b. How many passengers could fit in 78 cars?

   c. How many cars would be needed to fit 78 passengers?

(From Unit 5, Lesson 3.)

# Lesson 7: Tax and Tip

Let's learn about sales tax and tips.

## 7.1: Notice and Wonder: The Price of Sunglasses

You are on vacation and want to buy a pair of sunglasses for $10 or less. You find a pair with a price tag of $10. The cashier says the total cost will be $10.45.

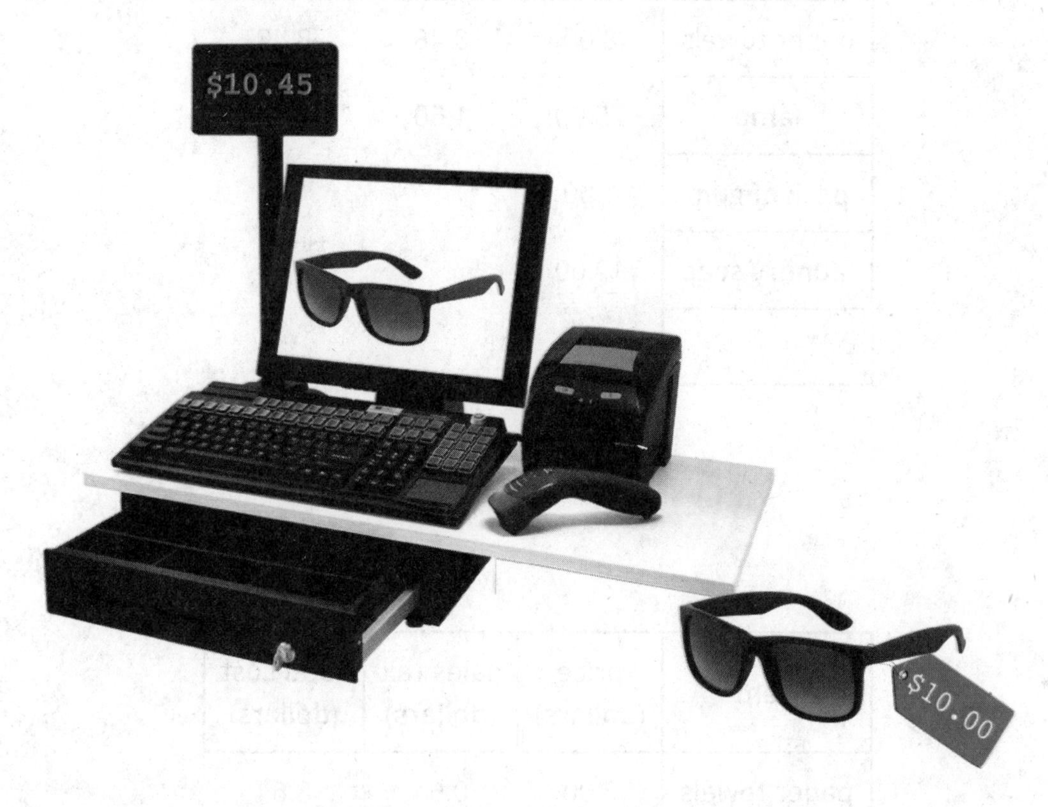


What do you notice? What do you wonder?

## 7.2: Shopping in Two Different Cities

Different cities have different sales tax rates. Here are the sales tax charges on the same items in two different cities. Complete the tables.

City 1

| item | price (dollars) | sales tax (dollars) | total cost (dollars) |
|---|---|---|---|
| paper towels | 8.00 | 0.48 | 8.48 |
| lamp | 25.00 | 1.50 | |
| pack of gum | 1.00 | | |
| laundry soap | 12.00 | | |
| | $x$ | | |

City 2

| item | price (dollars) | sales tax (dollars) | total cost (dollars) |
|---|---|---|---|
| paper towels | 8.00 | 0.64 | 8.64 |
| lamp | 25.00 | 2.00 | |
| pack of gum | 1.00 | | |
| laundry soap | 12.00 | | |
| | $x$ | | |

## 7.3: Shopping in a Third City

Here is the sales tax on the same items in City 3.

| item | price (dollars) | sales tax (dollars) |
|---|---|---|
| paper towels | 8.00 | 0.58 |
| lamp | 25.00 | 1.83 |
| pack of gum | 1.00 | 0.07 |
| laundry soap | 12.00 | |

1. What is the tax rate in this city?

2. For the sales tax on the laundry soap, Kiran says it should be $0.84. Lin says it should be $0.87. Do you agree with either of them? Explain your reasoning.

## 7.4: Dining at a Restaurant

1. Jada has a meal in a restaurant. She adds up the prices listed on the menu for everything they ordered and gets a subtotal of $42.00.

Date: Sep. 12th
Time: 6:55 PM
Server: # 27

| | |
|---|---|
| Bread Stix | 9.50 |
| Chicken Parm | 15.50 |
| Chef Salad | 12.00 |
| Lemon Soda | 2.00 |
| Tea | 3.00 |
| Subtotal | 42.00 |
| Sales Tax | 3.99 |
| Total | 45.99 |

a. When the check comes, it says they also need to pay $3.99 in sales tax. What percentage of the subtotal is the sales tax?

b. After tax, the total is $45.99. What percentage of the subtotal is the total?

c. They actually pay $52.99. The additional $7 is a tip for the server. What percentage of the subtotal is the tip?

2. The tax rate at this restaurant is 9.5%.

| Date: Sep 12th<br>Time: 6:04 PM<br>Server: # 27 | |
|---|---|
| Bread Stix | 9.50 |
| Ravioli Bites | 10.50 |
| Cheesecake | 4.95 |
| Subtotal | 24.95 |
| Sales Tax | ____ |
| Total | ____ |

Another person's subtotal is $24.95. How much will their sales tax be?

| Date: Sep 12th<br>Time: 7:12 PM<br>Server: # 27 | |
|---|---|
| Garden Salad | ____ |
| Broccoli Bites | ____ |
| Subtotal | ____ |
| Sales Tax | 1.61 |
| Total | ____ |

Some other person's sales tax is $1.61. How much was their subtotal?

**Are you ready for more?**

Elena's cousins went to a restaurant. The part of the entire cost of the meal that was tax and tip together was 25% of the cost of the food alone. What could the tax rate and tip rate be?

## Lesson 7 Summary

Many places have *sales tax*. A sales tax is an amount of money that a government agency collects on the sale of certain items. For example, a state might charge a tax on all cars purchased in the state. Often the tax rate is given as a percentage of the cost. For example, a state's tax rate on car sales might be 2%, which means that for every car sold in that state, the buyer has to pay a tax that is 2% of the sales price of the car.

Fractional percentages often arise when a state or city charges a sales tax on a purchase. For example, the sales tax in Arizona is 7.5%. This means that when someone buys something, they have to add 0.075 times the amount on the price tag to determine the total cost of the item.

For example, if the price tag on a T-shirt in Arizona says $11.50, then the sales tax is $(0.075) \cdot 11.5 = 0.8625$, which rounds to 86 cents. The customer pays $11.50 + 0.86$, or $12.36 for the shirt.

The total cost to the customer is the item price plus the sales tax. We can think of this as a percent increase. For example, in Arizona, the total cost to a customer is 107.5% of the price listed on the tag.

A *tip*is an amount of money that a person gives someone who provides a service. It is customary in many restaurants to give a tip to the server that is between 10% and 20% of the cost of the meal. If a person plans to leave a 15% tip on a meal, then the total cost will be 115% of the cost of the meal.

## Lesson 7 Practice Problems

1. In a city in Ohio, the sales tax rate is 7.25%. Complete the table to show the sales tax and the total price including tax for each item.

| item | price before tax ($) | sales tax ($) | price including tax ($) |
|---|---|---|---|
| pillow | 8.00 | | |
| blanket | 22.00 | | |
| trash can | 14.50 | | |

2. The sales tax rate in New Mexico is 5.125%. Select **all** the equations that represent the sales tax, $t$, you would pay in New Mexico for an item of cost $c$?

   A. $t = 5.125c$

   B. $t = 0.5125c$

   C. $t = 0.05125c$

   D. $t = c \div 0.05125$

   E. $t = \frac{5.125}{100}c$

3. Here are the prices of some items and the amount of sales tax charged on each in Nevada.

| cost of item ($) | sales tax ($) |
|---|---|
| 10 | 0.46 |
| 50 | 2.30 |
| 5 | 0.23 |

   a. What is the sales tax rate in Nevada?

   b. Write an expression for the amount of sales tax charged, in dollars, on an item that costs $c$ dollars.

4. Find each amount:

   a. 3.8% of 25

   b. 0.2% of 50

   c. 180.5% of 99

(From Unit 6, Lesson 6.)

5. On Monday, the high was 60 degrees Fahrenheit. On Tuesday, the high was 18% more. How much did the high increase from Monday to Tuesday?

(From Unit 6, Lesson 5.)

6. Complete the table. Explain or show your reasoning.

| object | radius | circumference |
|---|---|---|
| ceiling fan | 2.8 ft | |
| water bottle cap | 13 mm | |
| bowl | | 56.5 cm |
| drum | | 75.4 in |

(From Unit 5, Lesson 13.)

# Lesson 8: Percentage Situations

Let's find unknown percentages.

## 8.1: Tax, Tip, and Discount

What percentage of the car price is the tax?

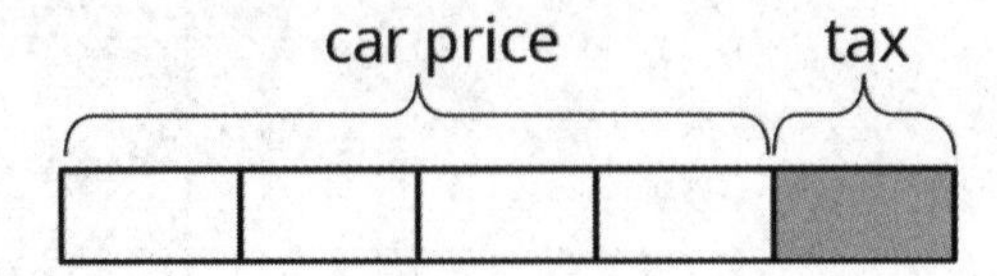


What percentage of the food cost is the tip?

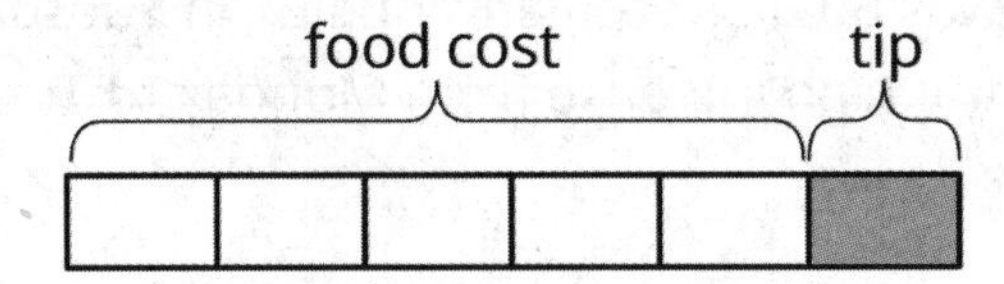


What percentage of the shirt cost is the discount?

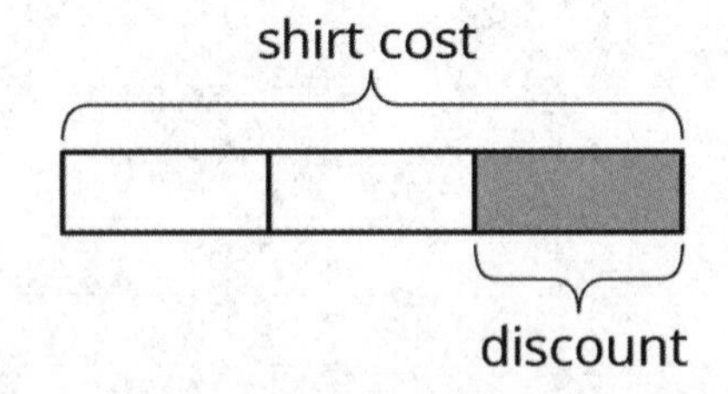

## 8.2: A Car Dealership

A car dealership pays a wholesale price of $12,000 to purchase a vehicle.

1. The car dealership wants to make a 32% profit.

   a. By how much will they mark up the price of the vehicle?

   b. After the markup, what is the retail price of the vehicle?

2. During a special sales event, the dealership offers a 10% discount off of the retail price. After the discount, how much will a customer pay for this vehicle?

### Are you ready for more?

This car dealership pays the salesperson a bonus for selling the car equal to 6.5% of the sale price. How much commission did the salesperson lose when they decided to offer a 10% discount on the price of the car?

## 8.3: Commission at a Gym

1. For each gym membership sold, the gym keeps $42 and the employee who sold it gets $8. What is the commission the employee earned as a percentage of the total cost of the gym membership?

2. If an employee sells a family pass for $135, what is the amount of the commission they get to keep?

## 8.4: Info Gap: Sporting Goods

Your teacher will give you either a *problem card* or a *data card*. Do not show or read your card to your partner.

If your teacher gives you the *problem card*:

1. Silently read your card and think about what information you need to be able to answer the question.
2. Ask your partner for the specific information that you need.
3. Explain how you are using the information to solve the problem.

   Continue to ask questions until you have enough information to solve the problem.
4. Share the *problem card* and solve the problem independently.
5. Read the *data card* and discuss your reasoning.

If your teacher gives you the *data card*:

1. Silently read your card.
2. Ask your partner *"What specific information do you need?"* and wait for them to *ask* for information.

   If your partner asks for information that is not on the card, do not do the calculations for them. Tell them you don't have that information.
3. Before sharing the information, ask "*Why do you need that information?*" Listen to your partner's reasoning and ask clarifying questions.
4. Read the *problem card* and solve the problem independently.
5. Share the *data card* and discuss your reasoning.

Pause here so your teacher can review your work. Ask your teacher for a new set of cards and repeat the activity, trading roles with your partner.

## 8.5: Card Sort: Percentage Situations

Your teacher will give you a set of cards. Take turns with your partner matching a situation with a descriptor. For each match, explain your reasoning to your partner. If you disagree, work to reach an agreement.

### Lesson 8 Summary

There are many everyday situations in which a percentage of an amount of money is added to or subtracted from that amount, in order to be paid to some other person or organization:

| | goes to | how it works |
|---|---|---|
| **sales tax** | the government | added to the price of the item |
| **gratuity (tip)** | the server | added to the cost of the meal |
| **interest** | the lender (or account holder) | added to the balance of the loan, credit card, or bank account |
| **markup** | the seller | added to the price of an item so the seller can make a profit |
| **markdown (discount)** | the customer | subtracted from the price of an item to encourage the customer to buy it |
| **commission** | the salesperson | subtracted from the payment that is collected |

For example, if a restaurant bill is \$34 and the customer pays \$40, they left 6 dollars as a tip for the server. That is 18% of \$34, so they left an 18% tip. From the customer's perspective, we can think of this as an 18% increase of the restaurant bill. If we know the initial amount and the final amount, we can also find the percent increase or percent decrease. For example, a plant was 12 inches tall and grew to be 15 inches tall. What percent increase is this? Here are two ways to solve this problem:

The plant grew 3 inches, because $15 - 12 = 3$. We can divide this growth by the original height, $3 \div 12 = 0.25$. So the height of the plant increased by 25%.

The plant's new height is 125% of the original height, because $15 \div 12 = 1.25$. This means the height increased by 25%, because $1.25 - 1 = 0.25$.

Here are two ways to solve the problem: A rope was 2.4 meters long. Someone cut it down to 1.9 meters. What percent decrease is this?

The rope is now $2.4 - 1.9$, or 0.5 meters shorter. We can divide this decrease by the original length, $0.5 \div 2.4 = 0.208\overline{3}$. So the length of the rope decreased by approximately 20.8%.

The rope's new length is about 79.2% of the original length, because $1.9 \div 2.4 = 0.791\overline{6}$. The length decreased by approximately 20.8%, because $1 - 0.792 = 0.208$.

## Lesson 8 Practice Problems

1. A car dealership pays $8,350 for a car. They mark up the price by 17.4% to get the retail price. What is the retail price of the car at this dealership?

2. A store has a 20% off sale on pants. With this discount, the price of one pair of pants before tax is $15.20. What was the original price of the pants?

   A. $3.04

   B. $12.16

   C. $18.24

   D. $19.00

3. A music store marks up the instruments it sells by 30%.

   a. If the store bought a guitar for $45, what will be its store price?

   b. If the price tag on a trumpet says $104, how much did the store pay for it?

   c. If the store paid $75 for a clarinet and sold it for $100, did the store mark up the price by 30%?

4. A family eats at a restaurant. The bill is $42. The family leaves a tip and spends $49.77.

    a. How much was the tip in dollars?

    b. How much was the tip as a percentage of the bill?

5. The price of gold is often reported per ounce. At the end of 2005, this price was $513. At the end of 2015, it was $1060. By what percentage did the price per ounce of gold increase?

6. A phone keeps track of the number of steps taken and the distance traveled. Based on the information in the table, is there a proportional relationship between the two quantities? Explain your reasoning.

| number of steps | distance in kilometers |
|---|---|
| 950 | 1 |
| 2,852 | 3 |
| 4,845 | 5.1 |

(From Unit 5, Lesson 4.)

7. A college student takes out a $7,500 loan from a bank. What will the balance of the loan be after one year (assuming the student has not made any payments yet):

   a. if the bank charges 3.8% interest each year?

   b. if the bank charges 5.3% interest each year?

(From Unit 6, Lesson 6.)

8. Match the situations with the equations.

   a. Mai slept for $x$ hours, and Kiran slept for $\frac{1}{10}$ less than that.

   b. Kiran practiced the piano for $x$ hours, and Mai practiced for $\frac{2}{5}$ less than that.

   c. Mai drank $x$ oz of juice and Kiran drank $\frac{4}{3}$ more than that.

   d. Kiran spent $x$ dollars and Mai spent $\frac{1}{4}$ less than that.

   e. Mai ate $x$ grams of almonds and Kiran ate 1.5 times more than that.

   f. Kiran collected $x$ pounds of recycling and Mai collected $\frac{3}{10}$ less than that.

   g. Mai walked $x$ kilometers and Kiran walked $\frac{3}{8}$ more than that.

   h. Kiran completed $x$ puzzles and Mai completed $\frac{3}{5}$ more than that.

   $y = 2.33x$

   $y = 1.375x$

   $y = 0.6x$

   $y = 0.9x$

   $y = 0.75x$

   $y = 1.6x$

   $y = 0.7x$

   $y = 2.5x$

(From Unit 6, Lesson 2.)

# Lesson 9: Measurement Error

Let's use percentages to describe how accurately we can measure.

## 9.1: Measuring to the Nearest

Your teacher will give you two rulers and three line segments labeled A, B, and C.

1. Use the centimeter ruler to measure each line segment to the nearest centimeter. Record these lengths in the first column of the table.

2. Use the millimeter ruler to measure each line segment to the nearest tenth of a centimeter. Record these lengths in the second column of the table.

| line segment | length (cm) as measured with the first ruler | length (cm) as measured with the second ruler |
|---|---|---|
| A | | |
| B | | |
| C | | |

## 9.2: Measuring a Soccer Field

A soccer field is 120 yards long. Han measures the length of the field using a 30-foot-long tape measure and gets a measurement of 358 feet, 10 inches.

1. What is the amount of the error?

2. Express the error as a percentage of the actual length of the field.

## 9.3: Measuring Your Classroom

Your teacher will tell you which three items to measure. Keep using the paper rulers from the earlier activity.

1. Between you and your partner, decide who will use which ruler.
2. Measure the three items assigned by your teacher and record your measurements in the first column of the appropriate table.

   Using the cm ruler:

| item | measured length (cm) | actual length (cm) | difference | percentage |
|---|---|---|---|---|
| | | | | |
| | | | | |
| | | | | |

   Using the mm ruler:

| item | measured length (cm) | actual length (cm) | difference | percentage |
|---|---|---|---|---|
| | | | | |
| | | | | |
| | | | | |

3. After you finish measuring the items, share your data with your partner. Next, ask your teacher for the actual lengths.
4. Calculate the difference between your measurements and the actual lengths in both tables.
5. For each difference, what percentage of the actual length is this amount? Record your answers in the last column of the tables.

**Are you ready for more?**

Before there were standard units of measurement, people often measured things using their hands or feet.

1. Measure the length of your foot to the nearest centimeter with your shoe on.

2. How many foot-lengths long is your classroom? Try to determine this as precisely as possible by carefully placing your heel next to your toe as you pace off the room.

3. Use this information to estimate the length of your classroom in centimeters.

4. Use a tape measure to measure the length of your classroom. What is the difference between the two measurements? Which one do you think is more accurate?

## Lesson 9 Summary

When we are measuring a length using a ruler or measuring tape, we can get a measurement that is different from the actual length. This could be because we positioned the ruler incorrectly, or it could be because the ruler is not very precise. There is always at least a small difference between the actual length and a measured length, even if it is a microscopic difference!

Here are two rulers with different markings.

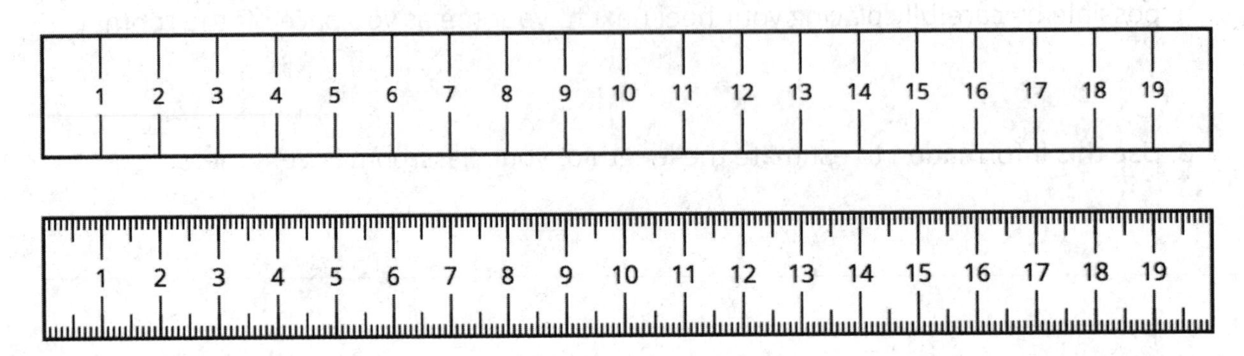


The second ruler is marked in millimeters, so it is easier to get a measurement to the nearest tenth of a centimeter with this ruler than with the first. For example, a line that is actually 6.2 cm long might be measured to be 6 cm long by the first ruler, because we measure to the nearest centimeter.

The **measurement error** is the positive difference between the measurement and the actual value. Measurement error is often expressed as a percentage of the actual value. We always use a positive number to express measurement error and, when appropriate, use words to describe whether the measurement is greater than or less than the actual value.

For example, if we get 6 cm when we measure a line that is actually 6.2 cm long, then the measurement error is 0.2 cm, or about 3.2%, because $0.2 \div 6.2 \approx 0.032$.

### Glossary

- measurement error

## Lesson 9 Practice Problems

1. The depth of a lake is 15.8 m.

    a. Jada accurately measured the depth of the lake to the nearest meter. What measurement did Jada get?

    b. By how many meters does the measured depth differ from the actual depth?

    c. Express the measurement error as a percentage of the actual depth.

2. A watermelon weighs 8,475 grams. A scale measured the weight with an error of 12% under the actual weight. What was the measured weight?

3. Noah's oven thermometer gives a reading that is 2% greater than the actual temperature.

    a. If the actual temperature is $325^{\circ}\mathrm{F}$, what will the thermometer reading be?

    b. If the thermometer reading is $76^{\circ}\mathrm{F}$, what is the actual temperature?

4. At the beginning of the month, there were 80 ounces of peanut butter in the pantry. Now, there is $\frac{1}{3}$ less than that. How many ounces of peanut butter are in the pantry now?

   A. $\frac{2}{3} \cdot 80$

   B. $\frac{1}{3} \cdot 80$

   C. $80 - \frac{1}{3}$

   D. $\left(1 + \frac{1}{3}\right) \cdot 80$

   (From Unit 6, Lesson 1.)

5. a. Fill in the table for side length and area of different squares.

| side length (cm) | area (cm$^2$) |
|---|---|
| 3 | |
| 100 | |
| 25 | |
| $s$ | |

   b. Is the relationship between the side length of a square and the area of a square proportional?

   (From Unit 5, Lesson 15.)

# Lesson 10: Percent Error

Let's use percentages to describe other situations that involve error.

## 10.1: Number Talk: Estimating a Percentage of a Number

Estimate.

25% of 15.8

9% of 38

1.2% of 127

0.53% of 6

0.06% of 202

## 10.2: Plants, Bicycles, and Crowds

1. Instructions to care for a plant say to water it with $\frac{3}{4}$ cup of water every day. The plant has been getting 25% too much water. How much water has the plant been getting?

2. The pressure on a bicycle tire is 63 psi. This is 5% higher than what the manual says is the correct pressure. What is the correct pressure?

3. The crowd at a sporting event is estimated to be 3,000 people. The exact attendance is 2,486 people. What is the **percent error**?

### Are you ready for more?

A micrometer is an instrument that can measure lengths to the nearest micron (a micron is a millionth of a meter). Would this instrument be useful for measuring any of the following things? If so, what would the largest percent error be?

1. The thickness of an eyelash, which is typically about 0.1 millimeters.

2. The diameter of a red blood cell, which is typically about 8 microns.

3. The diameter of a hydrogen atom, which is about 100 picometers (a picometer is a trillionth of a meter).

## 10.3: Measuring in the Heat

A metal measuring tape expands when the temperature goes above $50^\circ\text{F}$. For every degree Fahrenheit above 50, its length increases by 0.00064%.

1. The temperature is 100 degrees Fahrenheit. How much longer is a 30-foot measuring tape than its correct length?

2. What is the percent error?

### Lesson 10 Summary

**Percent error** can be used to describe any situation where there is a correct value and an incorrect value, and we want to describe the relative difference between them. For example, if a milk carton is supposed to contain 16 fluid ounces and it only contains 15 fluid ounces:

- the measurement error is 1 oz, and
- the percent error is 6.25% because $1 \div 16 = 0.0625$.

We can also use percent error when talking about estimates. For example, a teacher estimates there are about 600 students at their school. If there are actually 625 students, then the percent error for this estimate was 4%, because $625 - 600 = 25$ and $25 \div 625 = 0.04$.

### Glossary

- percent error

## Lesson 10 Practice Problems

1. A student estimated that it would take 3 hours to write a book report, but it actually took her 5 hours. What is the percent error for her estimate?

2. A radar gun measured the speed of a baseball at 103 miles per hour. If the baseball was actually going 102.8 miles per hour, what was the percent error in this measurement?

3. It took 48 minutes to drive downtown. An app estimated it would be less than that. If the error was 20%, what was the app's estimate?

4. A farmer estimated that there were 25 gallons of water left in a tank. If this is an underestimate by 16%, how much water was actually in the tank?

5. For each story, write an equation that describes the relationship between the two quantities.

   a. Diego collected $x$ kg of recycling. Lin collected $\frac{2}{5}$ more than that.

   b. Lin biked $x$ km. Diego biked $\frac{3}{10}$ less than that.

   c. Diego read for $x$ minutes. Lin read $\frac{4}{7}$ of that.

(From Unit 6, Lesson 1.)

6. For each diagram, decide if $y$ is an increase or a decrease of $x$. Then determine the percentage.

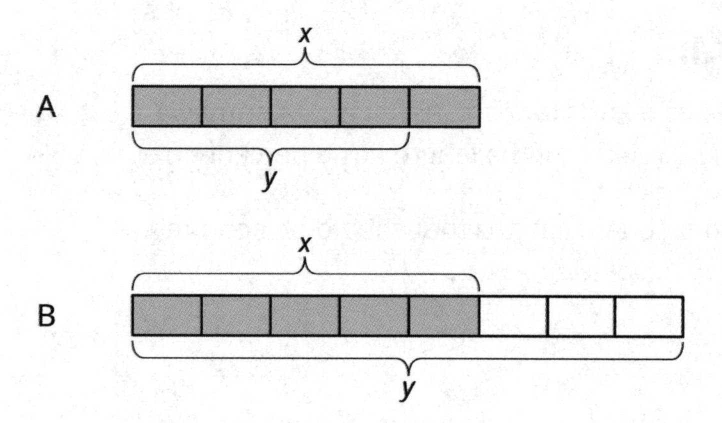


(From Unit 6, Lesson 8.)

7. Lin is making a window covering for a window that has the shape of a half circle on top of a square of side length 3 feet. How much fabric does she need?

(From Unit 5, Lesson 19.)

# Lesson 11: Error Intervals

Let's solve more problems about percent error.

## 11.1: A Lot of Iron Ore

An industrial scale is guaranteed by the manufacturer to have a percent error of no more than 1%. What is a possible reading on the scale if you put 500 kilograms of iron ore on it?

## 11.2: Saw Mill

1. A saw mill cuts boards that are 16 ft long. After they are cut, the boards are inspected and rejected if the length has a percent error of 1.5% or more.

    a. List some board lengths that should be accepted.

    b. List some board lengths that should be rejected.

2. The saw mill also cuts boards that are 10, 12, and 14 feet long. An inspector rejects a board that was 2.3 inches too long. What was the intended length of the board?

## 11.3: Info Gap: Quality Control

Your teacher will give you either a *problem card* or a *data card*. Do not show or read your card to your partner.

If your teacher gives you the *problem card*:

1. Silently read your card and think about what information you need to be able to answer the question.
2. Ask your partner for the specific information that you need.
3. Explain how you are using the information to solve the problem.

   Continue to ask questions until you have enough information to solve the problem.
4. Share the *problem card* and solve the problem independently.
5. Read the *data card* and discuss your reasoning.

If your teacher gives you the *data card*:

1. Silently read your card.
2. Ask your partner *"What specific information do you need?"* and wait for them to *ask* for information.

   If your partner asks for information that is not on the card, do not do the calculations for them. Tell them you don't have that information.
3. Before sharing the information, ask "*Why do you need that information?*" Listen to your partner's reasoning and ask clarifying questions.
4. Read the *problem card* and solve the problem independently.
5. Share the *data card* and discuss your reasoning.

Pause here so your teacher can review your work. Ask your teacher for a new set of cards and repeat the activity, trading roles with your partner.

## Lesson 11 Summary

Percent error is often used to express a range of possible values. For example, if a box of cereal is guaranteed to have 750 grams of cereal, with a margin of error of less than 5%, what are possible values for the actual number of grams of cereal in the box? The error could be as large as $(0.05) \cdot 750 = 37.5$ and could be either above or below than the correct amount.

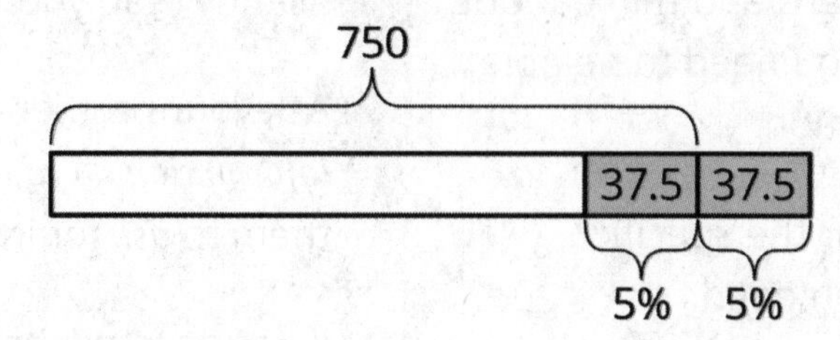


Therefore, the box can have anywhere between 712.5 and 787.5 grams of cereal in it, but it should not have 700 grams or 800 grams, because both of those are more than 37.5 grams away from 750 grams.

## Lesson 11 Practice Problems

1. Jada measured the height of a plant in a science experiment and finds that, to the nearest $\frac{1}{4}$ of an inch, it is $4\frac{3}{4}$ inches.

   a. What is the largest the actual height of the plant could be?

   b. What is the smallest the actual height of the plant could be?

   c. How large could the percent error in Jada's measurement be?

2. The reading on a car's speedometer has 1.6% maximum error. The speed limit on a road is 65 miles per hour.

   a. The speedometer reads 64 miles per hour. Is it possible that the car is going over the speed limit?

   b. The speedometer reads 66 miles per hour. Is the car definitely going over the speed limit?

3. Water is running into a bathtub at a constant rate. After 2 minutes, the tub is filled with 2.5 gallons of water. Write two equations for this proportional relationship. Use $w$ for the amount of water (gallons) and $t$ for time (minutes). In each case, what does the constant of proportionality tell you about the situation?

   (From Unit 5, Lesson 2.)

4. Noah picked 3 kg of cherries. Jada picked half as many cherries as Noah. How many total kg of cherries did Jada and Noah pick?

   A. $3 + 0.5$

   B. $3 - 0.5$

   C. $(1 + 0.5) \cdot 3$

   D. $1 + 0.5 \cdot 3$

(From Unit 6, Lesson 2.)

5. Here is a shape with some measurements in cm.

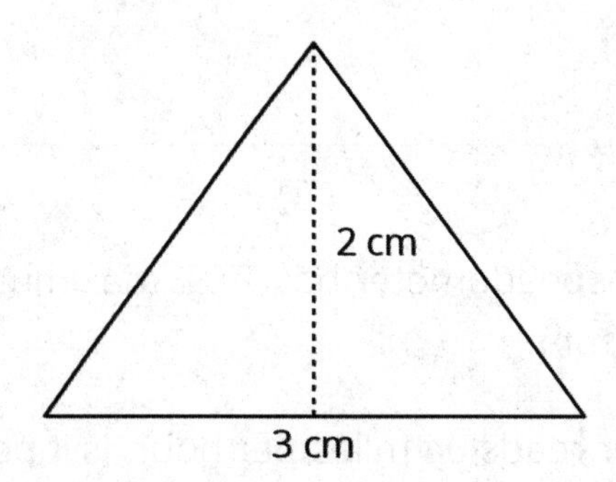


   a. Complete the table showing the area of different scaled copies of the triangle.

| scale factor | area ($cm^2$) |
| --- | --- |
| 1 | |
| 2 | |
| 5 | |
| $s$ | |

   b. Is the relationship between the scale factor and the area of the scaled copy proportional?

(From Unit 5, Lesson 15.)

# Lesson 12: Posing Percentage Problems

Let's explore how percentages are used in the news.

## 12.1: Sorting the News

Your teacher will give you a variety of news clippings that include percentages.

1. Sort the clippings into two piles: those that are about increases and those that are about decreases.

2. Were there any clippings that you had trouble deciding which pile they should go in?

## 12.2: Investigating

In the previous activity, you sorted news clippings into two piles.

1. For each pile, choose one example. Draw a diagram that shows how percentages are being used to describe the situation.

    a. Increase Example:

    b. Decrease Example:

2. For each example, write *two* questions that you can answer with the given information. Next, find the answers. Explain or show your reasoning.

## 12.3: Displaying the News

1. Choose the example that you find the most interesting. Create a visual display that includes:
    - a title that describes the situation
    - the news clipping
    - your diagram of the situation
    - the two questions you asked about the situation
    - the answers to each of your questions
    - an explanation of how you calculated each answer

   Pause here so your teacher can review your work.

2. Examine each display. Write one comment and one question for the group.

3. Next, read the comments and questions your classmates wrote for your group. Revise your display using the feedback from your classmates.

### Lesson 12 Summary

Statements about percentage increase or decrease need to specify what the whole is to be mathematically meaningful. Sometimes advertisements, media, etc. leave the whole ambiguous in order to make somewhat misleading claims. We should be careful to think critically about what mathematical claim is being made.

For example, if a disinfectant claims to "kill 99% of all bacteria," does it mean that

- It kills 99% of the number of bacteria on a surface?
- Or is it 99% of the types of bacteria commonly found inside the house?
- Or 99% of the total mass or volume of bacteria?
- Does it even matter if the remaining 1% are the most harmful bacteria?

Resolving questions of this type is an important step in making informed decisions.

# Learning Targets

### Lesson 1: Half as Much Again

- I can use the distributive property to rewrite an expression like $x + \frac{1}{2}x$ as $(1 + \frac{1}{2})x$.
- I understand that "half as much again" and "multiply by $\frac{3}{2}$" mean the same thing.

### Lesson 2: Say It with Decimals

- I can use the distributive property to rewrite an equation like $x + 0.5x = 1.5x$.
- I can write fractions as decimals.
- I understand that "half as much again" and "multiply by 1.5" mean the same thing.

### Lesson 3: Increasing and Decreasing

- I can draw a tape diagram that represents a percent increase or decrease.
- When I know a starting amount and the percent increase or decrease, I can find the new amount.

### Lesson 4: One Hundred Percent

- I can use a double number line diagram to help me solve percent increase and decrease problems.
- I understand that if I know how much a quantity has grown, then the original amount represents 100%.
- When I know the new amount and the percentage of increase or decrease, I can find the original amount.

### Lesson 5: Percent Increase and Decrease with Equations

- I can solve percent increase and decrease problems by writing an equation to represent the situation and solving it.

### Lesson 6: More and Less than 1%

- I can find percentages of quantities like 12.5% and 0.4%.
- I understand that to find 0.1% of an amount I have to multiply by 0.001.

### Lesson 7: Tax and Tip

- I understand and can solve problems about sales tax and tip.

### Lesson 8: Percentage Situations

- I can find the percentage increase or decrease when I know the original amount and the new amount.
- I understand and can solve problems about commission, interest, markups, and discounts.

### Lesson 9: Measurement Error

- I can represent measurement error as a percentage of the correct measurement.
- I understand that all measurements include some error.

### Lesson 10: Percent Error

- I can solve problems that involve percent error.

### Lesson 11: Error Intervals

- I can find a range of possible values for a quantity if I know the maximum percent error and the correct value.

### Lesson 12: Posing Percentage Problems

- I can write and solve problems about real-world situations that involve percent increase and decrease.

# Illustrative Mathematics

## Unit 7

STUDENT EDITION
Book 2

# Lesson 1: Positive and Negative Numbers

Let's explore how we represent temperatures and elevations.

## 1.1: Notice and Wonder: Memphis and Bangor

What do you notice? What do you wonder?

## 1.2: What's the Temperature?

1. Here are five thermometers. The first four thermometers show temperatures in Celsius. Write the temperatures in the blanks.

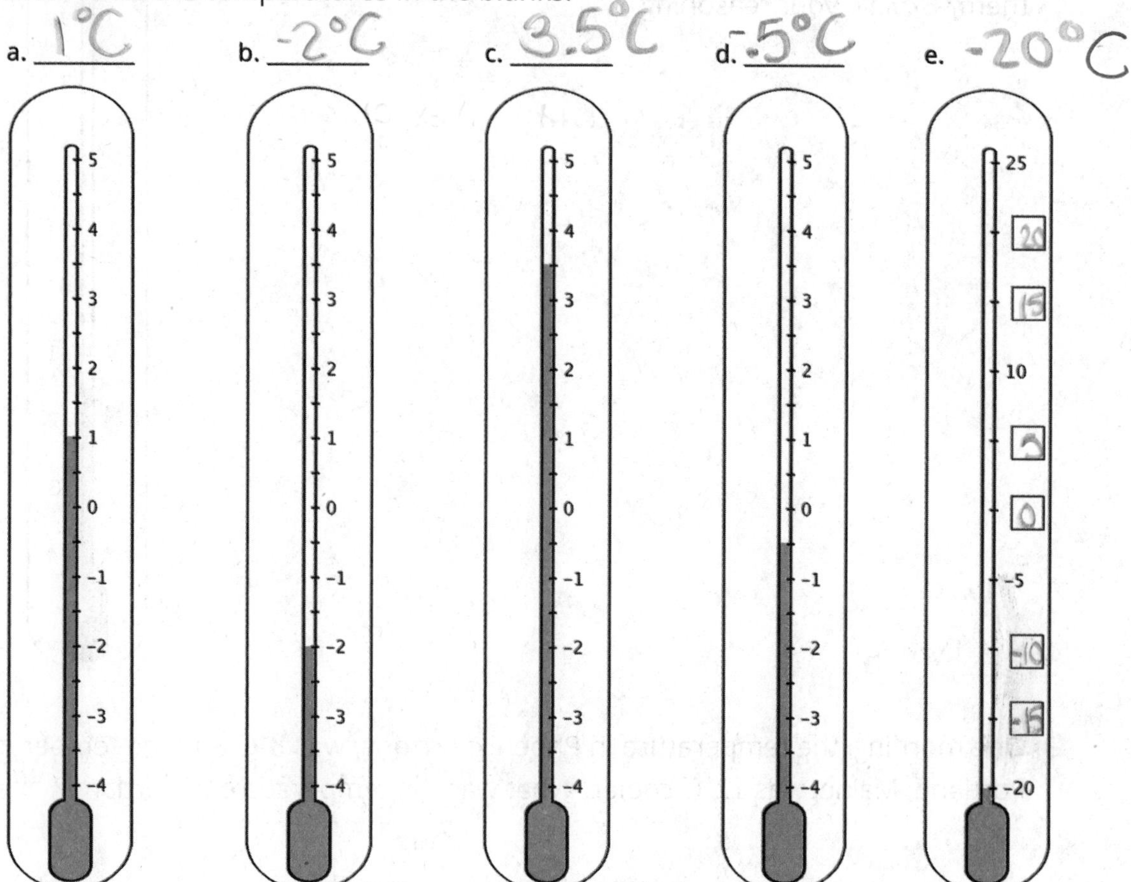


The last thermometer is missing some numbers. Write them in the boxes.

2. Elena says that the thermometer shown here reads -2.5°C because the line of the liquid is above -2°C. Jada says that it is -1.5°C. Do you agree with either one of them? Explain your reasoning.

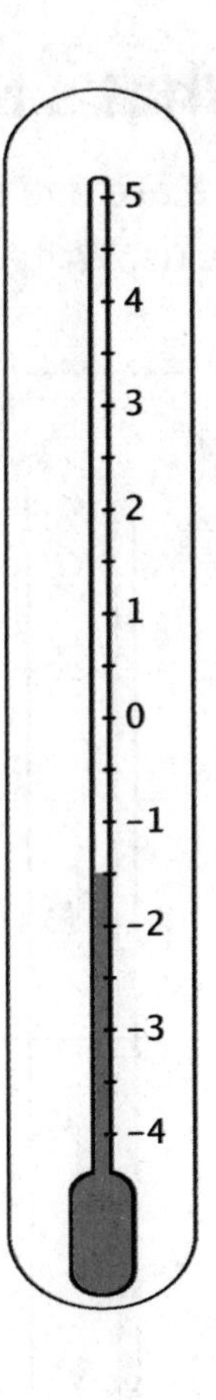


I agree with Jada because we are counting down so our number goes down.

3. One morning, the temperature in Phoenix, Arizona, was 8°C and the temperature in Portland, Maine, was 12°C cooler. What was the temperature in Portland?

$8 - 12$

$8 - 8 = 0$

$0 - 4 = -4$

The temp. in Portland is -4°C

## 1.3: Seagulls Soar, Sharks Swim

Here is a picture of some sea animals. The number line on the left shows the vertical position of each animal above or below sea level, in meters.

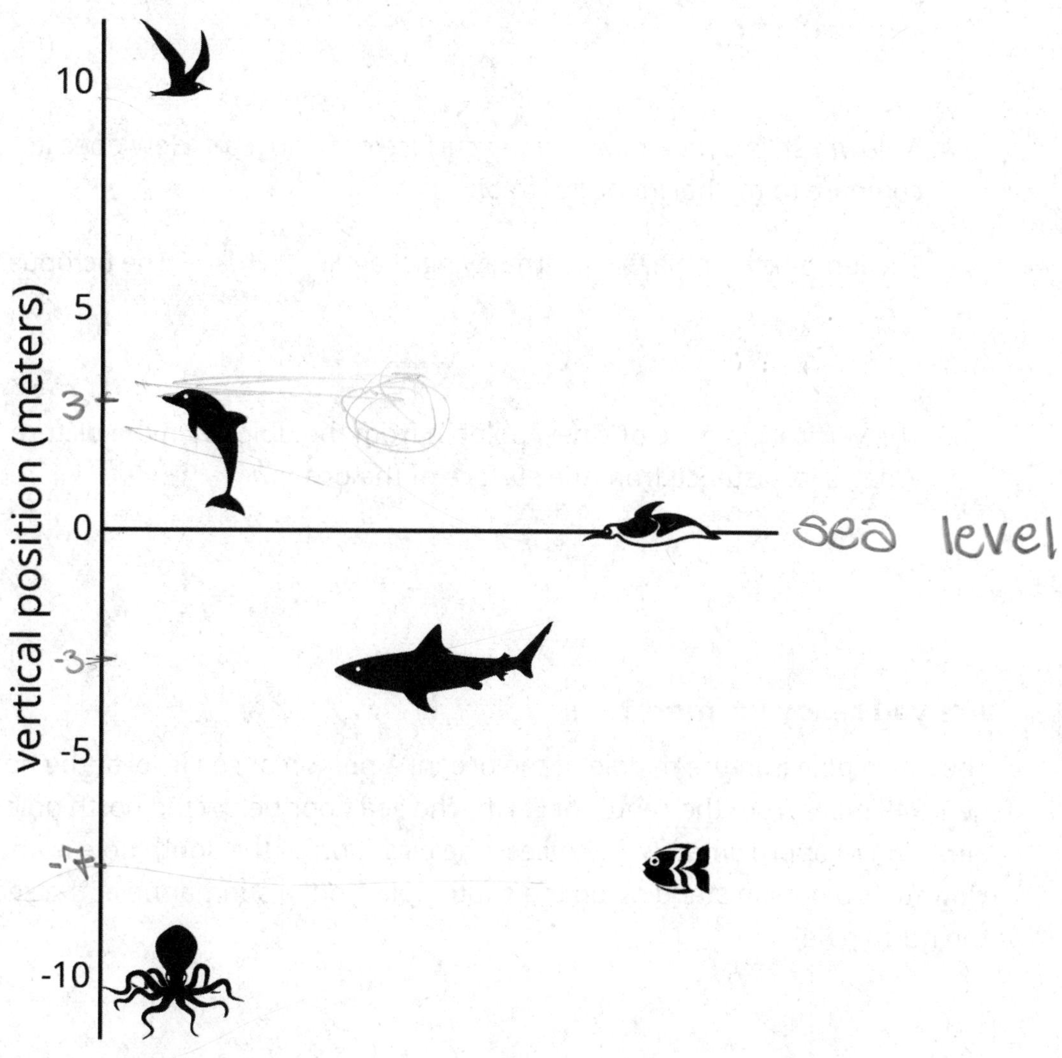


1. How far above or below sea level is each animal? Measure to their eye level.
2. A mobula ray is 3 meters above the surface of the ocean. How does its vertical position compare to the height or depth of:

The jumping dolphin? The flying seagull? The octopus?

some 0m 7m 13m

3. An albatross is 5 meters above the surface of the ocean. How does its vertical position compare to the height or depth of:

The jumping dolphin? The flying seagull? The octopus?

2m 5m 15m

4. A clownfish is 2 meters below the surface of the ocean. How does its vertical position compare to the height or depth of:

The jumping dolphin? The flying seagull? The octopus?

5 m. 12 m 8 m.

5. The vertical distance of a new dolphin from the dolphin in the picture is 3 meters. What is its distance from the surface of the ocean?

3m

**Are you ready for more?**

The north pole is in the middle of the ocean. A person at sea level at the north pole would be 3,949 miles from the center of Earth. The sea floor below the north pole is at an elevation of approximately -2.7 miles. The elevation of the south pole is about 1.7 miles. How far is a person standing on the south pole from a submarine at the sea floor below the north pole?

## 1.4: High Places, Low Places

1. Here is a table that shows elevations of various cities.

| city | elevation (feet) |
|---|---|
| Harrisburg, PA | 320 |
| Bethell, IN | 1,211 |
| Denver, CO | 5,280 |
| Coachella, CA | -22 |
| Death Valley, CA | -282 |
| New York City, NY | 33 |
| Miami, FL | 0 |

a. On the list of cities, which city has the second highest elevation?

b. How would you describe the elevation of Coachella, CA in relation to sea level?

c. How would you describe the elevation of Death Valley, CA in relation to sea level?

d. If you are standing on a beach right next to the ocean, what is your elevation?

e. How would you describe the elevation of Miami, FL?

f. A city has a higher elevation than Coachella, CA. Select all numbers that could represent the city's elevation. Be prepared to explain your reasoning.

- -11 feet
- -35 feet
- 4 feet
- -8 feet
- 0 feet

2. Here are two tables that show the elevations of highest points on land and lowest points in the ocean. Distances are measured from sea level.

| mountain | continent | elevation (meters) |
|---|---|---|
| Everest | Asia | 8,848 |
| Kilimanjaro | Africa | 5,895 |
| Denali | North America | 6,168 |
| Pikchu Pikchu | South America | 5,664 |

| trench | ocean | elevation (meters) |
|---|---|---|
| Mariana Trench | Pacific | -11,033 |
| Puerto Rico Trench | Atlantic | -8,600 |
| Tonga Trench | Pacific | -10,882 |
| Sunda Trench | Indian | -7,725 |

a. Which point in the ocean is the lowest in the world? What is its elevation?

b. Which mountain is the highest in the world? What is its elevation?

c. If you plot the elevations of the mountains and trenches on a vertical number line, what would 0 represent? What would points above 0 represent? What about points below 0?

d. Which is farther from sea level: the deepest point in the ocean, or the top of the highest mountain in the world? Explain.

### Are you ready for more?

A spider spins a web in the following way:

- It starts at sea level.
- It moves up one inch in the first minute.
- It moves down two inches in the second minute.
- It moves up three inches in the third minute.
- It moves down four inches in the fourth minute.

Assuming that the pattern continues, what will the spider's elevation be after an hour has passed?

### Lesson 1 Summary

**Positive numbers** are numbers that are greater than 0. **Negative numbers** are numbers that are less than zero. The meaning of a negative number in a context depends on the meaning of zero in that context.

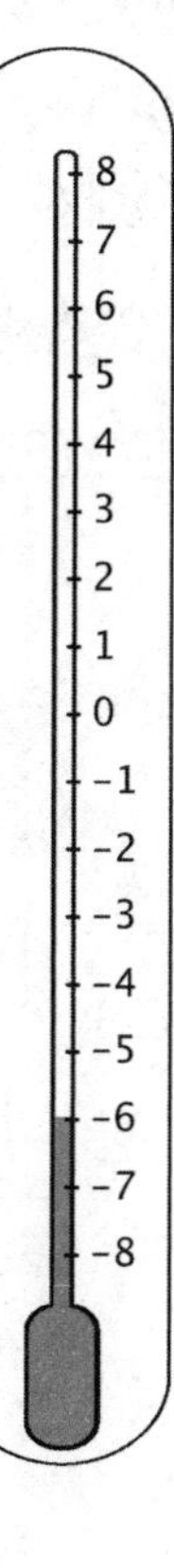


For example, if we measure temperatures in degrees Celsius, then 0 degrees Celsius corresponds to the temperature at which water freezes.

In this context, positive temperatures are warmer than the freezing point and negative temperatures are colder than the freezing point. A temperature of -6 degrees Celsius means that it is 6 degrees away from 0 and it is less than 0. This thermometer shows a temperature of -6 degrees Celsius.

If the temperature rises a few degrees and gets very close to 0 degrees without reaching it, the temperature is still a negative number.

Another example is elevation, which is a distance above or below sea level. An elevation of 0 refers to the sea level. Positive elevations are higher than sea level, and negative elevations are lower than sea level.

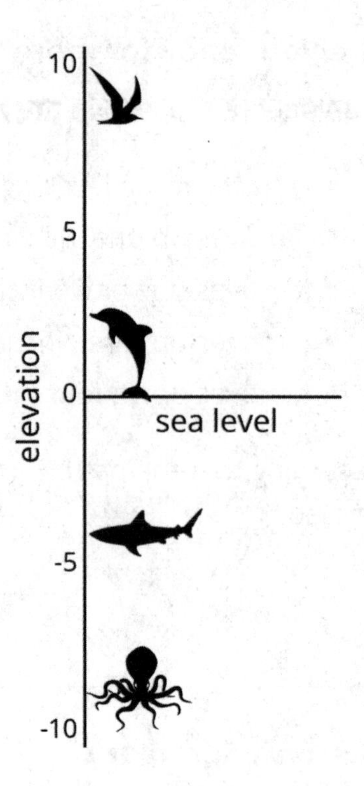


## Glossary

- negative number
- positive number

# Lesson 2: Comparing Positive and Negative Numbers

Let's compare numbers on the number line.

## 2.1: Which One Doesn't Belong: Inequalities

Which inequality doesn't belong?

- $\frac{5}{4} < 2$
- $8.5 > 0.95$
- $8.5 < 7$
- $10.00 < 100$

## 2.2: Comparing Temperatures

Here are the low temperatures, in degrees Celsius, for a week in Anchorage, Alaska.

| day | Mon | Tues | Weds | Thurs | Fri | Sat | Sun |
|---|---|---|---|---|---|---|---|
| temperature | 5 | -1 | -5.5 | -2 | 3 | 4 | 0 |

1. Plot the temperatures on a number line. Which day of the week had the lowest low temperature?

2. The lowest temperature ever recorded in the United States was -62 degrees Celsius, in Prospect Creek Camp, Alaska. The average temperature on Mars is about -55 degrees Celsius.

   a. Which is warmer, the coldest temperature recorded in the USA, or the average temperature on Mars? Explain how you know.

   MARS temp

   $-62 < -55$

   b. Write an inequality to show your answer.

3. On a winter day the low temperature in Anchorage, Alaska, was -21 degrees Celsius and the low temperature in Minneapolis, Minnesota, was -14 degrees Celsius.

   Jada said, "I know that 14 is less than 21, so -14 is also less than -21. This means that it was colder in Minneapolis than in Anchorage."

   Do you agree? Explain your reasoning.

   I Disagree because in negative numbers, the number farther away from 0 is the smaller number. The number closer to 0 is the bigger number.

## Are you ready for more?

Another temperature scale frequently used in science is the *Kelvin scale*. In this scale, 0 is the lowest possible temperature of anything in the universe, and it is -273.15 degrees in the Celsius scale. Each $1$ K is the same as $1^\circ$C, so $10$ K is the same as $-263.15^\circ$C.

1. Water boils at $100^\circ$C. What is this temperature in K?

2. Ammonia boils at $-35.5^\circ$C. What is the boiling point of ammonia in K?

3. Explain why only positive numbers (and 0) are needed to record temperature in K.

## 2.3: Rational Numbers on a Number Line

1. Plot the numbers -2, 4, -7, and 10 on the number line. Label each point with its numeric value.

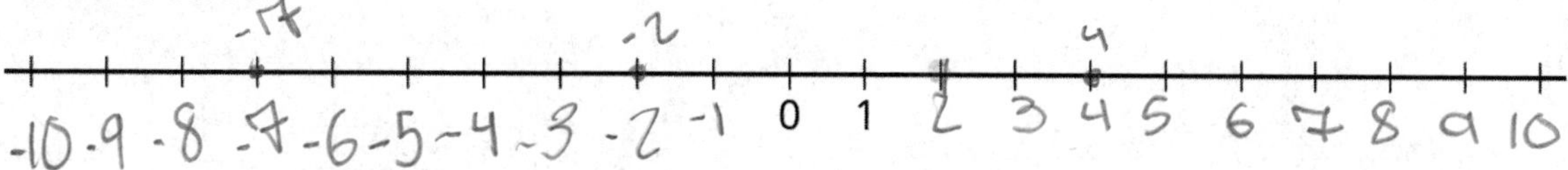


2. Decide whether each inequality statement is true or false. Be prepared to explain your reasoning.

   a. $-2 < 4$ ✓

   b. $-2 < -7$ ✗

   c. $4 > -7$ ✓

   d. $-7 > 10$ ✗

3. Andre says that $\frac{1}{4}$ is less than $-\frac{3}{4}$ because, of the two numbers, $\frac{1}{4}$ is closer to 0. Do you agree? Explain your reasoning. I do not agree because $\frac{1}{4}$ is a positive number. When you are in the positives, the farther away you are from 0, the larger you are. Positive numbers are larger than negative numbers.

4. Answer each question. Be prepared to explain how you know.

   a. Which number is greater: $\frac{1}{4}$ or $\frac{5}{4}$?

   $\frac{5}{4}$

   b. Which is farther from 0: $\frac{1}{4}$ or $\frac{5}{4}$?

   $\frac{5}{4}$

   c. Which number is greater: $-\frac{3}{4}$ or $\frac{5}{8}$?

   $\frac{5}{8}$

   d. Which is farther from 0: $-\frac{3}{4}$ or $\frac{5}{8}$?

   $-\frac{3}{4}$

e. Is the number that is farther from 0 always the greater number? Explain your reasoning. No. A distance cannot be negative

## Lesson 2 Summary

Here is a number line labeled with positive and negative numbers. The number 4 is positive, so its location is 4 units to the right of 0 on the number line. The number -1.1 is negative, so its location is 1.1 units to the left of 0 on the number line.

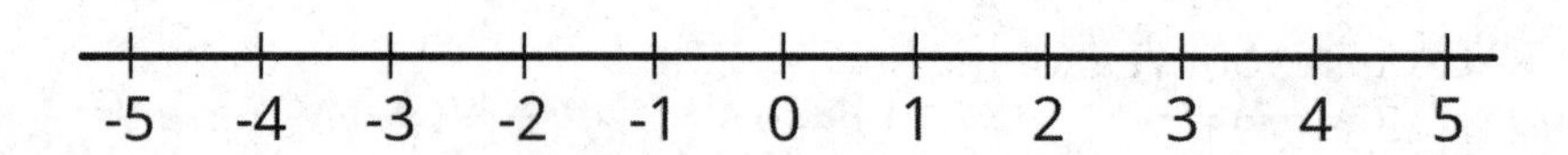


We say that the *opposite* of 8.3 is -8.3, and that the *opposite* of $\frac{-3}{2}$ is $\frac{3}{2}$. Any pair of numbers that are equally far from 0 are called **opposites**.

Points $A$ and $B$ are opposites because they are both 2.5 units away from 0, even though $A$ is to the left of 0 and $B$ is to the right of 0.

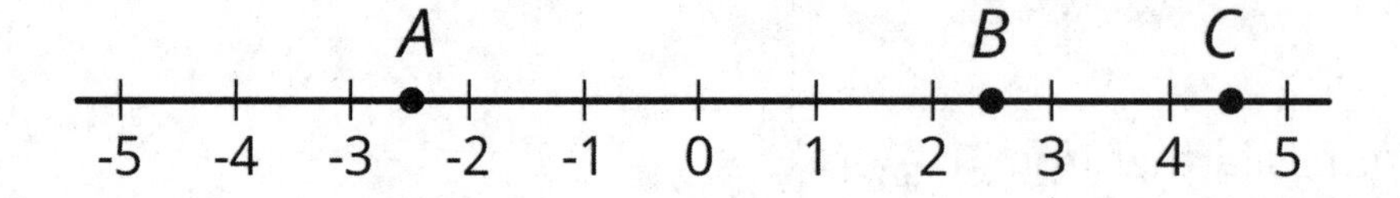


A positive number has a negative number for its opposite. A negative number has a positive number for its opposite. The opposite of 0 is itself.

You have worked with positive numbers for many years. All of the positive numbers you have seen—whole and non-whole numbers—can be thought of as fractions and can be located on a the number line.

To locate a non-whole number on a number line, we can divide the distance between two whole numbers into fractional parts and then count the number of parts. For example, 2.7

## 3.4: Drinks for Sale

A vending machine in an office building sells bottled beverages. The machine keeps track of all changes in the number of bottles from sales and from machine refills and maintenance. This record shows the changes for every 5-minute period over one hour.

| time | number of bottles |
|---|---|
| 8:00–8:04 | -1 |
| 8:05–8:09 | +12 |
| 8:10–8:14 | -4 |
| 8:15–8:19 | -1 |
| 8:20–8:24 | -5 |
| 8:25–8:29 | -12 |
| 8:30–8:34 | -2 |
| 8:35–8:39 | 0 |
| 8:40–8:40 | 0 |
| 8:45–8:49 | -6 |
| 8:50–8:54 | +24 |
| 8:55–8:59 | 0 |
| service | |

1. What might a positive number mean in this context? What about a negative number?

2. What would a "0" in the second column mean in this context?

3. Which numbers—positive or negative—result in fewer bottles in the machine?

4. At what time was there the greatest change to the number of bottles in the machine? How did that change affect the number of remaining bottles in the machine?

5. At which time period, 8:05–8:09 or 8:25–8:29, was there a greater change to the number of bottles in the machine? Explain your reasoning.

6. The machine must be emptied to be serviced. If there are 40 bottles in the machine when it is to be serviced, what number will go in the second column in the table?

### Are you ready for more?

Priya, Mai, and Lin went to a cafe on a weekend. Their shared bill came to $25. Each student gave the server a $10 bill. The server took this $30 and brought back five $1 bills in change. Each student took $1 back, leaving the rest, $2, as a tip for the server.

As she walked away from the cafe, Lin thought, "Wait—this doesn't make sense. Since I put in $10 and got $1 back, I wound up paying $9. So did Mai and Priya. Together, we paid $27. Then we left a $2 tip. That makes $29 total. And yet we originally gave the waiter $30. Where did the extra dollar go?"

Think about the situation and about Lin's question. Do you agree that the numbers didn't add up properly? Explain your reasoning.

### Lesson 3 Summary

To order rational numbers from least to greatest, we list them in the order they appear on the number line from left to right. For example, we can see that the numbers

-2.7, -1.3, 0.8

are listed from least to greatest because of the order they appear on the number line.

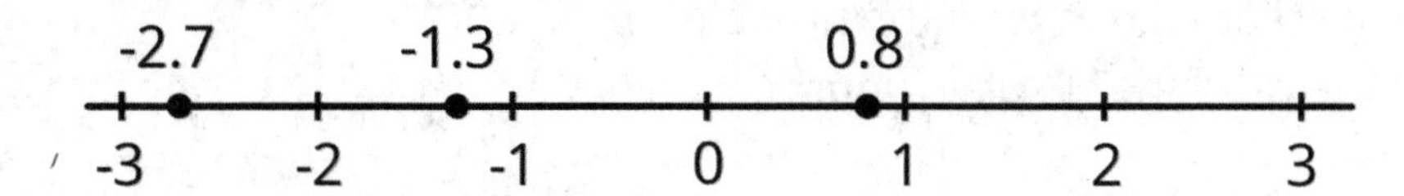

# Lesson 4: Absolute Value of Numbers

Let's explore distances from zero more closely.

## 4.1: Number Talk: Closer to Zero

For each pair of expressions, decide mentally which one has a value that is closer to 0.

$\frac{9}{11}$ or $\frac{15}{11}$

$\frac{1}{5}$ or $\frac{1}{9}$

1.25 or $\frac{5}{4}$

0.01 or 0.001

## 4.2: Jumping Flea

1. A flea is jumping around on a number line.

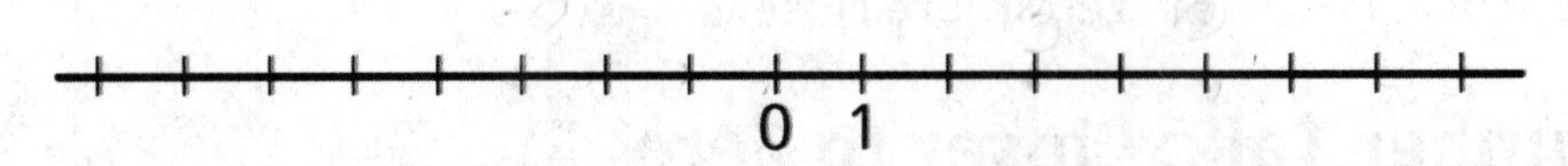


a. If the flea starts at 1 and jumps 4 units to the right, where does it end up? How far away from 0 is this?

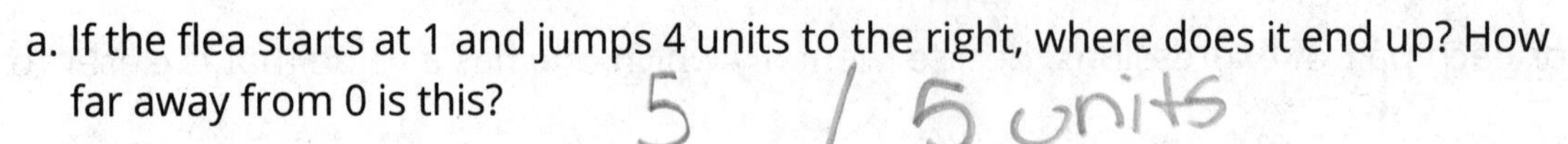


b. If the flea starts at 1 and jumps 4 units to the left, where does it end up? How far away from 0 is this?

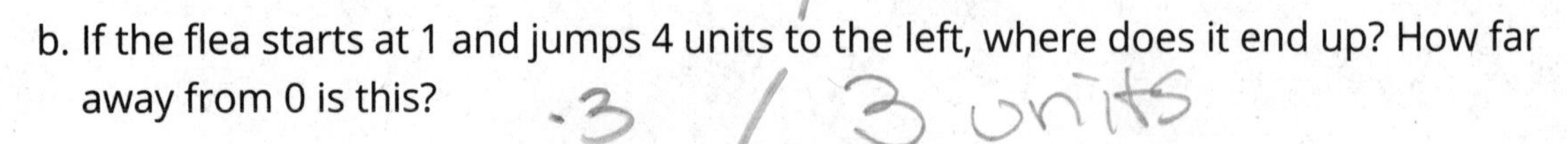


c. If the flea starts at 0 and jumps 3 units away, where might it land?

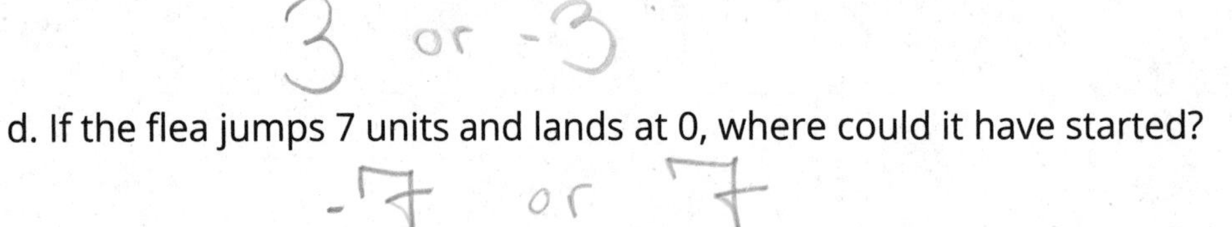


d. If the flea jumps 7 units and lands at 0, where could it have started?

e. The **absolute value** of a number is the distance it is from 0. The flea is currently to the left of 0 and the absolute value of its location is 4. Where on the number line is it?

-4

f. If the flea is to the left of 0 and the absolute value of its location is 5, where on the number line is it?

-5

g. If the flea is to the right of 0 and the absolute value of its location is 2.5, where on the number line is it?

2.5

2. We use the notation |-2| to say "the absolute value of -2," which means "the distance of -2 from 0 on the number line."

a. What does |-7| mean and what is its value?

b. What does |1.8| mean and what is its value?

## Lesson 4 Practice Problems

1. The temperature at dawn is $6^\circ\text{C}$ away from 0. Select **all** the temperatures that are possible.

   A. $-12^\circ\text{C}$

   B. $-6^\circ\text{C}$

   C. $0^\circ\text{C}$

   D. $6^\circ\text{C}$

   E. $12^\circ\text{C}$

2. On the number line, plot and label all numbers with an absolute value of $\frac{3}{2}$.

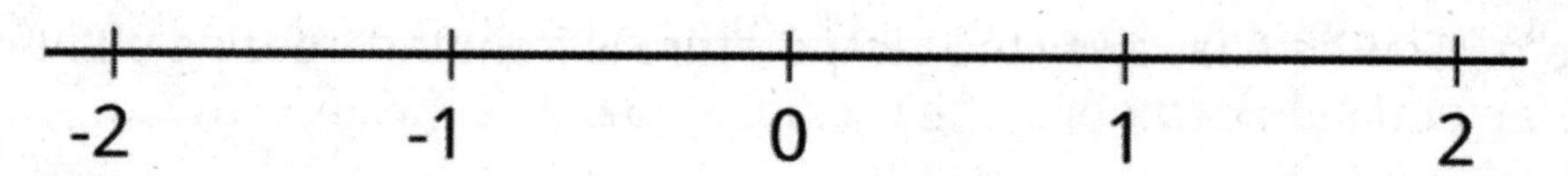


3. Put these numbers in order, from least to greatest.

   $|-2.7|$ 0 1.3 $|-1|$ 2

4. Lin's family needs to travel 325 miles to reach her grandmother's house.

   a. At 26 miles, what percentage of the trip's distance have they completed?

   b. How far have they traveled when they have completed 72% of the trip's distance?

   c. At 377 miles, what percentage of the trip's distance have they completed?

   (From Unit 3, Lesson 19.)

5. Elena donates some money to charity whenever she earns money as a babysitter. The table shows how much money, $d$, she donates for different amounts of money, $m$, that she earns.

| $d$ | 4.44 | 1.80 | 3.12 | 3.60 | 2.16 |
|---|---|---|---|---|---|
| $m$ | 37 | 15 | 26 | 30 | 18 |

a. What percent of her income does Elena donate to charity? Explain or show your work.

b. Which quantity, $m$ or $d$, would be the better choice for the dependent variable in an equation describing the relationship between $m$ and $d$? Explain your reasoning.

c. Use your choice from the second question to write an equation that relates $m$ and $d$.

(From Unit 4, Lesson 17.)

6. How many times larger is the first number in the pair than the second?

a. $3^4$ is ____ times larger than $3^3$.

b. $5^3$ is ____ times larger than $5^2$.

c. $7^{10}$ is ____ times larger than $7^8$.

d. $17^6$ is ____ times larger than $17^4$.

e. $5^{10}$ is ____ times larger than $5^4$.

(From Unit 4, Lesson 13.)

# Lesson 5: Comparing Numbers and Distance from Zero

Let's use absolute value and negative numbers to think about elevation.

## 5.1: Opposites

1. $a$ is a rational number. Choose a value for $a$ and plot it on the number line.

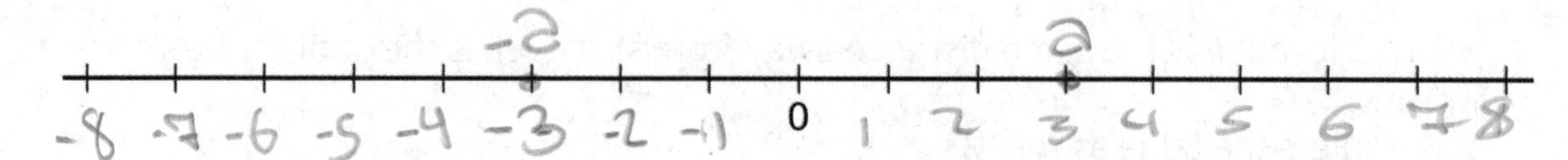


2. a. Based on where you plotted $a$, plot $-a$ on the same number line.

   b. What is the value of $-a$ that you plotted?

   the oppisite of a

3. Noah said, "If $a$ is a rational number, $-a$ will always be a negative number." Do you agree with Noah? Explain your reasoning.

   No, because if a was a negative number, -a would not be a negative number

LUCIA

## 5.2: Submarine

A submarine is at an elevation of -100 feet (100 feet below sea level). Let's compare the elevations of these four people to that of the submarine:

- Clare's elevation is greater than the elevation of the submarine. Clare is farther from sea level than the submarine.
- Andre's elevation is less than the elevation of the submarine. Andre is farther away from sea level than the submarine.
- Han's elevation is greater than the elevation of the submarine. Han is closer to sea level than is the submarine.
- Lin's elevation is the same distance away from sea level as the submarine's.

1. Complete the table as follows.

    a. Write a possible elevation for each person.

    b. Use $<$, $>$, or $=$ to compare the elevation of that person to that of the submarine.

    c. Use absolute value to tell how far away the person is from sea level (elevation 0).

    As an example, the first row has been filled with a possible elevation for Clare.

| | possible elevation | compare to submarine | distance from sea level |
|---|---|---|---|
| **Clare** | 150 feet | $150 > -100$ | $\lvert 150 \rvert$ or 150 feet |
| **Andre** | -150 | -175 < -100 | $\lvert 150 \rvert$ or 200 ft |
| **Han** | -50 | 50 > -100 | $\lvert -50 \rvert$ or 50 ft |
| **Lin** | 100 | 100 > -100 | $\lvert 100 \rvert$ or 100 ft. |

2. Priya says her elevation is less than the submarine's and she is closer to sea level. Is this possible? Explain your reasoning. No because if she is in negatives, the number closest to 0 is >.



iM

## 5.3: Info Gap: Points on the Number Line

Your teacher will give you either a *problem card* or a *data card*. Do not show or read your card to your partner.

If your teacher gives you the *problem card*:

1. Silently read your card and think about what information you need to be able to answer the question.
2. Ask your partner for the specific information that you need.
3. Explain how you are using the information to solve the problem.

   Continue to ask questions until you have enough information to solve the problem.
4. Share the *problem card* and solve the problem independently.
5. Read the *data card* and discuss your reasoning.

If your teacher gives you the *data card*:

1. Silently read your card.
2. Ask your partner *"What specific information do you need?"* and wait for them to *ask* for information.

   If your partner asks for information that is not on the card, do not do the calculations for them. Tell them you don't have that information.
3. Before sharing the information, ask "*Why do you need that information?*" Listen to your partner's reasoning and ask clarifying questions.
4. Read the *problem card* and solve the problem independently.
5. Share the *data card* and discuss your reasoning.

## 5.4: Inequality Mix and Match

Here are some numbers and inequality symbols. Work with your partner to write true comparison statements.

| | | | | | |
|---|---|---|---|---|---|
| -0.7 | $-\frac{3}{5}$ | 1 | 4 | $\lvert -8 \rvert$ | $<$ |
| $-\frac{6}{3}$ | -2.5 | 2.5 | 8 | $\lvert 0.7 \rvert$ | $=$ |
| -4 | 0 | $\frac{7}{2}$ | $\lvert 3 \rvert$ | $\lvert -\frac{5}{2} \rvert$ | $>$ |

One partner should select two numbers and one comparison symbol and use them to write a true statement using symbols. The other partner should write a sentence in words with the same meaning, using the following phrases:

- is equal to
- is the absolute value of
- is greater than
- is less than

For example, one partner could write $4 < 8$ and the other would write, "4 is less than 8." Switch roles until each partner has three true mathematical statements and three sentences written down.

## Are you ready for more?

For each question, choose a value for each variable to make the whole statement true. (When the word *and* is used in math, both parts have to be true for the whole statement to be true.) Can you do it if one variable is negative and one is positive? Can you do it if both values are negative?

1. $x < y$ and $|x| < y$.

2. $a < b$ and $|a| < |b|$.

3. $c < d$ and $|c| > d$.

4. $t < u$ and $|t| > |u|$.

## Lesson 5 Summary

We can use elevation to help us compare two rational numbers or two absolute values.

- Suppose an anchor has an elevation of -10 meters and a house has an elevation of 12 meters. To describe the anchor having a lower elevation than the house, we can write $-10 < 12$ and say "-10 is less than 12."

- The anchor is closer to sea level than the house is to sea level (or elevation of 0). To describe this, we can write $|-10| < |12|$ and say "the distance between -10 and 0 is less than the distance between 12 and 0."

We can use similar descriptions to compare rational numbers and their absolute values outside of the context of elevation.

- To compare the distance of -47.5 and 5.2 from 0, we can say: $|-47.5|$ is 47.5 units away from 0, and $|5.2|$ is 5.2 units away from 0, so $|-47.5| > |5.2|$.

- $|-18| > 4$ means that the absolute value of -18 is greater than 4. This is true because 18 is greater than 4.

# Lesson 5 Practice Problems

1. In the context of elevation, what would |-7| feet mean?

2. Match the the statements written in English with the mathematical statements.

A. The number -4 is a distance of 4 units away from 0 on the number line.

B. The number -63 is more than 4 units away from 0 on the number line.

C. The number 4 is greater than the number -4.

D. The numbers 4 and -4 are the same distance away from 0 on the number line.

E. The number -63 is less than the number 4.

F. The number -63 is further away from 0 than the number 4 on the number line.

1. $|-63| > 4$
2. $-63 < 4$
3. $|-63| > |4|$
4. $|-4| = 4$
5. $4 > -4$
6. $|4| = |-4|$

3. Compare each pair of expressions using >, <, or =.

- -32 ____ 15
- |-32| ____ |15|
- 5 ____ -5
- |5| ____ |-5|
- 2 ____ -17
- 2 ____ |-17|
- |-27| ____ |-45|
- |-27| ____ -45

4. Mai received and spent money in the following ways last month. For each example, write a signed number to represent the change in money from her perspective.

   a. Her grandmother gave her $25 in a birthday card.

   b. She earned $14 dollars babysitting.

   c. She spent $10 on a ticket to the concert.

   d. She donated $3 to a local charity

   e. She got $2 interest on money that was in her savings account.

   (From Unit 7, Lesson 3.)

5. Here are the lowest temperatures recorded in the last 2 centuries for some US cities.

   - Death Valley, CA was $-45^\circ\text{F}$ in January of 1937.
   - Danbury, CT was $-37^\circ\text{F}$ in February of 1943.
   - Monticello, FL was $-2^\circ\text{F}$ in February of 1899.
   - East Saint Louis, IL was $-36^\circ\text{F}$ in January of 1999.
   - Greenville, GA was $-17^\circ\text{F}$ in January of 1940.

   a. Which of these states has the lowest record temperature?

   b. Which state has a lower record temperature, FL or GA?

   c. Which state has a lower record temperature, CT or IL?

   d. How many more degrees colder is the record temperature for GA than for FL?

   (From Unit 7, Lesson 1.)

6. Noah was assigned to make 64 cookies for the bake sale. He made 125% of that number. 90% of the cookies he made were sold. How many of Noah's cookies were left after the bake sale?

   (From Unit 6, Lesson 4.)

# Lesson 6: Changing Temperatures

Let's add signed numbers.

## 6.1: Which One Doesn't Belong: Arrows

Which pair of arrows doesn't belong?

1.

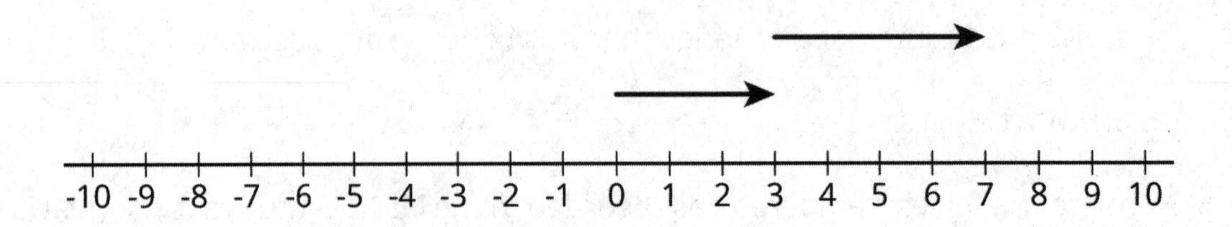


2.

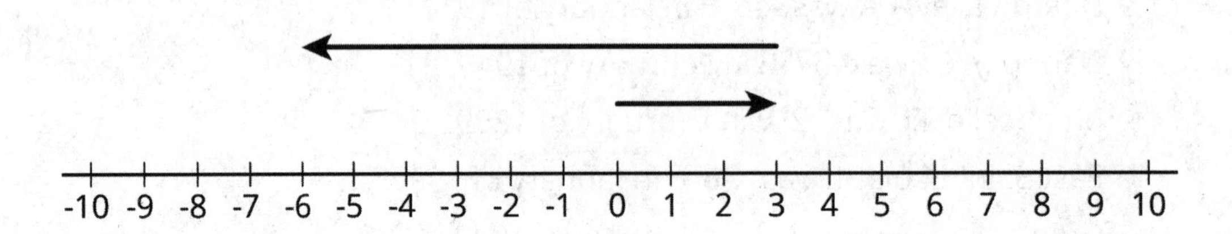


3.

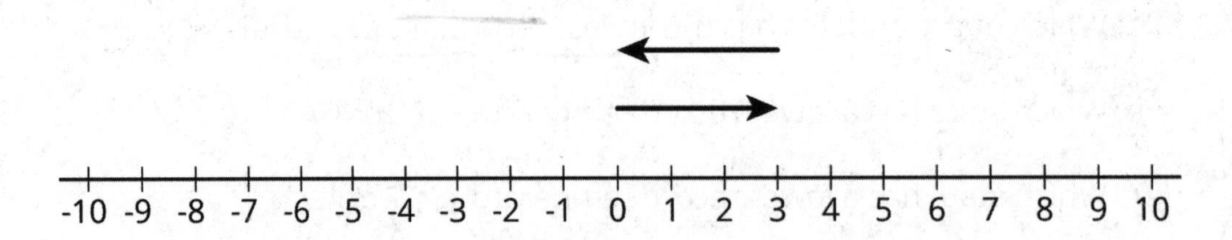


4.

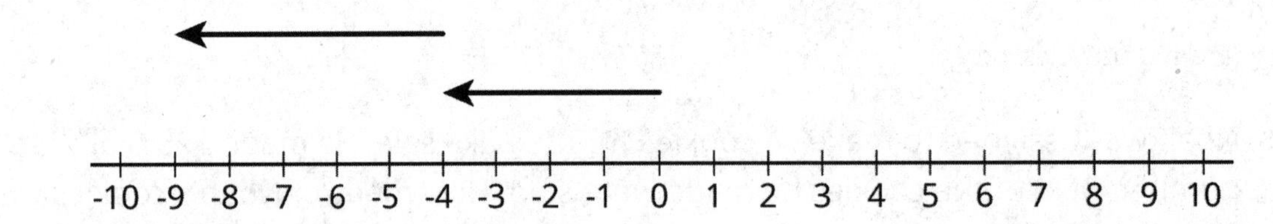

## 6.2: Warmer and Colder

1. Complete the table and draw a number line diagram for each situation.

| | start (°C) | change (°C) | final (°C) | addition equation |
|---|---|---|---|---|
| a | +40 | 10 degrees warmer | +50 | $40 + 10 = 50$ |
| b | +40 | 5 degrees colder | -35 | 40 + -5 = 35 |
| c | +40 | 30 degrees colder | 10 | 40 + -30 = 10 |
| d | +40 | 40 degrees colder | 0 | 40 + -40 = 0 |
| e | +40 | 50 degrees colder | -50 | 40 + -50 = -10 |

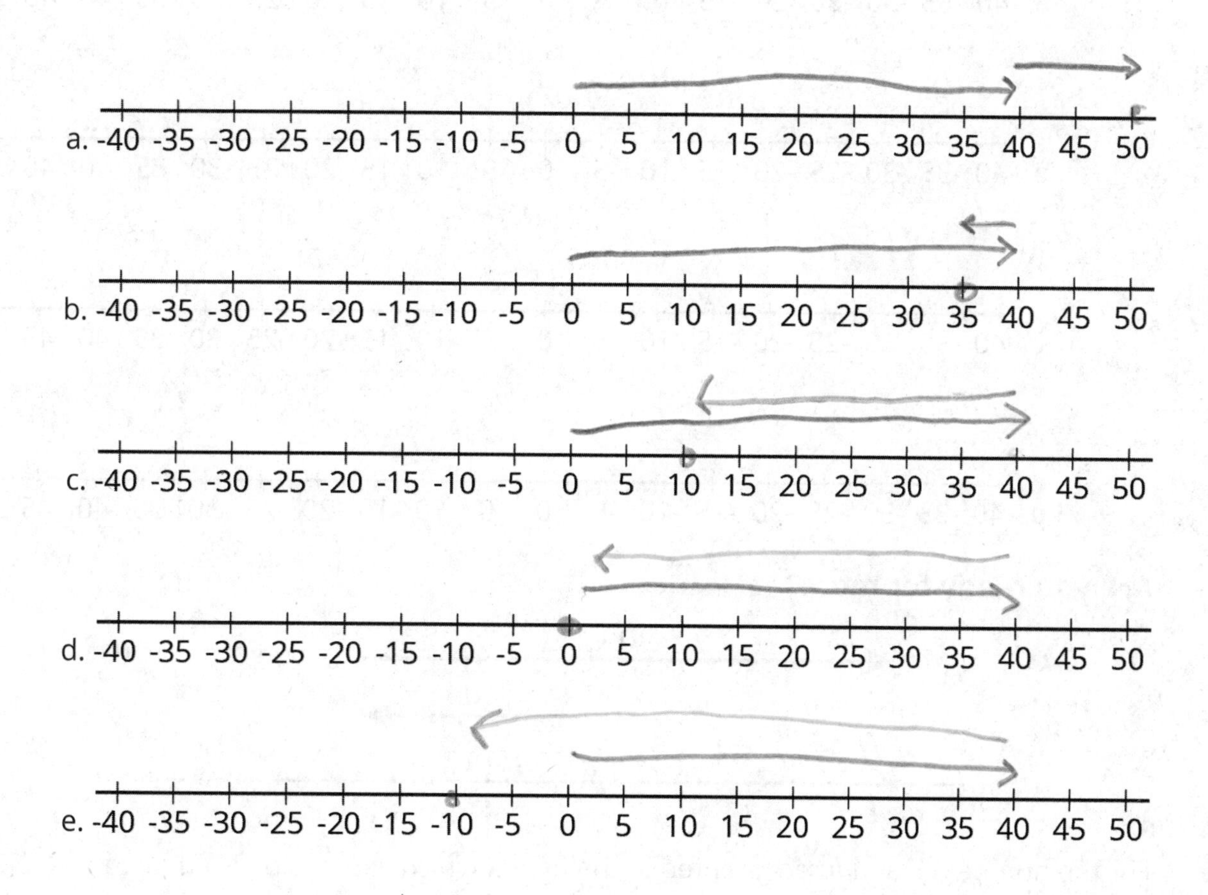

2. Complete the table and draw a number line diagram for each situation.

| | start (°C) | change (°C) | final (°C) | addition equation |
|---|---|---|---|---|
| a | -20 | 30 degrees warmer | 10 | -20 + 30 = 10 |
| b | -20 | 35 degrees warmer | 15 | -20 + 35 = 15 |
| c | -20 | 15 degrees warmer | -5 | -20 + 15 = -5 |
| d | -20 | 15 degrees colder | -35 | -20 + -15 = -35 |

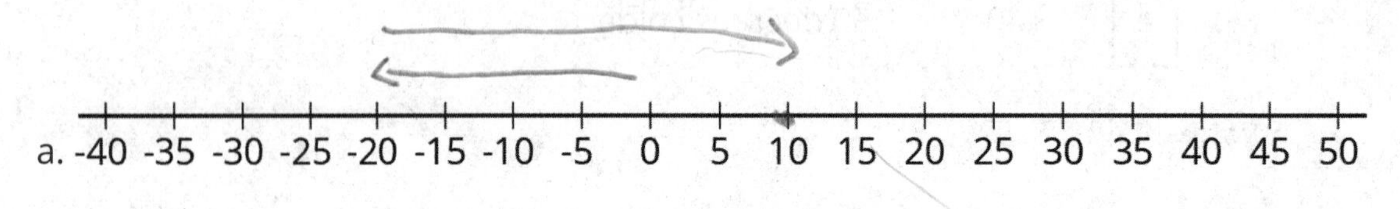


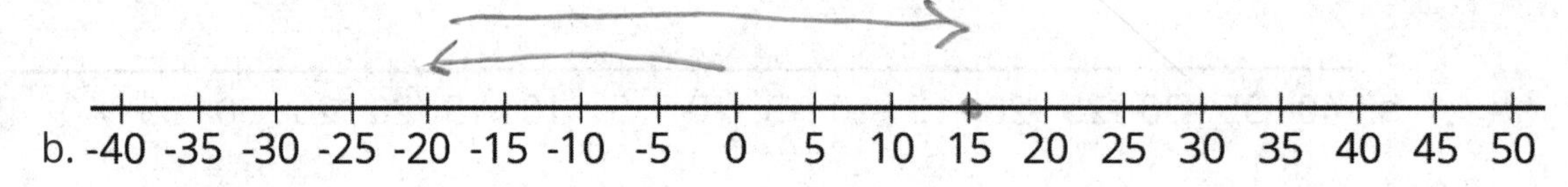


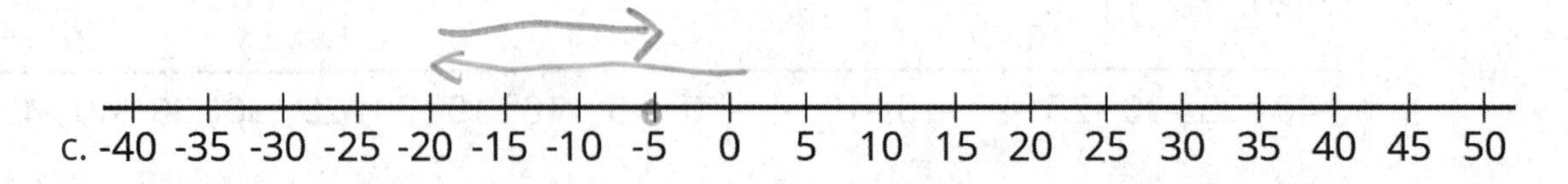


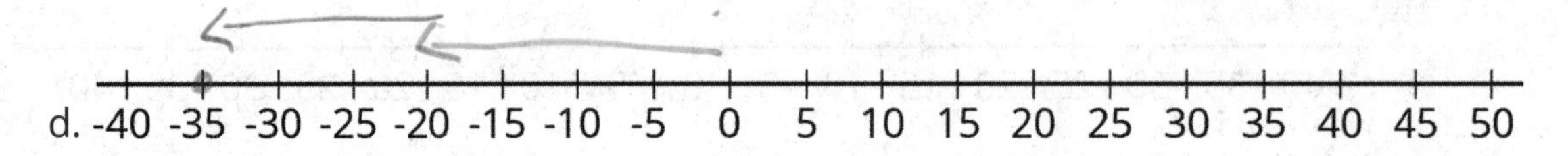


**Are you ready for more?**

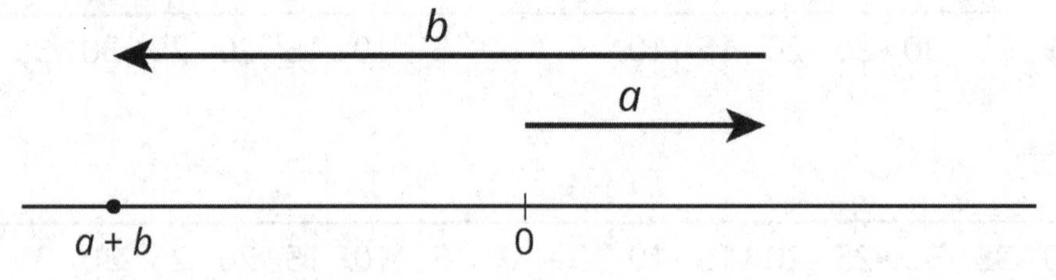


For the numbers $a$ and $b$ represented in the figure, which expression is equal to $|a + b|$?

$|a| + |b|$ $|a| - |b|$ $|b| - |a|$

## 6.3: Winter Temperatures

One winter day, the temperature in Houston is $8^\circ$ Celsius. Find the temperatures in these other cities. Explain or show your reasoning.

1. In Orlando, it is $10^\circ$ warmer than it is in Houston.

2. In Salt Lake City, it is $8^\circ$ colder than it is in Houston.

3. In Minneapolis, it is $20^\circ$ colder than it is in Houston.

4. In Fairbanks, it is $10^\circ$ colder than it is in *Minneapolis*.

5. Write an addition equation that represents the relationship between the temperature in Houston and the temperature in Fairbanks.

## Lesson 6 Summary

If it is $42^\circ$ outside and the temperature increases by $7^\circ$, then we can add the initial temperature and the change in temperature to find the final temperature.

$$42 + 7 = 49$$

If the temperature decreases by $7^\circ$, we can either subtract $42 - 7$ to find the final temperature, or we can think of the change as $-7^\circ$. Again, we can add to find the final temperature.

$$42 + (-7) = 35$$

In general, we can represent a change in temperature with a positive number if it increases and a negative number if it decreases. Then we can find the final temperature by adding the initial temperature and the change. If it is $3^\circ$ and the temperature decreases by $7^\circ$, then we can add to find the final temperature.

$$3 + (-7) = -4$$

We can represent signed numbers with arrows on a number line. We can represent positive numbers with arrows that start at 0 and points to the right. For example, this arrow represents +10 because it is 10 units long and it points to the right.

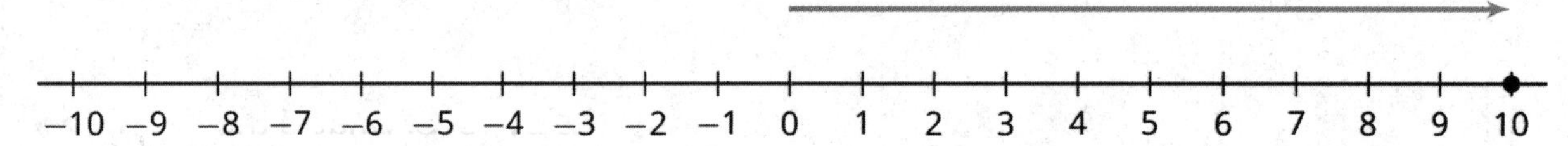


We can represent negative numbers with arrows that start at 0 and point to the left. For example, this arrow represents -4 because it is 4 units long and it points to the left.

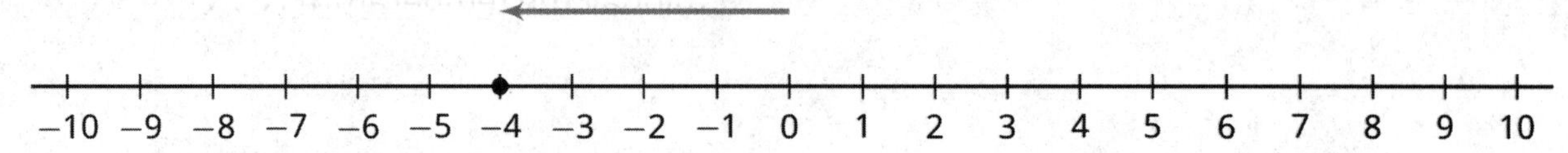


To represent addition, we put the arrows "tip to tail." So this diagram represents $3 + 5$:

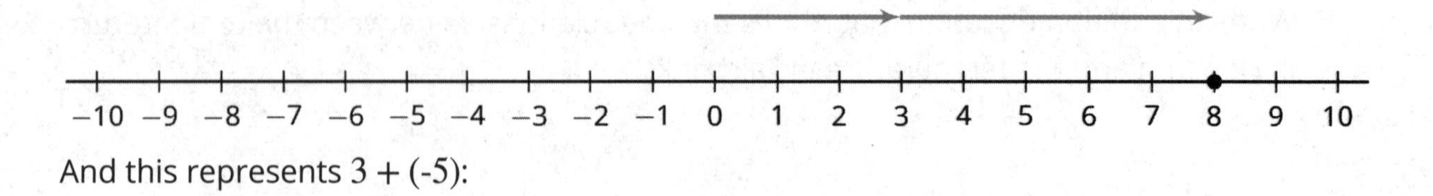


And this represents $3 + (-5)$:

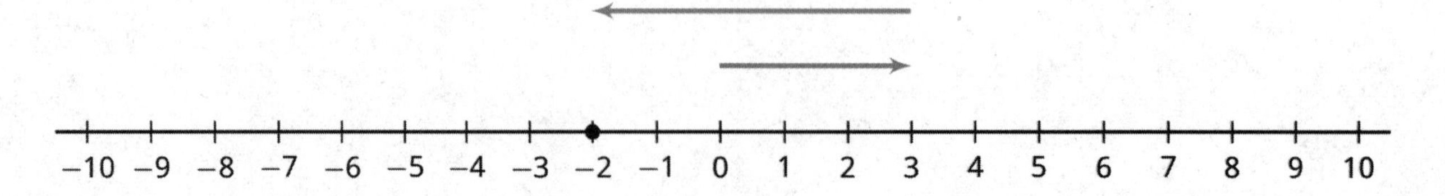

## Lesson 6 Practice Problems

1. Draw a diagram to represent each of these situations. Then write an addition expression that represents the final temperature.

   a. The temperature was $80^\circ F$ and then fell $20^\circ F$.

   b. The temperature was $-13^\circ F$ and then rose $9^\circ F$.

   c. The temperature was $-5^\circ F$ and then fell $8^\circ F$.

2. a. The temperature is $-2^\circ C$. If the temperature rises by $15^\circ C$, what is the new temperature?

   b. At midnight the temperature is $-6^\circ C$. At midday the temperature is $9^\circ C$. By how much did the temperature rise?

3. Complete each statement with a number that makes the statement true.

   a. ____ $< 7^\circ C$

   b. ____ $< -3^\circ C$

   c. $-0.8^\circ C <$ ____ $< -0.1^\circ C$

   d. ____ $> -2^\circ C$

   (From Unit 7, Lesson 1.)

4. Match the statements written in English with the mathematical statements. All of these statements are true.

A. The number -15 is further away from 0 than the number -12 on the number line.

B. The number -12 is a distance of 12 units away from 0 on the number line.

C. The distance between -12 and 0 on the number line is greater than -15.

D. The numbers 12 and -12 are the same distance away from 0 on the number line.

E. The number -15 is less than the number -12.

F. The number 12 is greater than the number -12.

1. $|-12| > -15$
2. $-15 < -12$
3. $|-15| > |-12|$
4. $|-12| = 12$
5. $12 > -12$
6. $|12| = |-12|$

(From Unit 7, Lesson 5.)

5. Evaluate each expression.

- $2^3 \cdot 3$
- $6^2 \div 4$
- $\frac{4^2}{2}$
- $2^3 - 2$
- $3^1$
- $10^2 + 5^2$

(From Unit 4, Lesson 13.)

6. Decide whether each table could represent a proportional relationship. If the relationship could be proportional, what would be the constant of proportionality?

a. The number of wheels on a group of buses.

| number of buses | number of wheels | wheels per bus |
|---|---|---|
| 5 | 30 | |
| 8 | 48 | |
| 10 | 60 | |
| 15 | 90 | |

b. The number of wheels on a train.

| number of train cars | number of wheels | wheels per train car |
|---|---|---|
| 20 | 184 | |
| 30 | 264 | |
| 40 | 344 | |
| 50 | 424 | |

(From Unit 5, Lesson 4.)

# Lesson 7: Changing Elevation

Let's solve problems about adding signed numbers.

## 7.1: That's the Opposite

1. Draw arrows on a number line to represents these situations:

    a. The temperature was -5 degrees. Then the temperature rose 5 degrees.

    b. A climber was 30 feet above sea level. Then she descended 30 feet.

2. What's the opposite?

    a. Running 150 feet east.

    b. Jumping down 10 steps.

    c. Pouring 8 gallons into a fish tank.

## Lesson 7 Summary

The opposite of a number is the same distance from 0 but on the other side of 0.

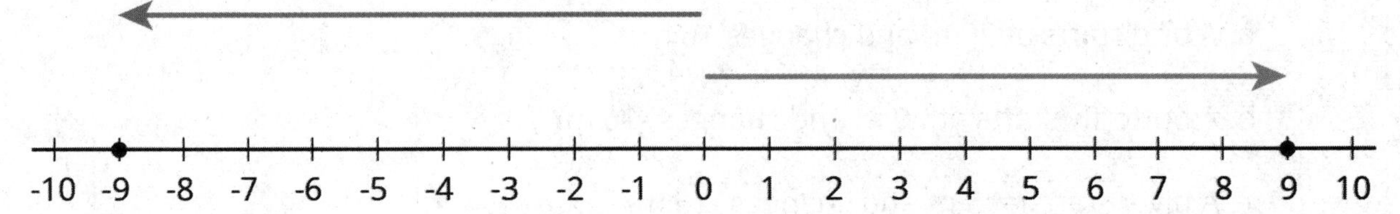


The opposite of -9 is 9. When we add opposites, we always get 0. This diagram shows that $9 + -9 = 0$.

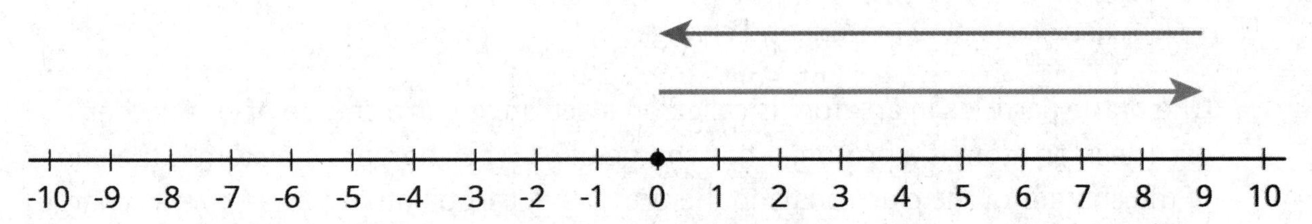


When we add two numbers with the same sign, the arrows that represent them point in the same direction. When we put the arrows tip to tail, we see the sum has the same sign.

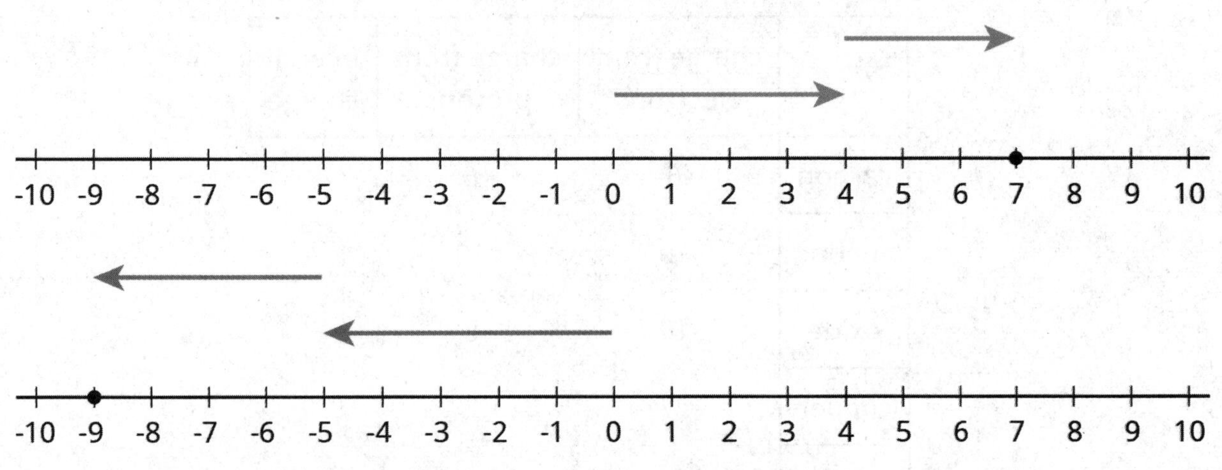


To find the sum, we add the magnitudes and give it the correct sign. For example, $(-5) + (-4) = -(5 + 4)$.

On the other hand, when we add two numbers with different signs, we subtract their magnitudes (because the arrows point in the opposite direction) and give it the sign of the number with the larger magnitude. For example, $(-5) + 12 = +(12 - 5)$.

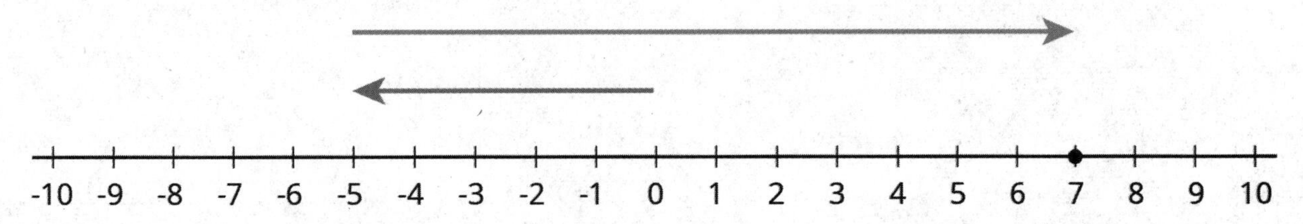

## Lesson 7 Practice Problems

1. What is the final elevation if

   a. A bird starts at 20 m and changes 16 m?

   b. A butterfly starts at 20 m and changes -16 m?

   c. A diver starts at 5 m and changes -16 m?

   d. A whale starts at -9 m and changes 11 m?

   e. A fish starts at -9 meters and changes -11 meters?

2. One of the particles in an atom is called an electron. It has a charge of -1. Another particle in an atom is a proton. It has charge of +1. The charge of an atom is the sum of the charges of the electrons and the protons. A carbon atom has an overall charge of 0, because it has 6 electrons and 6 protons and $-6 + 6 = 0$. Find the overall charge for the rest of the elements on the list.

| | charge from electrons | charge from protons | overall charge |
|---|---|---|---|
| carbon | -6 | +6 | 0 |
| neon | -10 | +10 | |
| oxide | -10 | +8 | |
| copper | -27 | +29 | |
| tin | -50 | +50 | |

3. Add.

$14.7 + 28.9$ $\quad$ $-9.2 + 4.4$ $\quad$ $-81.4 + (-12)$ $\quad$ $51.8 + (-0.8)$

4. Select **all** the true statements.

   A. $-5 < |-5|$

   B. $|-6| < -5$

   C. $|-6| < 3$

   D. $4 < |-7|$

   E. $|-7| < |-8|$

   (From Unit 7, Lesson 5.)

5. Last week, the price, in dollars, of a gallon of gasoline was $g$. This week, the price of gasoline per gallon increased by 5%. Which expressions represent this week's price, in dollars, of a gallon of gasoline? Select **all** that apply.

   A. $g + 0.05$

   B. $g + 0.05g$

   C. $1.05g$

   D. $0.05g$

   E. $(1 + 0.05)g$

   (From Unit 6, Lesson 5.)

6. A family goes to a restaurant. When the bill comes, this is printed at the bottom of it:

   Gratuity Guide For Your Convenience:
   15% would be $4.89
   18% would be $5.87
   20% would be $6.52

   How much was the price of the meal? Explain your reasoning.

   (From Unit 6, Lesson 7.)

# Lesson 8: Money and Debts

Let's apply what we know about signed numbers to money.

## 8.1: Concert Tickets

Priya wants to buy three tickets for a concert. She has earned $135 and each ticket costs $50. She borrows the rest of the money she needs from a bank and buys the tickets.

1. How can you represent the amount of money that Priya has after buying the tickets?

2. How much more money will Priya need to earn to pay back the money she borrowed from the bank?

3. How much money will she have after she pays back the money she borrowed from the bank?

### Are you ready for more?

The *national debt* of a country is the total amount of money the government of that country owes. Imagine everyone in the United States was asked to help pay off the national debt. How much would each person have to pay?

### Lesson 8 Summary

Banks use positive numbers to represent money that gets put into an account and negative numbers to represent money that gets taken out of an account. When you put money into an account, it is called a **deposit**. When you take money out of an account, it is called a **withdrawal**.

People also use negative numbers to represent debt. If you take out more money from your account than you put in, then you owe the bank money, and your account balance will be a negative number to represent that debt. For example, if you have $200 in your bank account, and then you write a check for $300, you will owe the bank $100 and your account balance will be -$100.

| starting balance | deposits and withdrawals | new balance |
|---|---|---|
| 0 | 50 | $0 + 50$ |
| 50 | 150 | $50 + 150$ |
| 200 | -300 | $200 + (-300)$ |
| -100 | | |

In general, you can find a new account balance by adding the value of the deposit or withdrawal to it. You can also tell quickly how much money is needed to repay a debt using the fact that to get to zero from a negative value you need to add its opposite.

### Glossary

- deposit
- withdrawal

## Lesson 8 Practice Problems

1. The table shows five transactions and the resulting account balance in a bank account, except some numbers are missing. Fill in the missing numbers.

| | transaction amount | account balance |
|---|---|---|
| transaction 1 | 200 | 200 |
| transaction 2 | -147 | 53 |
| transaction 3 | 90 | |
| transaction 4 | -229 | |
| transaction 5 | | 0 |

2. a. Clare has $54 in her bank account. A store credits her account with a $10 refund. How much does she now have in the bank?

   b. Mai's bank account is overdrawn by $60, which means her balance is -$60. She gets $85 for her birthday and deposits it into her account. How much does she now have in the bank?

   c. Tyler is overdrawn at the bank by $180. He gets $70 for his birthday and deposits it. What is his account balance now?

   d. Andre has $37 in his bank account and writes a check for $87. After the check has been cashed, what will the bank balance show?

3. Add.

a. $5\frac{3}{4} + (-\frac{1}{4})$

b. $-\frac{2}{3} + \frac{1}{6}$

c. $-\frac{8}{5} + (-\frac{3}{4})$

(From Unit 7, Lesson 7.)

4. Which is greater, $\frac{-9}{20}$ or -0.5? Explain how you know. If you get stuck, consider plotting the numbers on a number line.

(From Unit 7, Lesson 2.)

5. Decide whether or not each equation represents a proportional relationship.

a. Volume measured in cups ($c$) vs. the same volume measured in ounces ($z$): $c = \frac{1}{8}z$

b. Area of a square ($A$) vs. the side length of the square ($s$): $A = s^2$

c. Perimeter of an equilateral triangle ($P$) vs. the side length of the triangle ($s$): $3s = P$

d. Length ($L$) vs. width ($w$) for a rectangle whose area is 60 square units: $L = \frac{60}{w}$

(From Unit 5, Lesson 5.)

# Lesson 9: Representing Subtraction

Let's subtract signed numbers.

## 9.1: Equivalent Equations

Consider the equation $2 + 3 = 5$. Here are some more equations, using the same numbers, that express the same relationship in a different way:

$3 + 2 = 5$ $\quad$ $5 - 3 = 2$ $\quad$ $5 - 2 = 3$

For each equation, write two more equations, using the same numbers, that express the same relationship in a different way.

1. $9 + (-1) = 8$

2. $-11 + x = 7$

## 9.2: Subtraction with Number Lines

1. Here is an unfinished number line diagram that represents a sum of 8.

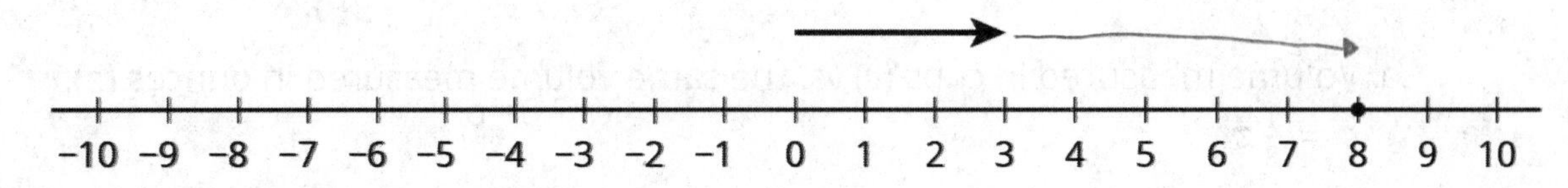


a. How long should the other arrow be?

5

b. For an equation that goes with this diagram, Mai writes $3 + ? = 8$. Tyler writes $8 - 3 = ?$. Do you agree with either of them?

c. What is the unknown number? How do you know?

2. Here are two more unfinished diagrams that represent sums.

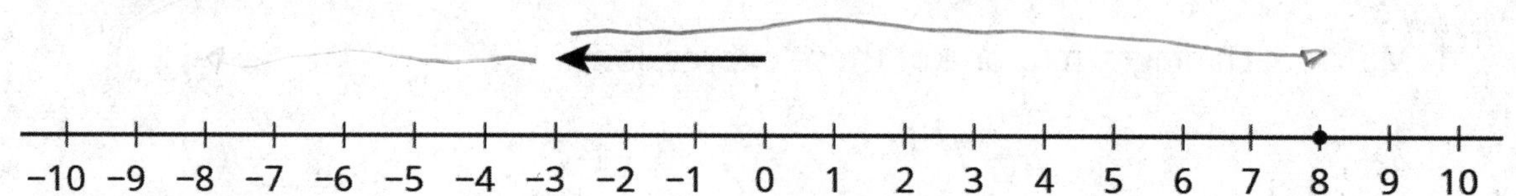

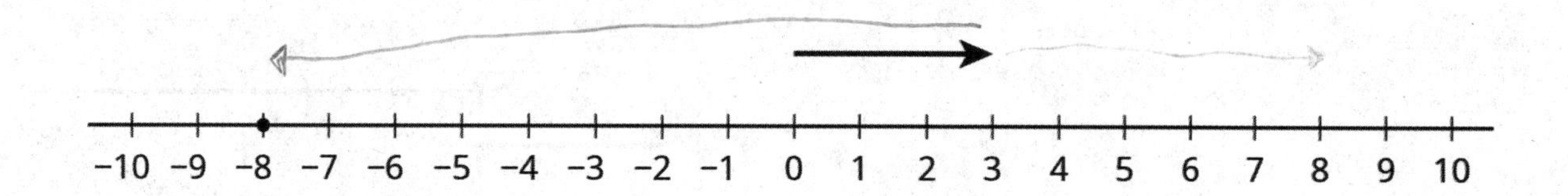

For each diagram:

a. What equation would Mai write if she used the same reasoning as before?

-3 + ? = 8 | 3 + ? = -8

b. What equation would Tyler write if he used the same reasoning as before?

8 - ? = -3 | ? - 3 = 8

c. How long should the other arrow be?

11

d. What number would complete each equation? Be prepared to explain your reasoning.

11

3. Draw a number line diagram for (-8) – (-3) = ? What is the unknown number? How do you know?

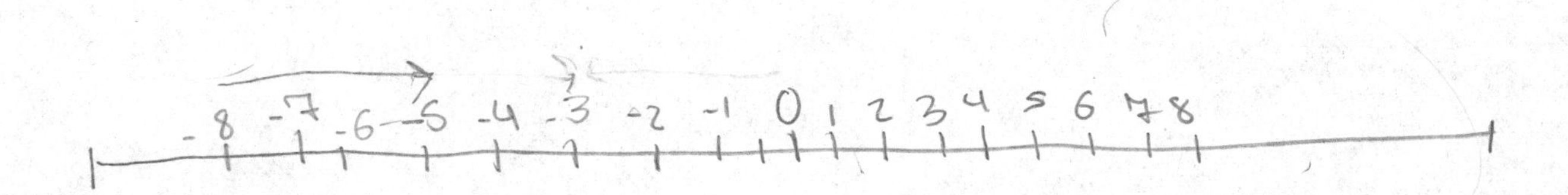

## 9.3: We Can Add Instead

1. Match each diagram to one of these expressions:

$3 + 7$ a

$3 - 7$ c

$3 + (-7)$ 

$3 - (-7)$ 

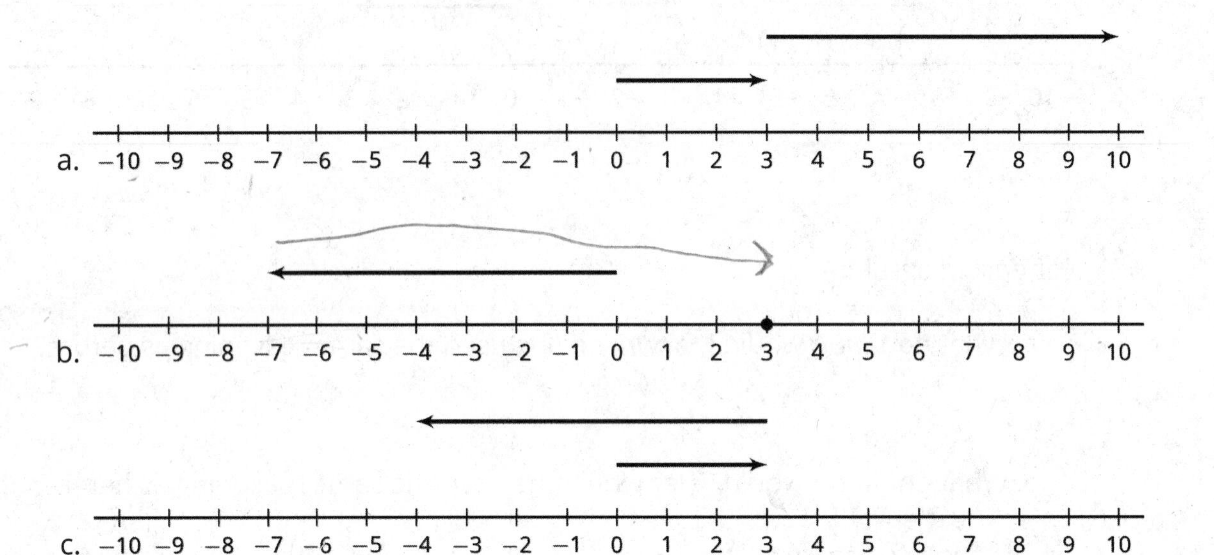


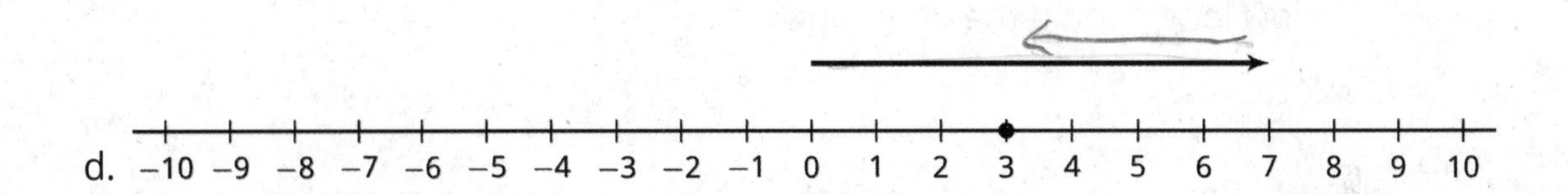


2. Which expressions in the first question have the same value? What do you notice?

3. Complete each of these tables. What do you notice?

| expression | value |
|---|---|
| $8 + (-8)$ | 0 |
| $8 - 8$ | 0 |
| $8 + (-5)$ | 3 |
| $8 - 5$ | 3 |
| $8 + (-12)$ | -4 |
| $8 - 12$ | -4 |

| expression | value |
|---|---|
| $-5 + 5$ | 0 |
| $-5 - (-5)$ | 0 |
| $-5 + 9$ | 4 |
| $-5 - (-9)$ | 4 |
| $-5 + 2$ | -3 |
| $-5 - (-2)$ | 3 |

**Are you ready for more?**

It is possible to make a new number system using *only* the numbers 0, 1, 2, and 3. We will write the symbols for adding and subtracting in this system like this: $2 \oplus 1 = 3$ and $2 \ominus 1 = 1$. The table shows some of the sums.

| $\oplus$ | 0 | 1 | 2 | 3 |
|---|---|---|---|---|
| 0 | 0 | 1 | 2 | 3 |
| 1 | 1 | 2 | 3 | 0 |
| 2 | 2 | 3 | 0 | 1 |
| 3 | | | | |

1. In this system, $1 \oplus 2 = 3$ and $2 \oplus 3 = 1$. How can you see that in the table?
2. What do you think $3 \oplus 1$ should be?
3. What about $3 \oplus 3$?
4. What do you think $3 \ominus 1$ should be?
5. What about $2 \ominus 3$?
6. Can you think of any uses for this number system?

## Lesson 9 Summary

The equation $7 - 5 = ?$ is equivalent to $? + 5 = 7$. The diagram illustrates the second equation.

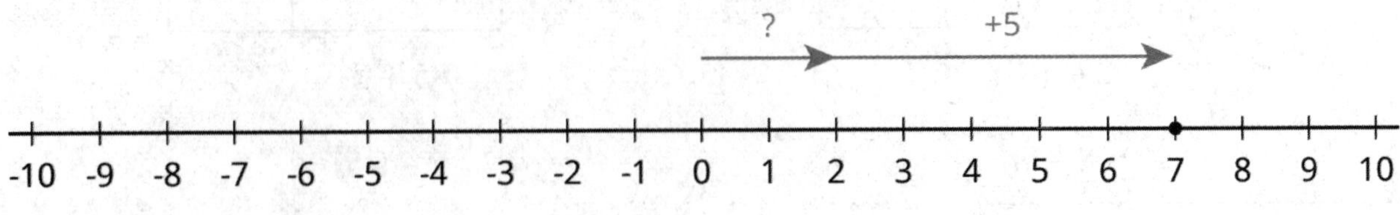


Notice that the value of $7 + (-5)$ is 2.

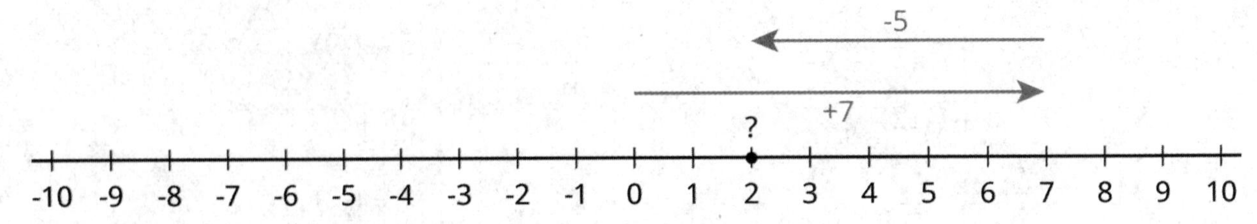


We can solve the equation $? + 5 = 7$ by adding -5 to both sides. This shows that $7 - 5 = 7 + (-5)$

Likewise, $3 - 5 = ?$ is equivalent to $? + 5 = 3$.

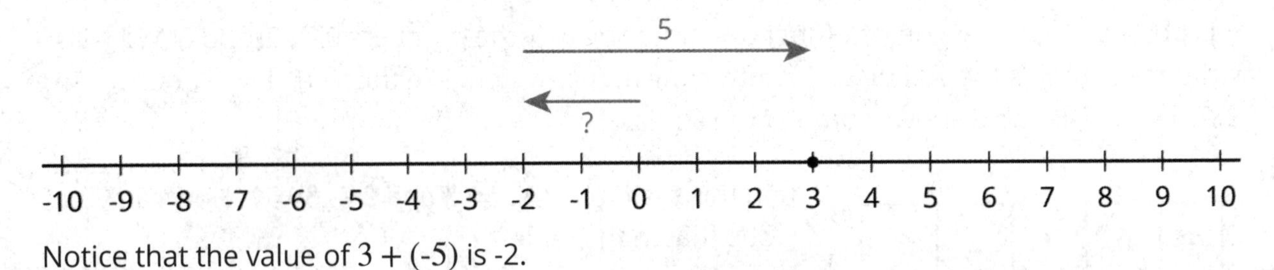


Notice that the value of $3 + (-5)$ is -2.

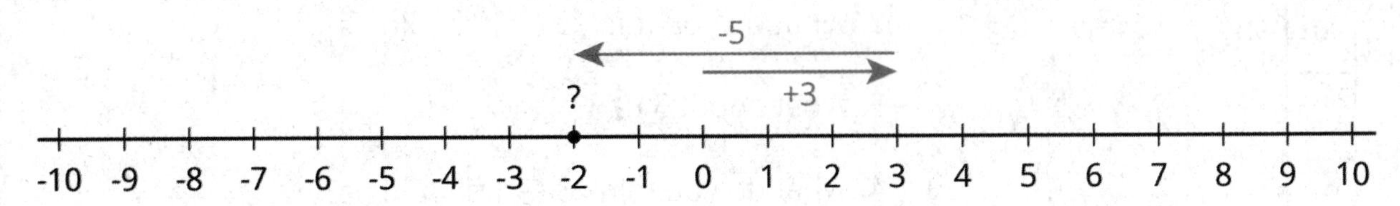


We can solve the equation $? + 5 = 3$ by adding -5 to both sides. This shows that $3 - 5 = 3 + (-5)$

In general:

$$a - b = a + (\text{-}b)$$

If $a - b = x$, then $x + b = a$. We can add $\text{-}b$ to both sides of this second equation to get that $x = a + (\text{-}b)$

## Lesson 9 Practice Problems

1. Write each subtraction equation as an addition equation.

   a. $a - 9 = 6$

   b. $p - 20 = -30$

   c. $z - (-12) = 15$

   d. $x - (-7) = -10$

2. Find each difference. If you get stuck, consider drawing a number line diagram.

   a. $9 - 4$

   b. $4 - 9$

   c. $9 - (-4)$

   d. $-9 - (-4)$

   e. $-9 - 4$

   f. $4 - (-9)$

   g. $-4 - (-9)$

   h. $-4 - 9$

3. Find the solution to each equation mentally.

   a. $30 + a = 40$

   b. $500 + b = 200$

   c. $-1 + c = -2$

   d. $d + 3{,}567 = 0$

4. A restaurant bill is $59 and you pay $72. What percentage gratuity did you pay?

(From Unit 6, Lesson 7.)

5. One kilogram is 2.2 pounds. Complete the tables. What is the interpretation of the constant of proportionality in each case?

______ kilogram per pound

| pounds | kilograms |
|---|---|
| 2.2 | 1 |
| 11 | |
| 5.5 | |
| 1 | |

______ pounds per kilogram

| kilograms | pounds |
|---|---|
| 1 | 2.2 |
| 7 | |
| 30 | |
| 0.5 | |

(From Unit 5, Lesson 1.)

# Lesson 10: Subtracting Rational Numbers

Let's bring addition and subtraction together.

## 10.1: Number Talk: Missing Addend

Solve each equation mentally. Rewrite each addition equation as a subtraction equation.

$247 + c = 458$

$c + 43.87 = 58.92$

$\frac{15}{8} + c = \frac{51}{8}$

## 10.2: Expressions with Altitude

A mountaineer is changing elevations. Write an expression that represents the difference between the final elevation and beginning elevation. Then write the value of the change. The first one is done for you.

| beginning elevation (feet) | final elevation (feet) | difference between final and beginning | change |
|---|---|---|---|
| +400 | +900 | $900 - 400$ | +500 |
| +400 | +50 | 50 - 400 | -350 |
| +400 | -120 | -120 - 400 | -520 |
| -200 | +610 | 610 - -200 | 810 |
| -200 | -50 | -50 - -200 | 150 |
| -200 | -500 | -200 + -500 | -300 |
| -200 | 0 | 0 + 200 | 200 |

**Are you ready for more?**

Fill in the table so that every row and every column sums to 0. Can you find another way to solve this puzzle?

| | -12 | 0 | | 5 |
|---|---|---|---|---|
| 0 | | | -18 | 25 |
| 25 | | -18 | 5 | -12 |
| -12 | | | | -18 |
| | -18 | 25 | -12 | |

| | -12 | 0 | | 5 |
|---|---|---|---|---|
| 0 | | | -18 | 25 |
| 25 | | -18 | 5 | -12 |
| -12 | | | | -18 |
| | -18 | 25 | -12 | |

## 10.3: Does the Order Matter?

1. Find the value of each subtraction expression.

| A | | B | |
|---|---|---|---|
| $3 - 2$ | 1 | $2 - 3$ | -1 |
| $5 - (-9)$ | 14 | $(-9) - 5$ | -14 |
| $(-11) - 2$ | -13 | $2 - (-11)$ | 13 |
| $(-6) - (-3)$ | -3 | $(-3) - (-6)$ | 3 |
| $(-1.2) - (-3.6)$ | 2.4 | $(-3.6) - (-1.2)$ | -2.4 |
| $(-2\frac{1}{2}) - (-3\frac{1}{2})$ | 1 | $(-3\frac{1}{2}) - (-2\frac{1}{2})$ | -1 |

2. What do you notice about the expressions in Column A compared to Column B?

They're flipped

3. What do you notice about their values?

They're opposites

## 10.4: Phone Inventory

A store tracks the number of cell phones it has in stock and how many phones it sells.

The table shows the inventory for one phone model at the beginning of each day last week. The inventory changes when they sell phones or get shipments of phones into the store.

| | inventory | change |
|---|---|---|
| Monday | 18 | -2 |
| Tuesday | 16 | -5 |
| Wednesday | 11 | -7 |
| Thursday | 4 | -6 |
| Friday | -2 | 20 |

1. What do you think it means when the change is positive? Negative?

pos. - they got more in stock

neg. - they sold them

2. What do you think it means when the inventory is positive? Negative?

positive - that is how many they have in stock

negative - they owe that many phones to (minimum)

3. Based on the information in the table, what do you think the inventory will be at on Saturday morning? Explain your reasoning.

I think it will be 18 because they bought 20 phones and they gave two away

4. What is the difference between the greatest inventory and the least inventory?

18 - |-2| = 20

## Lesson 10 Summary

When we talk about the difference of two numbers, we mean, "subtract them." Usually, we subtract them in the order they are named. For example, the difference of +8 and -6 is $8 - (-6)$.

The difference of two numbers tells you how far apart they are on the number line. 8 and -6 are 14 units apart, because $8 - (-6) = 14$:

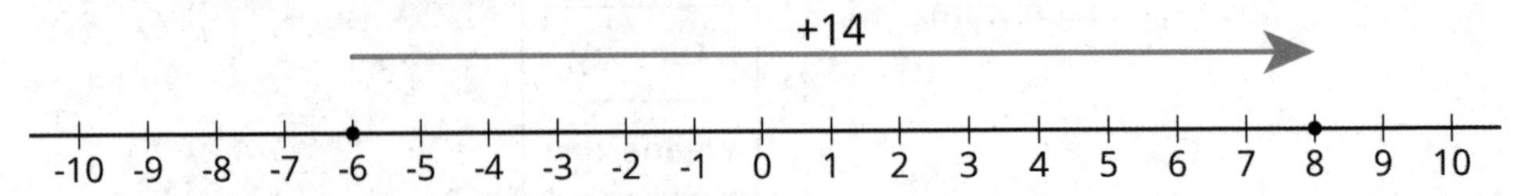


Notice that if you subtract them in the opposite order, you get the opposite number:

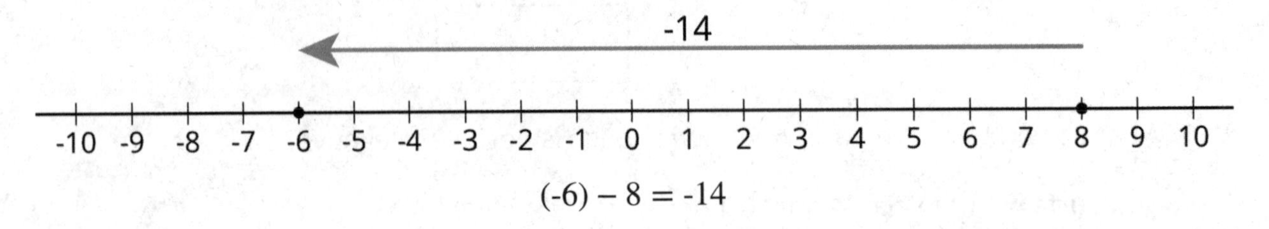


$(-6) - 8 = -14$

In general, the distance between two numbers $a$ and $b$ on the number line is $|a - b|$. Note that the *distance* between two numbers is always positive, no matter the order. But the *difference* can be positive or negative, depending on the order.

Sometimes we use positive and negative numbers to represent quantities in context. Here are some contexts we have studied that can be represented with positive and negative numbers:

- temperature
- elevation
- inventory
- an account balance
- electricity flowing in and flowing out

In these situations, using positive and negative numbers, and operations on positive and negative numbers, helps us understand and analyze them. To solve problems in these situations, we just have to understand what it means when the quantity is positive, when it is negative, and what it means to add and subtract them.

## Lesson 10 Practice Problems

1. Write a sentence to answer each question:

   a. How much warmer is 82 than 40?

   b. How much warmer is 82 than -40?

2. a. What is the difference in height between 30 m up a cliff and 87 m up a cliff? What is the distance between these positions?

   b. What is the difference in height between an albatross flying at 100 m above the surface of the ocean and a shark swimming 30 m below the surface? What is the distance between them if the shark is right below the albatross?

3. The table shows four transactions and the resulting account balance in a bank account, except some numbers are missing. Fill in the missing numbers.

| | transaction amount | account balance |
|---|---|---|
| transaction 1 | 360 | 360 |
| transaction 2 | -22.50 | 337.50 |
| transaction 3 | | 182.35 |
| transaction 4 | | -41.40 |

4. Find each difference.

   - $(-5) - 6$
   - $\frac{2}{5} - \frac{3}{5}$
   - $35 - (-8)$
   - $-4\frac{3}{8} - (-1\frac{1}{4})$

5. Last week, it rained $g$ inches. This week, the amount of rain decreased by 5%. Which expressions represent the amount of rain that fell this week? Select **all** that apply.

   A. $g - 0.05$

   B. $g - 0.05g$

   C. $0.95g$

   D. $0.05g$

   E. $(1 - 0.05)g$

   (From Unit 6, Lesson 5.)

6. A company produces screens of different sizes. Based on the table, could there be a relationship between the number of pixels and the area of the screen? If so, write an equation representing the relationship. If not, explain your reasoning.

| square inches of screen | number of pixels |
|---|---|
| 6 | 31,104 |
| 72 | 373,248 |
| 105 | 544,320 |
| 300 | 1,555,200 |

(From Unit 5, Lesson 5.)

# Lesson 11: Constructing the Coordinate Plane

Let's explore and extend the coordinate plane.

## 11.1: Guess My Line

1. Choose a horizontal or a vertical line on the grid. Draw 4 points on the line and label each point with its coordinates.

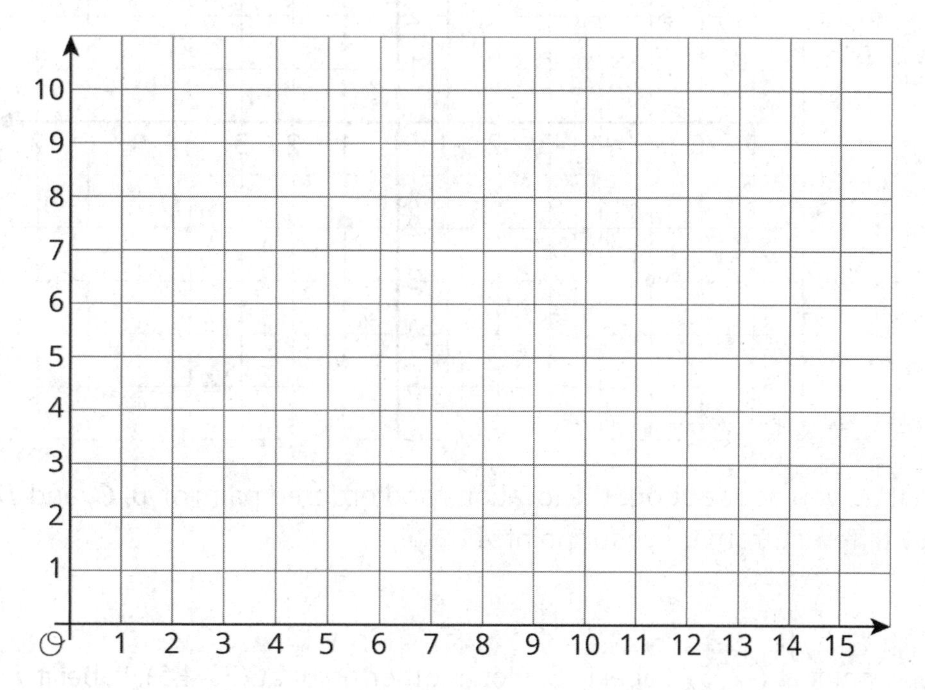


2. Tell your partner whether your line is horizontal or vertical, and have your partner guess the locations of your points by naming coordinates.

   If a guess is correct, put an X through the point. If your partner guessed a point that is on your line but not the point that you plotted, say, "That point is on my line, but is not one of my points."

   Take turns guessing each other's points, 3 guesses per turn.

## 11.2: The Coordinate Plane

1. Label each point on the coordinate plane with an ordered pair.

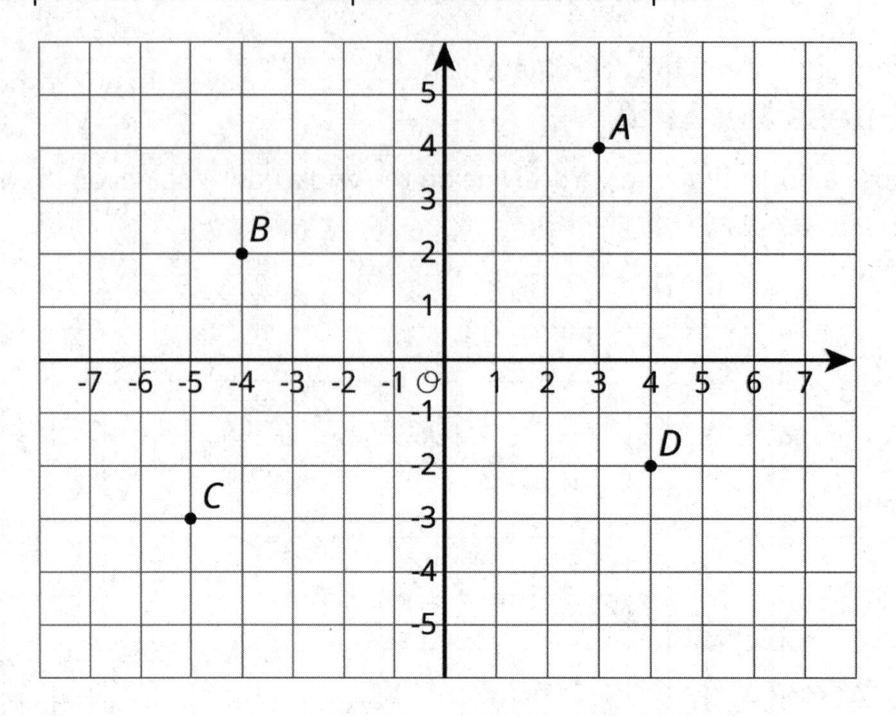


2. What do you notice about the locations and ordered pairs of $B$, $C$, and $D$? How are they different from those for point $A$?

3. Plot a point at $(-2, 5)$. Label it $E$. Plot another point at $(3, -4.5)$. Label it $F$.

4. The coordinate plane is divided into four **quadrants**, I, II, III, and IV, as shown here.

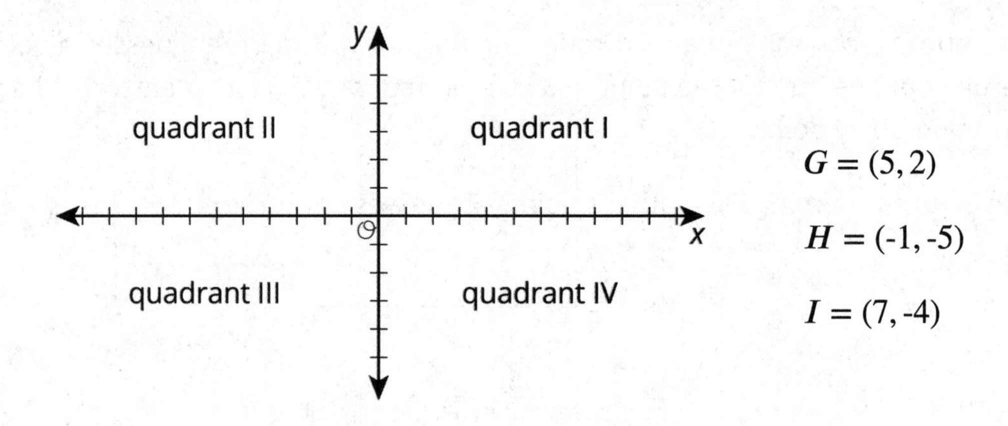


$G = (5, 2)$

$H = (-1, -5)$

$I = (7, -4)$

5. In which quadrant is point $G$ located? Point $H$? Point $I$?

6. A point has a positive $y$-coordinate. In which quadrant could it be?

## 11.3: Axes Drawing Decisions

1. Here are three sets of coordinates. For each set, draw and label an appropriate pair of axes and plot the points.

   a. (1, 2), (3, -4), (-5, -2), (0, 2.5)

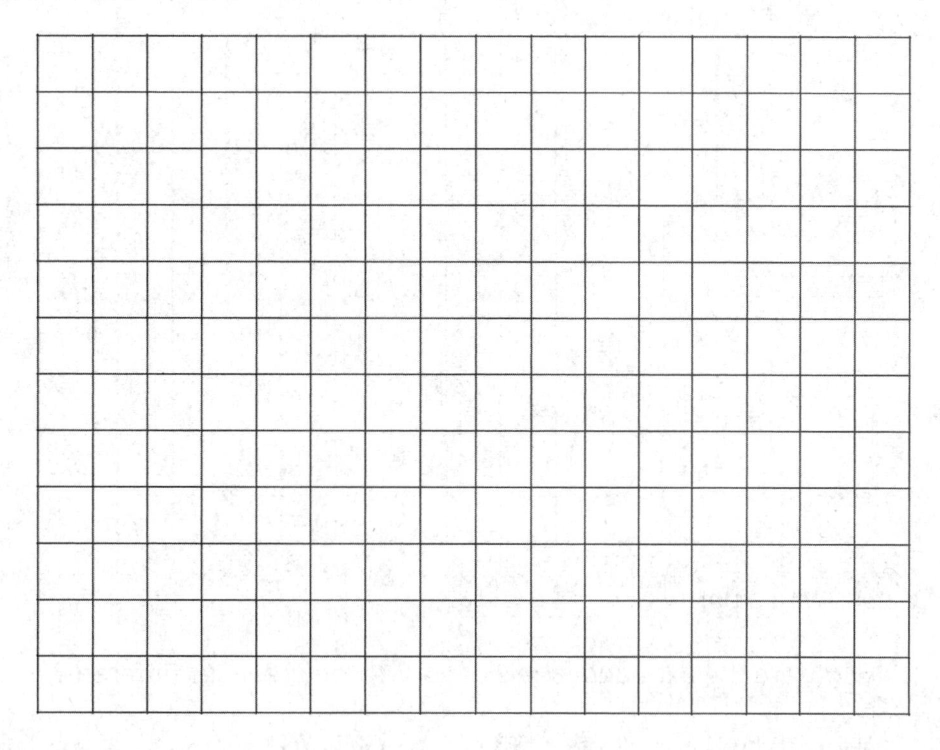

   b. (50, 50), (0, 0), (-10, -30), (-35, 40)

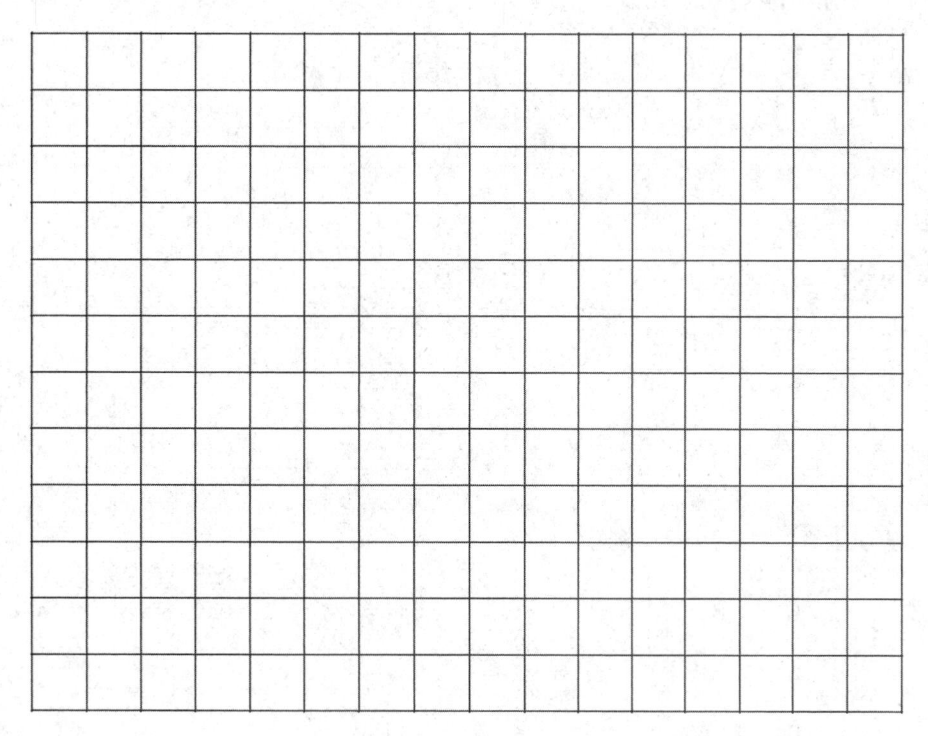

c. $\left(\frac{1}{4}, \frac{3}{4}\right)$, $\left(\frac{-5}{4}, \frac{1}{2}\right)$, $\left(-1\frac{1}{4}, \frac{-3}{4}\right)$, $\left(\frac{1}{4}, \frac{-1}{2}\right)$

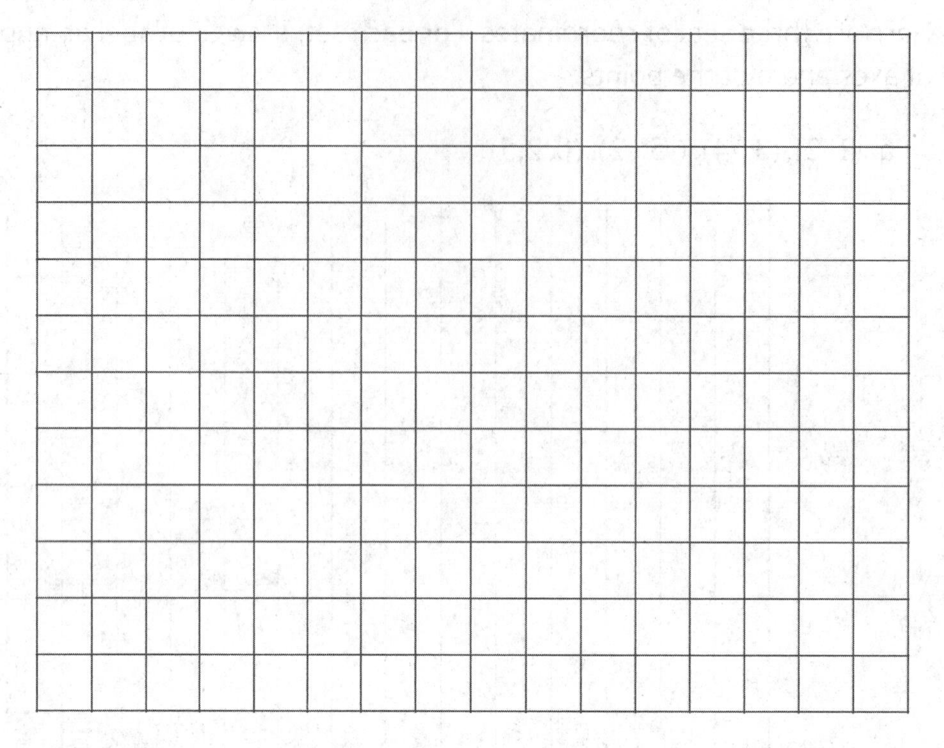

2. Discuss with a partner:

    - How are the axes and labels of your three drawings different?
    - How did the coordinates affect the way you drew the axes and label the numbers?

## Lesson 11 Summary

Just as the number line can be extended to the left to include negative numbers, the $x$- and $y$-axis of a coordinate plane can also be extended to include negative values.

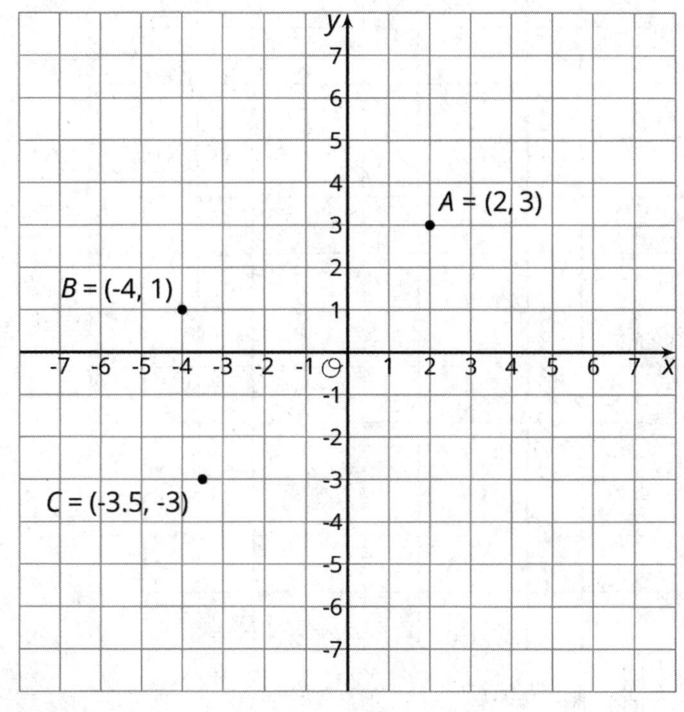


The ordered pair $(x, y)$ can have negative $x$- and $y$-values. For $B = (-4, 1)$, the $x$-value of -4 tells us that the point is 4 units to the left of the $y$-axis. The $y$-value of 1 tells us that the point is one unit above the $x$-axis.

The same reasoning applies to the points $A$ and $C$. The $x$- and $y$-coordinates for point $A$ are positive, so $A$ is to the right of the $y$-axis and above the $x$-axis. The $x$- and $y$-coordinates for point $C$ are negative, so $C$ is to the left of the $y$-axis and below the $x$-axis.

## Glossary

- quadrant

## Lesson 11 Practice Problems

1. a. Graph these points in the coordinate plane: $(-2, 3)$, $(2, 3)$, $(-2, -3)$, $(2, -3)$.

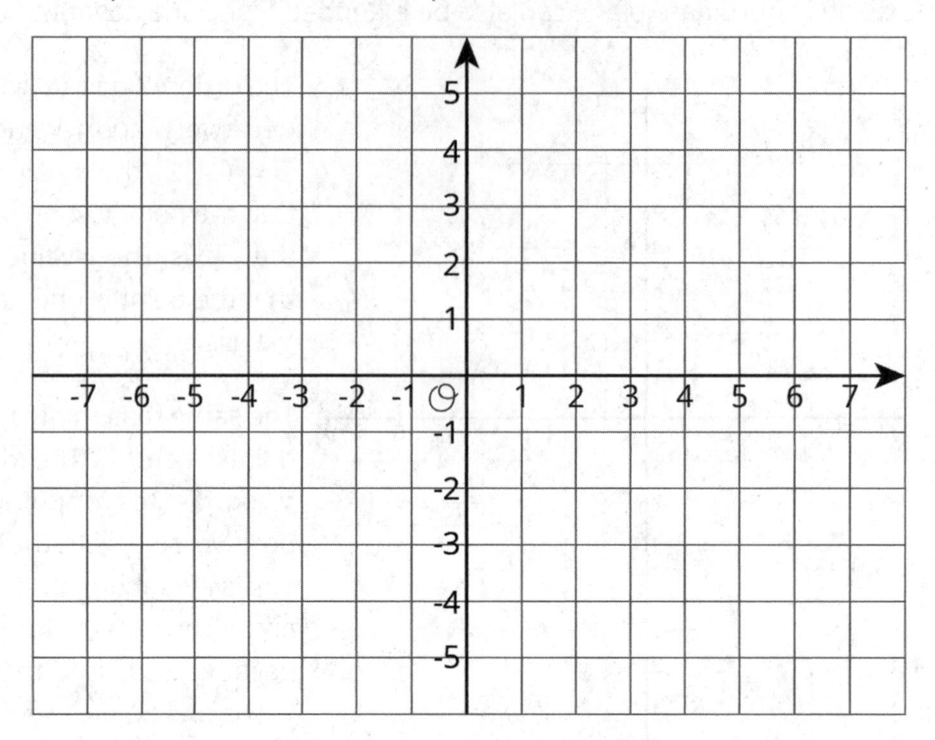


   b. Connect all of the points. Describe the figure.

2. Diego was asked to plot these points: $(-50, 0)$, $(150, 100)$, $(200, -100)$, $(350, 50)$, $(-250, 0)$. What interval could he use for each axis? Explain your reasoning.

3. Write the coordinates of each point.

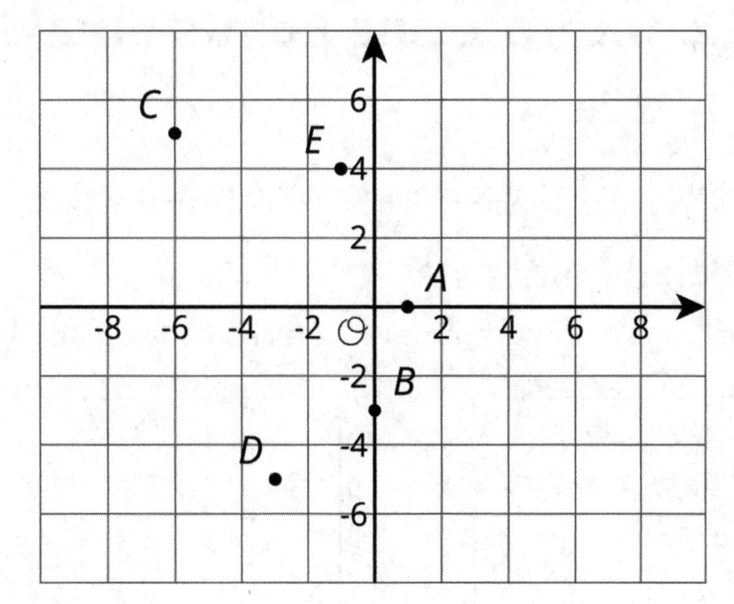


4. Find the value of each expression.

a. $16.2 + -8.4$

b. $\frac{2}{5} - \frac{3}{5}$

c. $-9.2 + -7$

d. $-4\frac{3}{8} - (-1\frac{1}{4})$

(From Unit 7, Lesson 10.)

5. Lin and Tyler are drawing circles. Tyler's circle has twice the diameter of Lin's circle. Tyler thinks that his circle will have twice the area of Lin's circle as well. Do you agree with Tyler?

(From Unit 5, Lesson 15.)

# Lesson 12: Interpreting Points on a Coordinate Plane

Let's examine what points on the coordinate plane can tell us.

## 12.1: Unlabeled Points

Label each point on the coordinate plane with the appropriate letter and ordered pair.

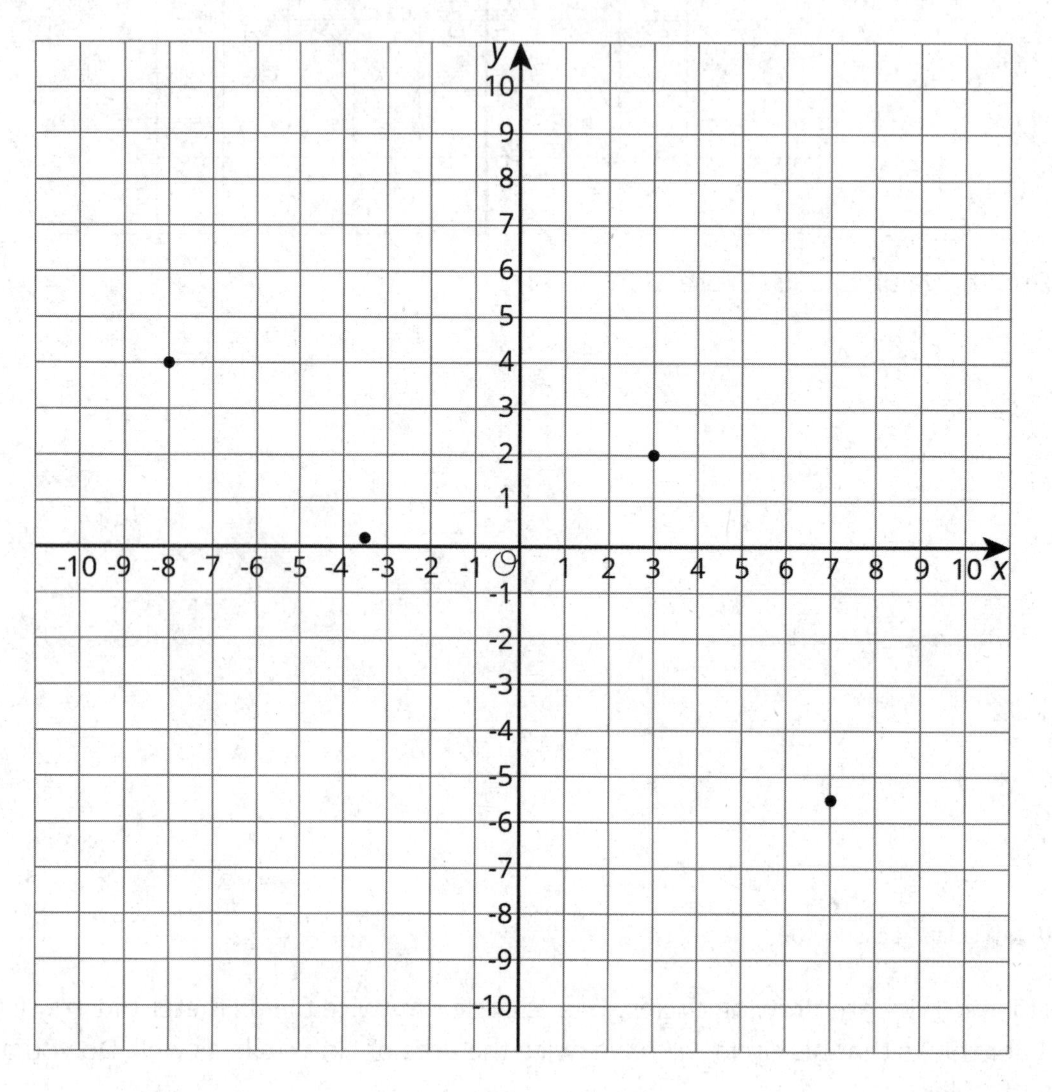


$A = (7, \text{-}5.5)$ $B = (\text{-}8, 4)$ $C = (3, 2)$ $D = (\text{-}3.5, 0.2)$

## 12.2: Account Balance

The graph shows the balance in a bank account over a period of 14 days. The axis labeled $b$ represents account balance in dollars. The axis labeled $d$ represents the day.

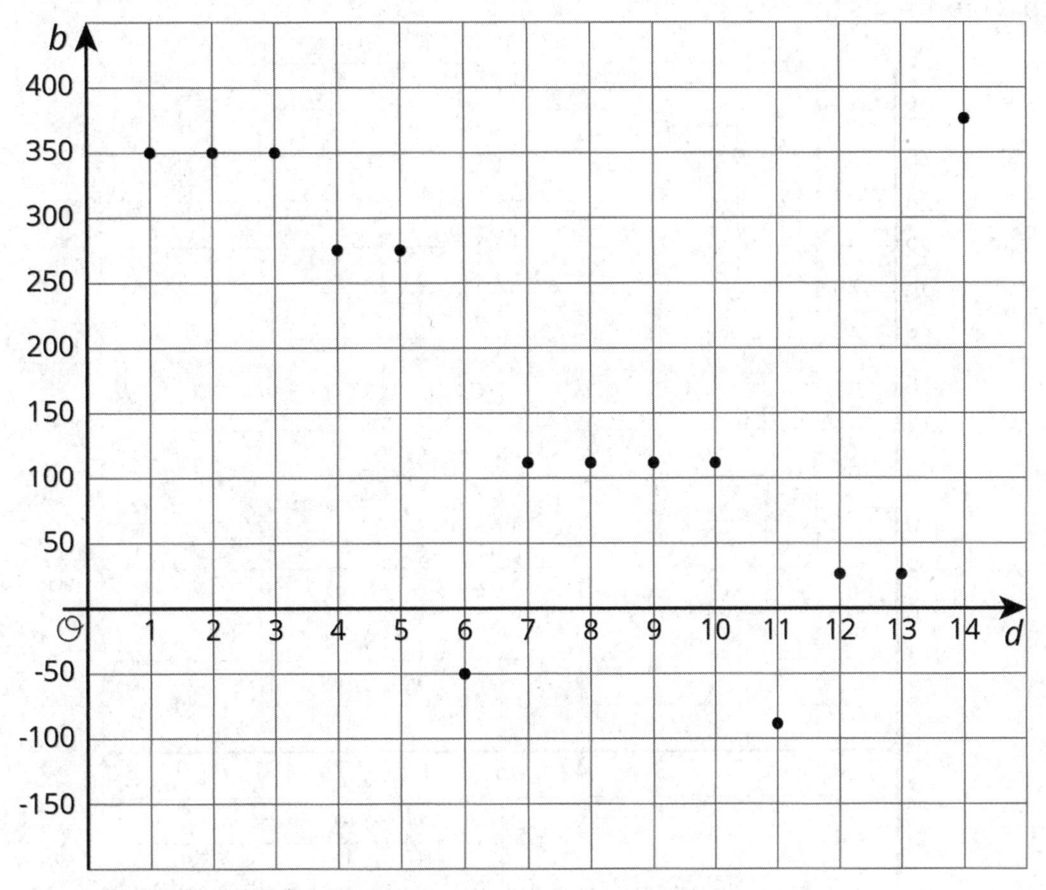


1. Estimate the greatest account balance. On which day did it occur?

2. Estimate the least account balance. On which day did it occur?

3. What does the point $(6, -50)$ tell you about the account balance?

4. How can we interpret $|-50|$ in the context?

## 12.3: High and Low Temperatures

The coordinate plane shows the high and low temperatures in Nome, Alaska over a period of 8 days. The axis labeled $T$ represents temperatures in degrees Fahrenheit. The axis labeled $d$ represents the day.

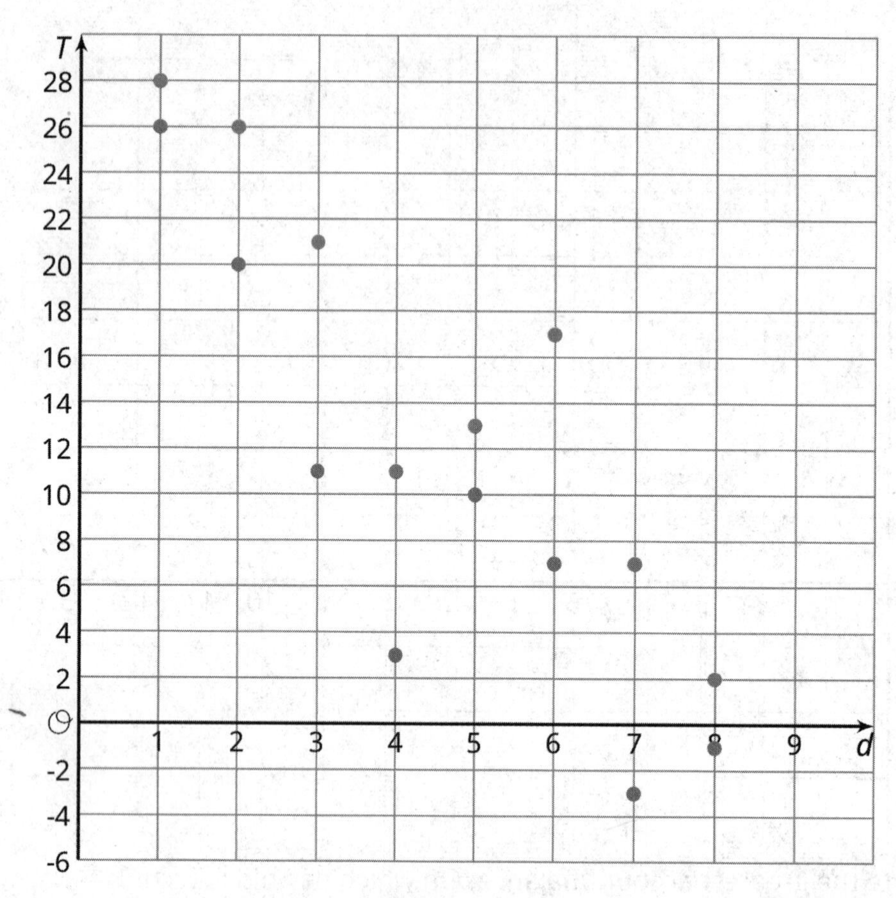


1. a. What was the warmest high temperature?

   (1,28) 28°

   b. What was the coldest high temperature?

   (1,26) 26°

   c. Write an inequality to compare the warmest and coldest high temperatures.

   28° > 26°

2. a. What was the coldest low temperature?

   (7,-3) -3°

   b. What was the warmest low temperature?

   (8-2) 2°

   c. Write an inequality to compare the warmest and coldest low temperatures.

   -3° < 2°

3. a. On which day(s) did the *largest* difference between the high and low temperatures occur? Write down this difference.

3, 6, 7 5

b. On which day(s) did the *smallest* difference between the high and low temperatures occur? Write down this difference.

1 1

**Are you ready for more?**

To get from the point $(2, 1)$ to $(-4, 3)$ you can go two units up and six units to the left, for a total distance of eight units. This is called the "taxicab distance," because a taxi driver would have to drive eight blocks to get between those two points on a map.

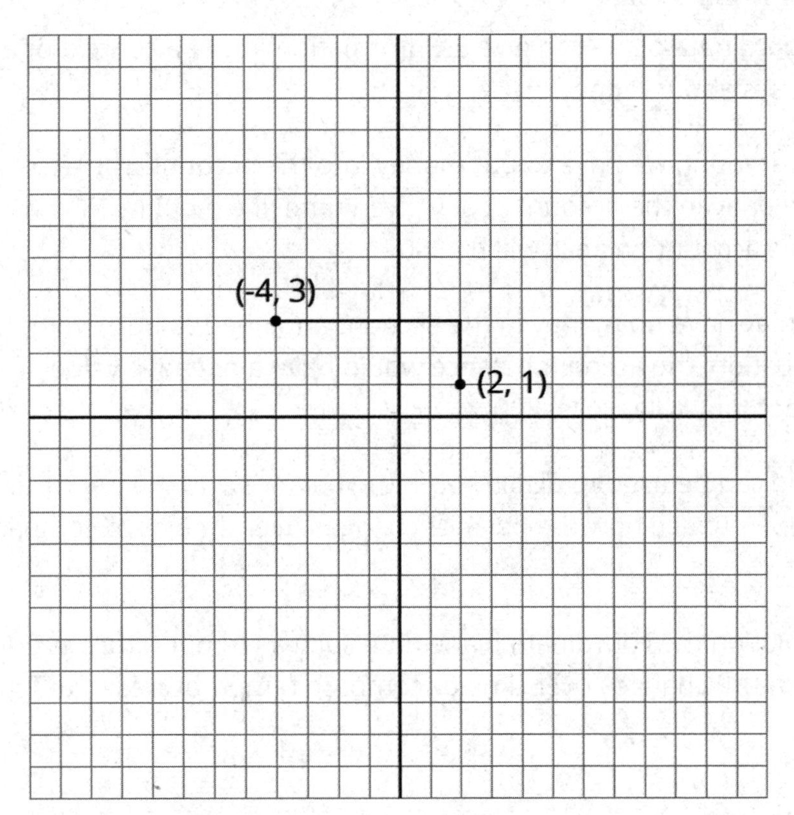


1. Find as many points as you can that have a taxicab distance of eight units away from $(2, 1)$. What shape do these points make?

2. The point $(0, 3)$ is 4 taxicab units away from $(-4, 3)$ and 4 taxicab units away from $(2, 1)$.

   a. Find as many other points as you can that are 4 taxicab units away from *both* $(-4, 3)$ and $(2, 1)$.

   b. Are there any points that are 3 taxicab units away from both points?

### Lesson 12 Summary

Points on the coordinate plane can give us information about a context or a situation. One of those contexts is about money.

To open a bank account, we have to put money into the account. The account balance is the amount of money in the account at any given time. If we put in $350 when opening the account, then the account balance will be 350.

Sometimes we may have no money in the account and need to borrow money from the bank. In that situation, the account balance would have a negative value. If we borrow $200, then the account balance is -200.

A coordinate grid can be used to display both the balance and the day or time for any balance. This allows to see how the balance changes over time or to compare the balances of different days.

Similarly, if we plot on the coordinate plane data such as temperature over time, we can see how temperature changes over time or compare temperatures of different times.

# Lesson 12 Practice Problems

1. The elevation of a submarine is shown in the table. Draw and label coordinate axes with an appropriate scale and plot the points.

| time after noon (hours) | elevation (meters) |
|---|---|
| 0 | -567 |
| 1 | -892 |
| 2 | -1,606 |
| 3 | -1,289 |
| 4 | -990 |
| 5 | -702 |
| 6 | -365 |

2. $30 + -30 = 0$.

   a. Write another sum of two numbers that equals 0.

   b. Write a sum of three numbers that equals 0.

   c. Write a sum of four numbers that equals 0, none of which are opposites.

(From Unit 7, Lesson 7.)

3. The $x$-axis represents the number of hours before or after noon, and the $y$-axis represents the temperature in degrees Celsius.

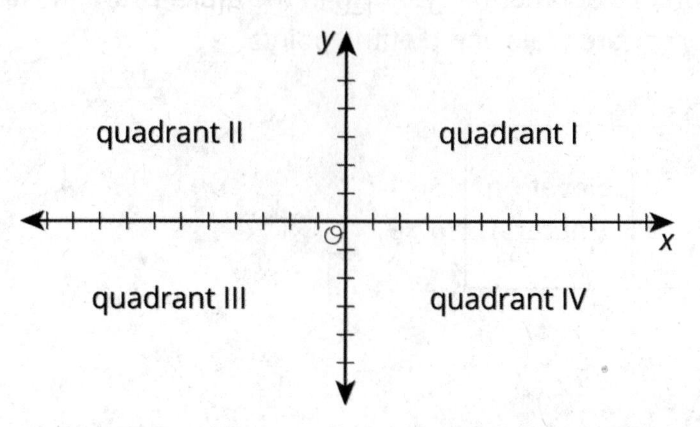


a. At 9 a.m., it was below freezing. In what quadrant would this point be plotted?

b. At 11 a.m., it was $10^\circ$C. In what quadrant would this point be plotted?

c. Choose another time and temperature. Then tell the quadrant where the point should be plotted.

d. What does the point $(0, 0)$ represent in this context?

4. Solve each equation.

$3a = 12$

$b + 3.3 = 8.9$

$1 = \frac{1}{4}c$

$5\frac{1}{2} = d + \frac{1}{4}$

$2e = 6.4$

(From Unit 4, Lesson 4.)

5. Crater Lake in Oregon is shaped like a circle with a diameter of about 5.5 miles.

   a. How far is it around the perimeter of Crater Lake?

   b. What is the area of the surface of Crater Lake?

(From Unit 5, Lesson 19.)

6. A type of green paint is made by mixing 2 cups of yellow with 3.5 cups of blue.

   a. Find a mixture that will make the same shade of green but a smaller amount.

   b. Find a mixture that will make the same shade of green but a larger amount.

   c. Find a mixture that will make a different shade of green that is bluer.

   d. Find a mixture that will make a different shade of green that is more yellow.

(From Unit 5, Lesson 1.)

# Lesson 13: Distances and Shapes on the Coordinate Plane

Let's explore distance on the coordinate plane.

## 13.1: Coordinate Patterns

Plot points in your assigned quadrant and label them with their coordinates.

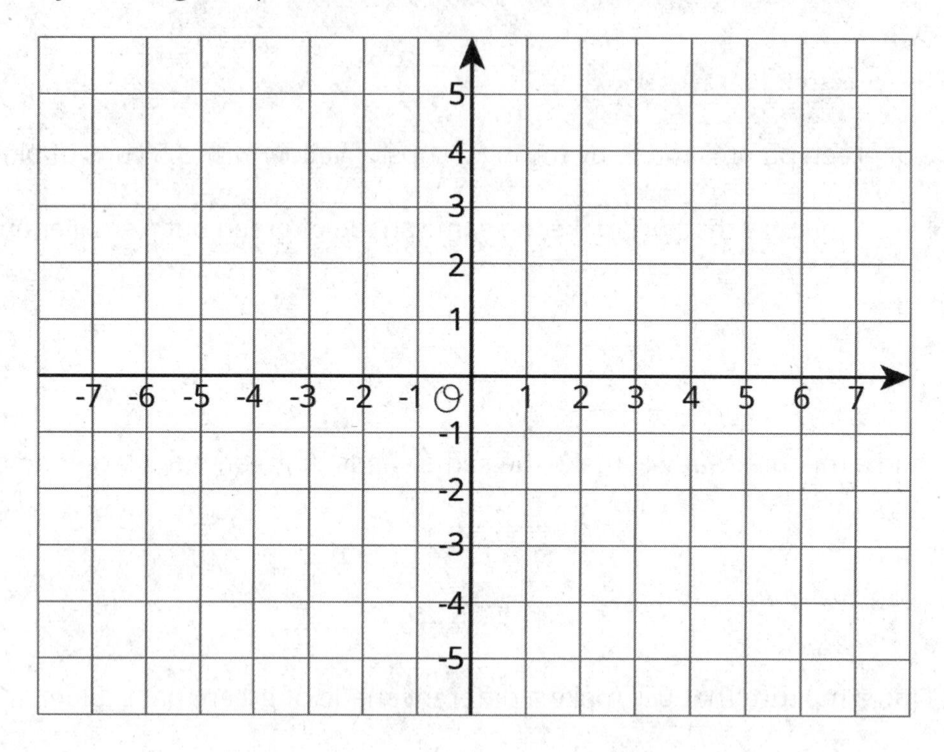

## 13.2: Signs of Numbers in Coordinates

1. Write the coordinates of each point.

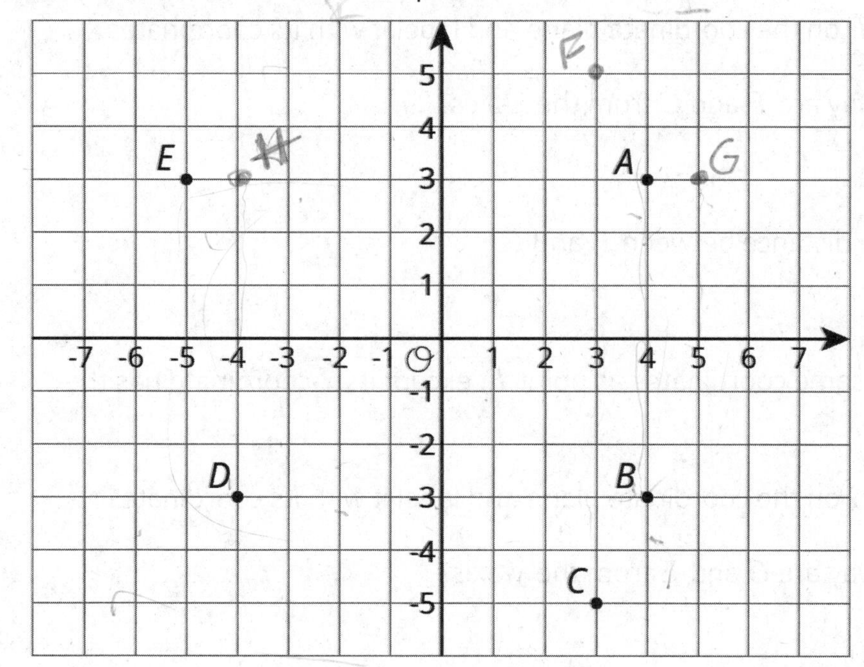


$A =$ (4,3)

$B =$ (4, -3)

$C =$ (3, -5)

$D =$ (-4, -3)

$E =$ (-5, 3)

F = (3, 5)

G = (5, 3)

H = (-4, 3)

2. Answer these questions for each pair of points.

- How are the coordinates the same? How are they different?
- How far away are they from the y-axis? To the left or to the right of it?
- How far away are they from the x-axis? Above or below it?

a. $A$ and $B$

same = x
dif. = y

4 units away
R

A = 3 up
B = 3 down

b. $B$ and $D$

same = y
dif = x

4 units away
L and R

B = 3 down
D = 3 down

c. $A$ and $D$

A = 3 up
D = 3 down

same = absolute distance
dif. = y and x are oppisites

4 units away
L and R

Pause here for a class discussion.

3. Point $F$ has the same coordinates as point $C$, except its $y$-coordinate has the opposite sign.

    a. Plot point $F$ on the coordinate plane and label it with its coordinates.

    b. How far away are $F$ and $C$ from the $x$-axis?

    5

    c. What is the distance between $F$ and $C$?

    10

4. Point $G$ has the same coordinates as point $E$, except its $x$-coordinate has the opposite sign.

    a. Plot point $G$ on the coordinate plane and label it with its coordinates.

    b. How far away are $G$ and $E$ from the $y$-axis?

    5

    c. What is the distance between $G$ and $E$?

    10

5. Point $H$ has the same coordinates as point $B$, except its *both* coordinates have the opposite sign. In which quadrant is point $H$?

    II

## Lesson 13 Summary

The points $A = (5, 2)$, $B = (-5, 2)$, $C = (-5, -2)$, and $D = (5, -2)$ are shown in the plane. Notice that they all have almost the same coordinates, except the signs are different. They are all the same distance from each axis but are in different quadrants.

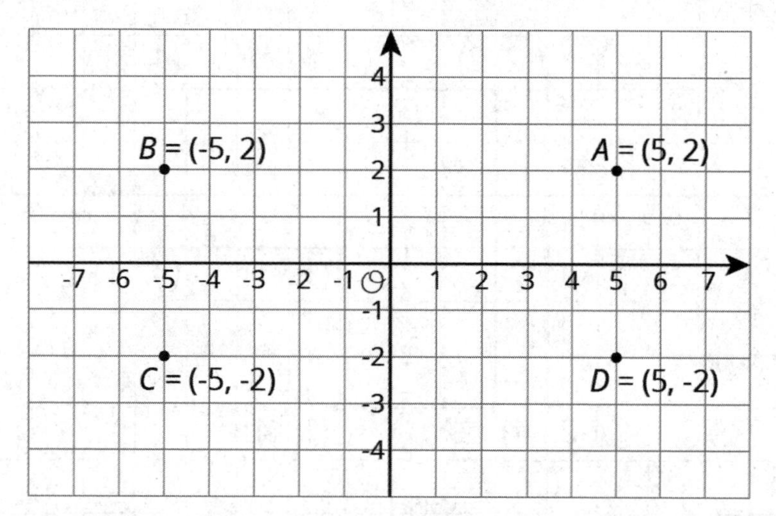


Notice that the vertical distance between points $A$ and $D$ is 4 units, because point $A$ is 2 units above the horizontal axis and point $D$ is 2 units below the horizontal axis. The horizontal distance between points $A$ and $B$ is 10 units, because point $B$ is 5 units to the left of the vertical axis and point $A$ is 5 units to the right of the vertical axis.

We can always tell which quadrant a point is located in by the signs of its coordinates.

| $x$ | $y$ | quadrant |
|---|---|---|
| positive | positive | I |
| negative | positive | II |
| negative | negative | III |
| positive | negative | IV |

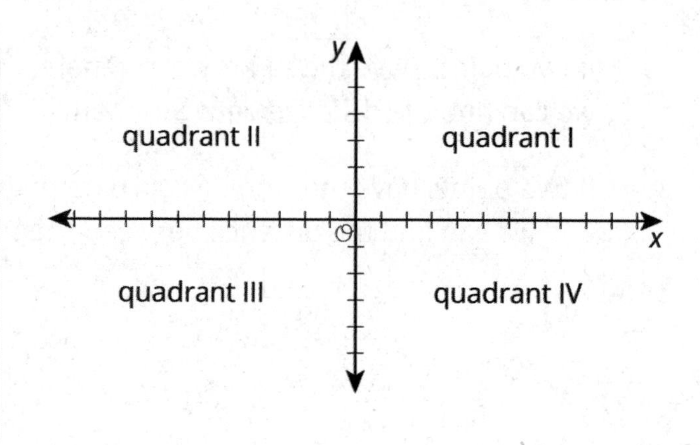


In general:

- If two points have $x$-coordinates that are opposites (like 5 and -5), they are the same distance away from the vertical axis, but one is to the left and the other to the right.
- If two points have $y$-coordinates that are opposites (like 2 and -2), they are the same distance away from the horizontal axis, but one is above and the other below.

When two points have the same value for the first or second coordinate, we can find the distance between them by subtracting the coordinates that are different.

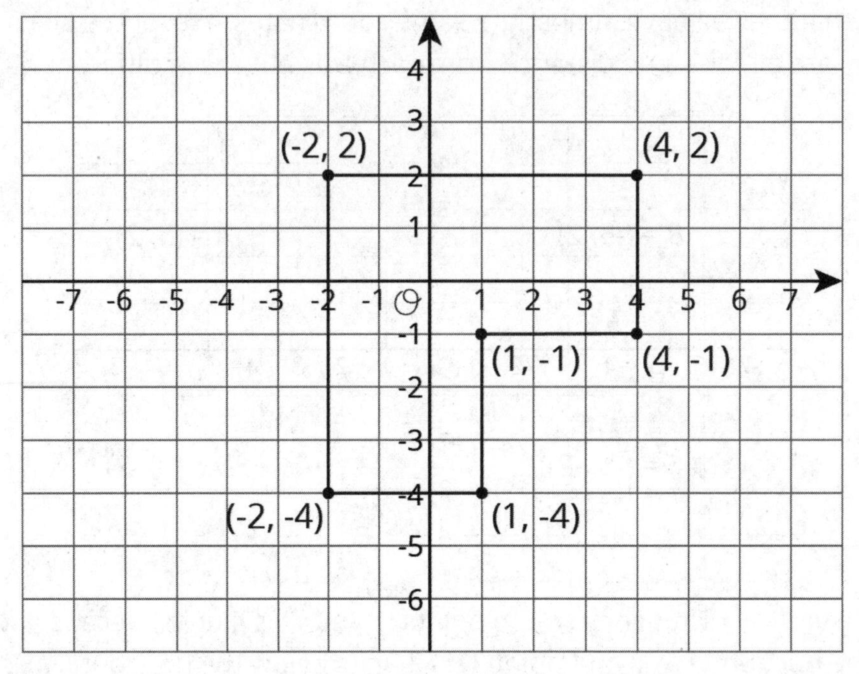


For example, we can find the perimeter of this polygon by finding the sum of its side lengths. Starting from $(-2, 2)$ and moving clockwise, we can see that the lengths of the segments are 6, 3, 3, 3, 3, and 6 units. The perimeter is therefore 24 units.

In general:

- If two points have the same $x$-coordinate, they will be on the same vertical line, and we can find the distance between them.
- If two points have the same $y$-coordinate, they will be on the same horizontal line, and we can find the distance between them.

# Lesson 14: Position, Speed, Direction

Let's use signed numbers to represent movement.

## 14.1: Distance, Rate, Time

1. An airplane moves at a constant speed of 120 miles per hour for 3 hours. How far does it go?

2. A train moves at constant speed and travels 6 miles in 4 minutes. What is its speed in miles per minute?

3. A car moves at a constant speed of 50 miles per hour. How long does it take the car to go 200 miles?

## 14.2: Velocity

A traffic safety engineer was studying travel patterns along a highway. She set up a camera and recorded the speed and direction of cars and trucks that passed by the camera. Positions to the east of the camera are positive, and to the west are negative.

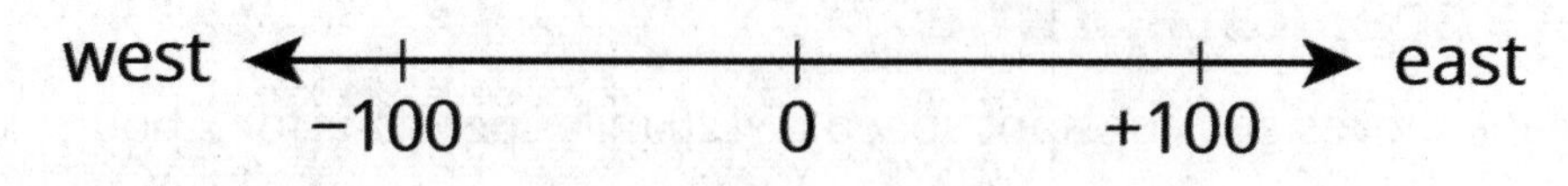


Vehicles that are traveling towards the east have a positive velocity, and vehicles that are traveling towards the west have a negative velocity.

1. Complete the table with the position of each vehicle if the vehicle is traveling at a constant speed for the indicated time period. Then write an equation.

| velocity (meters per second) | time after passing the camera (seconds) | ending position (meters) | equation describing the position |
|---|---|---|---|
| +25 | +10 | +250 | $25 \cdot 10 = 250$ |
| -20 | +30 | | |
| +32 | +40 | | |
| -35 | +20 | | |
| +28 | 0 | | |

2. If a car is traveling east when it passes the camera, will its position be positive or negative 60 seconds after it passes the camera? If we multiply two positive numbers, is the result positive or negative?

3. If a car is traveling west when it passes the camera, will its position be positive or negative 60 seconds after it passes the camera? If we multiply a negative and a positive number, is the result positive or negative?

### Are you ready for more?

In many contexts we can interpret negative rates as "rates in the opposite direction." For example, a car that is traveling -35 miles per hour is traveling in the opposite direction of a car that is traveling 40 miles per hour.

1. What could it mean if we say that water is flowing at a rate of -5 gallons per minute?

2. Make up another situation with a negative rate, and explain what it could mean.

## 14.3: Before and After

Where was the girl:

1. 5 seconds *after* this picture was taken? Mark her approximate location on the picture.

2. 5 seconds *before* this picture was taken? Mark her approximate location on the picture.

## 14.4: Backwards in Time

A traffic safety engineer was studying travel patterns along a highway. She set up a camera and recorded the speed and direction of cars and trucks that passed by the camera. Positions to the east of the camera are positive, and to the west are negative.

1. Here are some positions and times for one car:

| position (feet) | -180 | -120 | -60 | 0 | 60 | 120 |
|---|---|---|---|---|---|---|
| time (seconds) | -3 | -2 | -1 | 0 | 1 | 2 |

   a. In what direction is this car traveling?

   b. What is its velocity?

2. a. What does it mean when the time is zero?

   b. What could it mean to have a negative time?

3. Here are the positions and times for a different car whose velocity is -50 feet per second:

| position (feet) | | | | 0 | -50 | -100 |
|---|---|---|---|---|---|---|
| time (seconds) | -3 | -2 | -1 | 0 | 1 | 2 |

   a. Complete the table with the rest of the positions.

   b. In what direction is this car traveling? Explain how you know.

4. Complete the table for several different cars passing the camera.

| | velocity (meters per second) | time after passing the camera (seconds) | ending position (meters) | equation |
|---|---|---|---|---|
| car C | +25 | +10 | +250 | $25 \cdot 10 = 250$ |
| car D | -20 | +30 | | |
| car E | +32 | -40 | | |
| car F | -35 | -20 | | |
| car G | -15 | -8 | | |

5. a. If a car is traveling east when it passes the camera, will its position be positive or negative 60 seconds *before* it passes the camera?

   b. If we multiply a positive number and a negative number, is the result positive or negative?

6. a. If a car is traveling west when it passes the camera, will its position be positive or negative 60 seconds *before* it passes the camera?

   b. If we multiply two negative numbers, is the result positive or negative?

### Lesson 14 Summary

We can use signed numbers to represent the position of an object along a line. We pick a point to be the reference point, and call it zero. Positions to the right of zero are positive. Positions to the left of zero are negative.

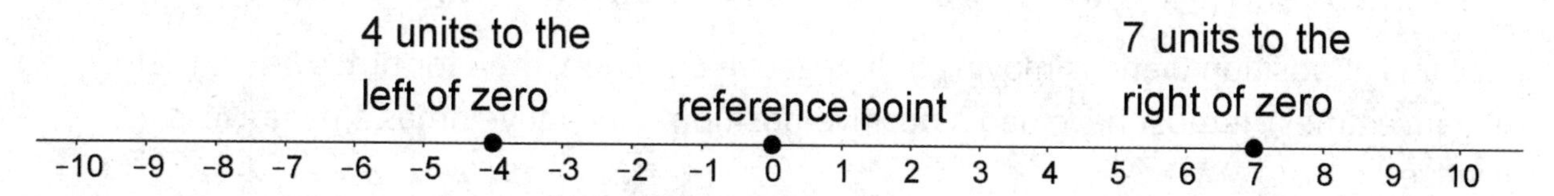


When we combine speed with direction indicated by the sign of the number, it is called *velocity*. For example, if you are moving 5 meters per second to the right, then your velocity is +5 meters per second. If you are moving 5 meters per second to the left, then your velocity is -5 meters per second.

If you start at zero and move 5 meters per second for 10 seconds, you will be $5 \cdot 10 = 50$ meters to the right of zero. In other words, $5 \cdot 10 = 50$.

If you start at zero and move -5 meters per second for 10 seconds, you will be $5 \cdot 10 = 50$ meters to the *left* of zero. In other words,

$$-5 \cdot 10 = -50$$

We can also use signed numbers to represent time relative to a chosen point in time. We can think of this as starting a stopwatch. The positive times are after the watch starts, and negative times are times before the watch starts.

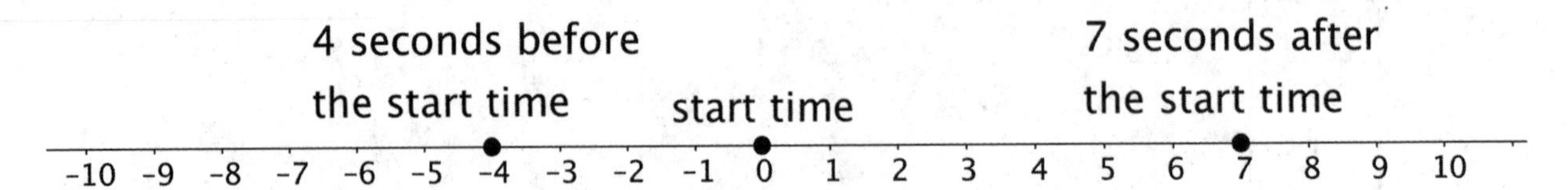


If a car is at position 0 and is moving in a positive direction, then for times after that (positive times), it will have a positive position. A positive times a positive is positive.

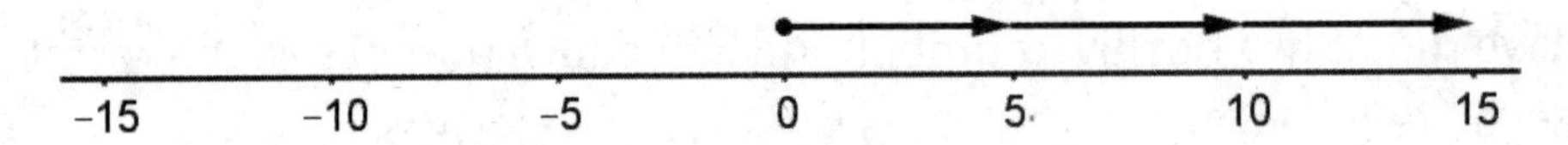


If a car is at position 0 and is moving in a negative direction, then for times after that (positive times), it will have a negative position. A negative times a positive is negative.

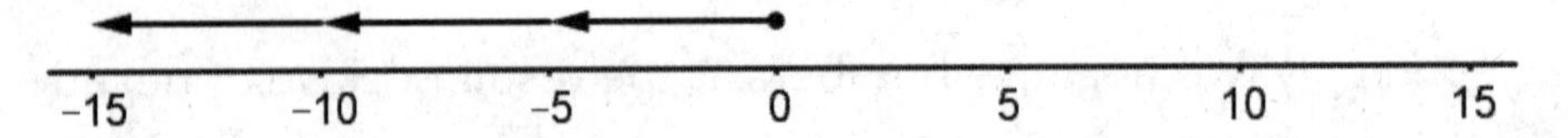


If a car is at position 0 and is moving in a positive direction, then for times *before* that (negative times), it must have had a negative position. A positive times a negative is negative.

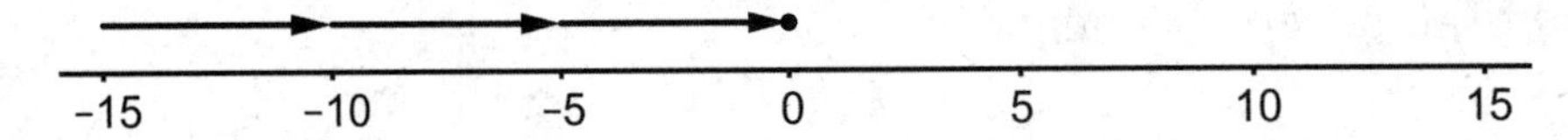


If a car is at position 0 and is moving in a negative direction, then for times *before* that (negative times), it must have had a positive position. A negative times a negative is positive.

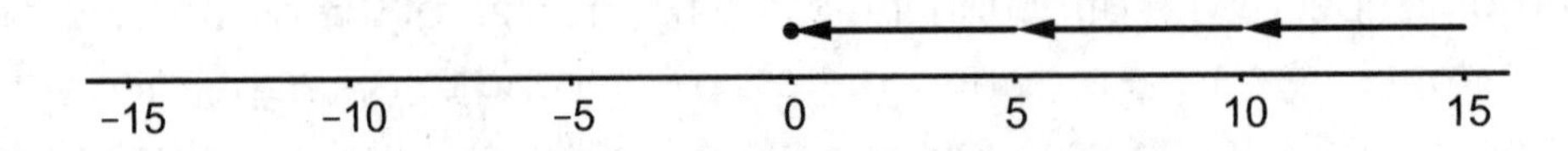

## Lesson 14 Practice Problems

1. a. How could you distinguish between traveling west at 5 miles per hour and traveling east at 5 miles per hour without using the words "east" and "west"?

   b. Four people are cycling. They each start at the same point. (0 represents their starting point.) Plot their finish points after five seconds of cycling on a number line

      - Lin cycles at 5 meters per second
      - Diego cycles at -4 meters per second
      - Elena cycles at 3 meters per second
      - Noah cycles at -6 meters per second

2. A weather station on the top of a mountain reports that the temperature is currently $0^\circ C$ and has been falling at a constant rate of $3^\circ C$ per hour. If it continues to fall at this rate, find each indicated temperature. Explain or show your reasoning.

   a. What will the temperature be in 2 hours?

   b. What will the temperature be in 5 hours?

   c. What will the temperature be in half an hour?

   d. What was the temperature 1 hour ago?

   e. What was the temperature 3 hours ago?

   f. What was the temperature 4.5 hours ago?

3. Fill in the missing numbers in these equations

   a. $-2 \cdot (-4.5) = ?$

   b. $(-8.7) \cdot (-10) = ?$

   c. $(-7) \cdot ? = 14$

   d. $? \cdot (-10) = 90$

4. a. Here are the vertices of rectangle $FROG$: $(-2, 5), (-2, 1), (6, 5), (6, 1)$. Find the perimeter of this rectangle. If you get stuck, try plotting the points on a coordinate plane.

   b. Find the area of the rectangle $FROG$.

   c. Here are the coordinates of rectangle $PLAY$: $(-11, 20), (-11, -3), (-1, 20), (-1, -3)$. Find the perimeter and area of this rectangle. See if you can figure out its side lengths without plotting the points.

(From Unit 7, Lesson 10.)

5. Tyler orders a meal that costs \$15.

   a. If the tax rate is 6.6%, how much will the sales tax be on Tyler's meal?

   b. Tyler also wants to leave a tip for the server. How much do you think he should pay in all? Explain your reasoning.

(From Unit 6, Lesson 7.)

# Lesson 15: Multiplying Rational Numbers

Let's get more practice multiplying signed numbers.

## 15.1: Which One Doesn't Belong: Expressions

Which expression doesn't belong?

$7.9x$

$7.9 + x$

$7.9 \cdot (-10)$

$-79$

## 15.2: Rational Numbers Multiplication Grid

1. Complete the *shaded* boxes in the multiplication square.

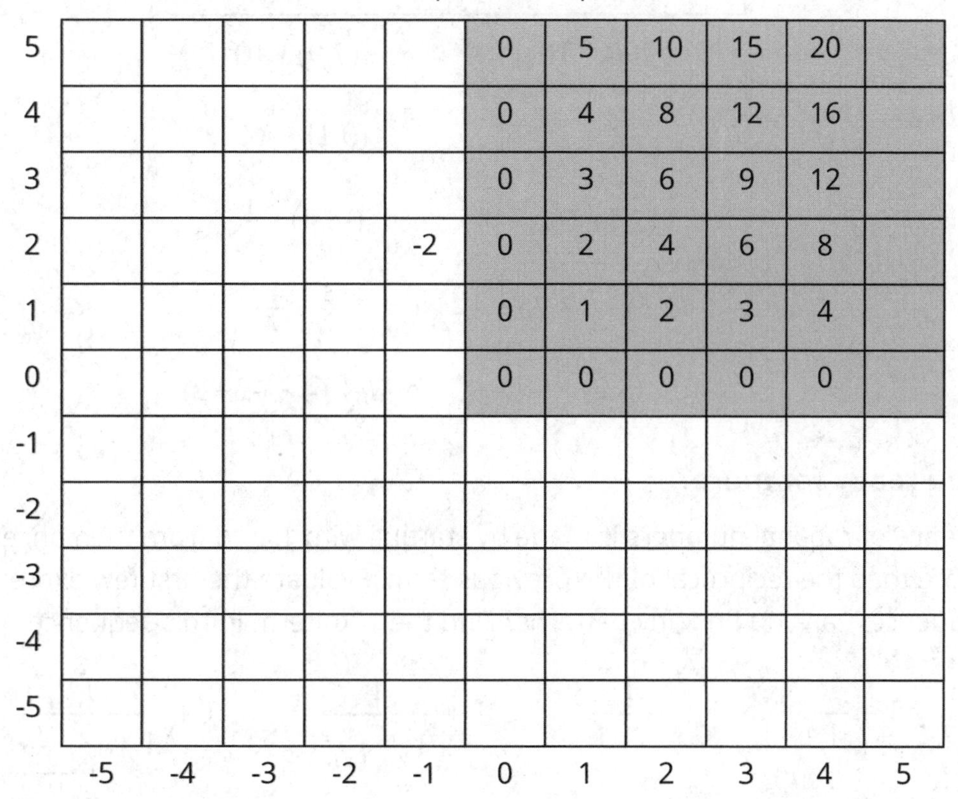


2. Look at the patterns along the rows and columns. Continue those patterns into the unshaded boxes.

3. Complete the whole table.

4. What does this tell you about multiplication with negative numbers?

## 15.3: Card Sort: Matching Expressions

Your teacher will give you cards with multiplication expressions on them. Match the expressions that are equal to each other. There will be 3 cards in each group.

## 15.4: Row Game: Multiplying Rational Numbers

Evaluate the expressions in one of the columns. Your partner will work on the other column. Check in with your partner after you finish each row. Your answers in each row should be the same. If your answers aren't the same, work together to find the error.

| column A | column B |
|---|---|
| $790 \div 10$ | $(7.9) \cdot 10$ |
| $-\frac{6}{7} \cdot 7$ | $(0.1) \cdot -60$ |
| $(2.1) \cdot -2$ | $(-8.4) \cdot \frac{1}{2}$ |
| $(2.5) \cdot (-3.25)$ | $-\frac{5}{2} \cdot \frac{13}{4}$ |
| $-10 \cdot (3.2) \cdot (-7.3)$ | $5 \cdot (-1.6) \cdot (-29.2)$ |

### Are you ready for more?

A sequence of rational numbers is made by starting with 1, and from then on, each term is one more than the reciprocal of the previous term. Evaluate the first few expressions in the sequence. Can you find any patterns? Find the 10th term in this sequence.

$$1 \qquad 1+\frac{1}{1} \qquad 1+\frac{1}{1+1} \qquad 1+\frac{1}{1+\frac{1}{1+1}} \qquad 1+\frac{1}{1+\frac{1}{1+\frac{1}{1+1}}} \qquad \dots$$

## Lesson 15 Summary

We can think of $3 \cdot 5$ as $5 + 5 + 5$, which has a value of 15.

We can think of $3 \cdot (-5)$ as $-5 + -5 + -5$, which has a value of -15.

We know we can multiply positive numbers in any order: $3 \cdot 5 = 5 \cdot 3$

If we can multiply signed numbers in any order, then $(-5) \cdot 3$ would also equal -15.

Now let's think about multiplying two negatives.

We can find $-5 \cdot (3 + -3)$ in two ways:

- Applying the distributive property:

  $$-5 \cdot 3 + -5 \cdot (-3)$$

- Adding the numbers in parentheses:

  $$-5 \cdot (0) = 0$$

This means that these expressions must be equal.

$$-5 \cdot 3 + -5 \cdot (-3) = 0$$

Multiplying the first two numbers gives

$$-15 + -5 \cdot (-3) = 0$$

Which means that

$$-5 \cdot (-3) = 15$$

There was nothing special about these particular numbers. This always works!

- A positive times a positive is always positive.

  For example, $\frac{3}{5} \cdot \frac{7}{8} = \frac{21}{40}$.
- A negative times a negative is also positive.

  For example, $-\frac{3}{5} \cdot -\frac{7}{8} = \frac{21}{40}$.
- A negative times a positive or a positive times a negative is always negative.

  For example, $\frac{3}{5} \cdot -\frac{7}{8} = -\frac{3}{5} \cdot \frac{7}{8} = -\frac{21}{40}$.
- A negative times a negative times a negative is also negative.

  For example, $-3 \cdot -4 \cdot -5 = -60$.

## Lesson 15 Practice Problems

1. Evaluate each expression:

   a. $-1 \cdot 2 \cdot 3$

   b. $-1 \cdot (-2) \cdot 3$

   c. $-1 \cdot (-2) \cdot (-3)$

2. Find the value of each expression.

   a. $\frac{1}{4} \cdot (-12)$

   b. $-\frac{1}{3} \cdot 39$

   c. $(-\frac{4}{5}) \cdot (-75)$

   d. $-\frac{2}{5} \cdot (-\frac{3}{4})$

   e. $\frac{8}{3} \cdot -42$

3. Fill in the missing numbers in these equations

   a. $(-7) \cdot ? = -14$

   b. $? \cdot 3 = -15$

   c. $? \cdot 4 = 32$

   d. $-49 \cdot 3 = ?$

   (From Unit 7, Lesson 14.)

4. These three points form a horizontal line: $(-3.5, 4)$, $(0, 4)$, and $(6.2, 4)$. Name two additional points that fall on this line.

(From Unit 7, Lesson 11.)

5. Order each set of numbers from least to greatest.

   a. 4, 8, -2, -6, 0

   b. -5, -5.2, 5.5, $-5\frac{1}{2}$, $\frac{-5}{2}$

(From Unit 7, Lesson 1.)

6. Decide whether each table could represent a proportional relationship. If the relationship could be proportional, what would be the constant of proportionality?

   a. Annie's Attic is giving away $5 off coupons.

| original price | sale price |
| --- | --- |
| $15 | $10 |
| $25 | $20 |
| $35 | $30 |

   b. Bettie's Boutique is having a 20% off sale.

| original price | sale price |
| --- | --- |
| $15 | $12 |
| $25 | $20 |
| $35 | $28 |

(From Unit 5, Lesson 4.)

# Lesson 16: Dividing Rational Numbers

Let's divide signed numbers.

## 16.1: Tell Me Your Sign

Consider the equation: $-27x = -35$

Without computing:

1. Is the **solution** to this equation positive or negative?

2. Are either of these two numbers solutions to the equation?

$\frac{35}{27}$ $\quad$ $-\frac{35}{27}$

## 16.2: Multiplication and Division

1. Find the missing values in the equations

   a. $-3 \cdot 4 = ?$

   b. $-3 \cdot ? = 12$

   c. $3 \cdot ? = 12$

   d. $? \cdot -4 = 12$

   e. $? \cdot 4 = -12$

2. Rewrite the unknown factor problems as division problems.

3. Complete the sentences. Be prepared to explain your reasoning.

   a. The sign of a positive number divided by a positive number is always:

   b. The sign of a positive number divided by a negative number is always:

   c. The sign of a negative number divided by a positive number is always:

   d. The sign of a negative number divided by a negative number is always:

4. Han and Clare walk towards each other at a constant rate, meet up, and then continue past each other in opposite directions. We will call the position where they meet up 0 feet and the time when they meet up 0 seconds.

   - Han's velocity is 4 feet per second.
   - Clare's velocity is -5 feet per second.

   a. Where is each person 10 seconds before they meet up?

   b. When is each person at the position -10 feet from the meeting place?

### Are you ready for more?

It is possible to make a new number system using *only* the numbers 0, 1, 2, and 3. We will write the symbols for multiplying in this system like this: $1 \otimes 2 = 2$. The table shows some of the products.

| $\otimes$ | 0 | 1 | 2 | 3 |
|---|---|---|---|---|
| 0 | 0 | 0 | 0 | 0 |
| 1 | | 1 | 2 | 3 |
| 2 | | | 0 | 2 |
| 3 | | | | |

1. In this system, $1 \otimes 3 = 3$ and $2 \otimes 3 = 2$. How can you see that in the table?
2. What do you think $2 \otimes 1$ is?
3. What about $3 \otimes 3$?
4. What do you think the solution to $3 \otimes n = 2$ is?
5. What about $2 \otimes n = 3$?

## 16.3: Drilling Down

A water well drilling rig has dug to a height of -60 feet after one full day of continuous use.

1. Assuming the rig drilled at a constant rate, what was the height of the drill after 15 hours?

2. If the rig has been running constantly and is currently at a height of -147.5 feet, for how long has the rig been running?

3. Use the coordinate grid to show the drill's progress.

hours
0 10 20 30 40 50 60 70 80 90 100
-50
-100
-150
-200
-250
feet

4. At this rate, how many hours will it take until the drill reaches -250 feet?

## Lesson 16 Summary

Any division problem is actually a multiplication problem:

- $6 \div 2 = 3$ because $2 \cdot 3 = 6$
- $6 \div -2 = -3$ because $-2 \cdot -3 = 6$
- $-6 \div 2 = -3$ because $2 \cdot -3 = -6$
- $-6 \div -2 = 3$ because $-2 \cdot 3 = -6$

Because we know how to multiply signed numbers, that means we know how to divide them.

- The sign of a positive number divided by a negative number is always negative.
- The sign of a negative number divided by a positive number is always negative.
- The sign of a negative number divided by a negative number is always positive.

A number that can be used in place of the variable that makes the equation true is called a **solution** to the equation. For example, for the equation $x \div -2 = 5$, the solution is -10, because it is true that $-10 \div -2 = 5$.

## Glossary

- solution to an equation

# Lesson 16 Practice Problems

1. Find the quotients:

   a. $24 \div -6$

   b. $-15 \div 0.3$

   c. $-4 \div -20$

2. Find the quotients.

   a. $\frac{2}{5} \div \frac{3}{4}$

   b. $\frac{9}{4} \div \frac{-3}{4}$

   c. $\frac{-5}{7} \div \frac{-1}{3}$

   d. $\frac{-5}{3} \div \frac{1}{6}$

3. Is the solution positive or negative?

   a. $2 \cdot x = 6$

   b. $-2 \cdot x = 6.1$

   c. $2.9 \cdot x = -6.04$

   d. $-2.473 \cdot x = -6.859$

4. Find the solution mentally.

   a. $3 \cdot -4 = a$

   b. $b \cdot (-3) = -12$

   c. $-12 \cdot c = 12$

   d. $d \cdot 24 = -12$

5. Find the products.

   a. $(100) \cdot (-0.09)$

   b. $(-7) \cdot (-1.1)$

   c. $(-7.3) \cdot (5)$

   d. $(-0.2) \cdot (-0.3)$

   (From Unit 7, Lesson 14.)

6. Which graphs could not represent a proportional relationship? Explain how you decided.

A

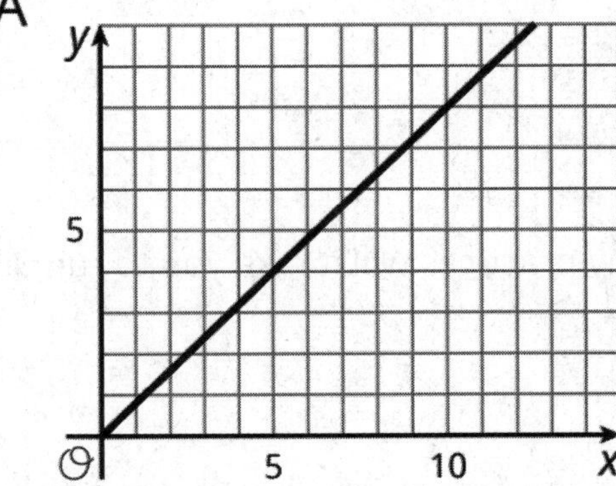


B

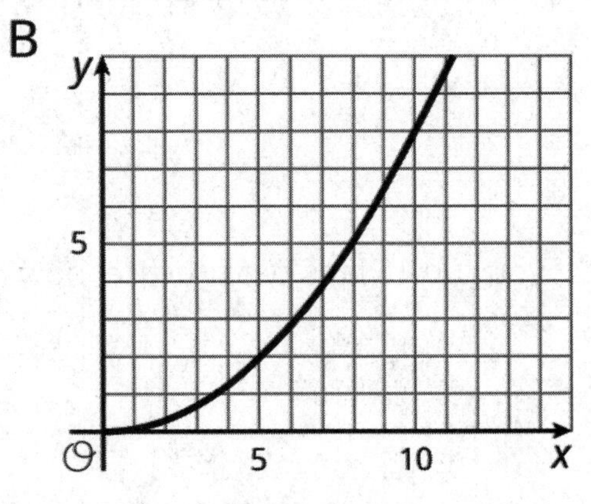


C

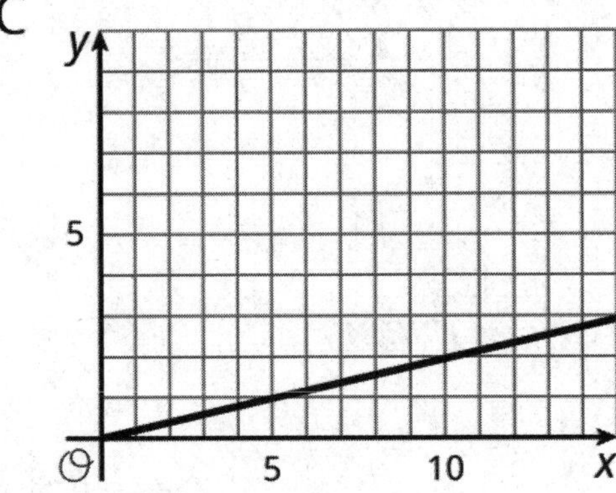


D

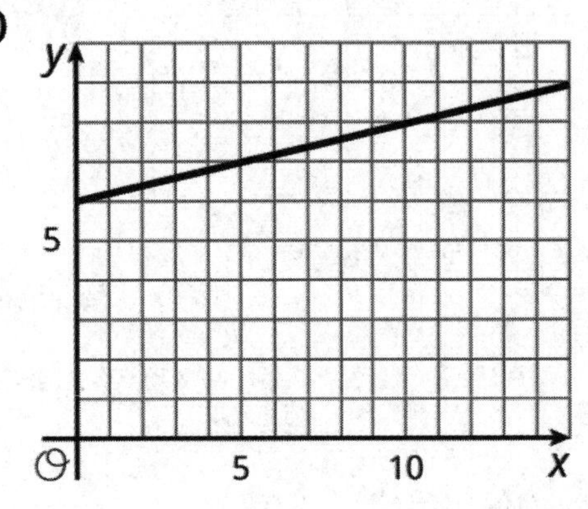


   (From Unit 5, Lesson 7.)

# Lesson 17: Negative Rates

Let's apply what we know about signed numbers.

## 17.1: Grapes per Minute

1. If you eat 5 grapes per minute for 8 minutes, how many grapes will you eat?

2. If you hear 9 new songs per day for 3 days, how many new songs will you hear?

3. If you run 15 laps per practice, how many practices will it take you to run 30 laps?

## 17.2: Water Level in the Aquarium

1. A large aquarium should contain 10,000 liters of water when it is filled correctly. It will overflow if it gets up to 12,000 liters. The fish will get sick if it gets down to 4,000 liters. The aquarium has an automatic system to help keep the correct water level. If the water level is too low, a faucet fills it. If the water level is too high, a drain opens.

   One day, the system stops working correctly. The faucet starts to fill the aquarium at a rate of 30 liters per minute, and the drain opens at the same time, draining the water at a rate of 20 liters per minute.

   a. Is the water level rising or falling? How do you know?

   b. How long will it take until the tank starts overflowing or the fish get sick?

2. A different aquarium should contain 15,000 liters of water when filled correctly. It will overflow if it gets to 17,600 liters.

   One day there is an accident and the tank cracks in 4 places. Water flows out of each crack at a rate of $\frac{1}{2}$ liter per hour. An emergency pump can re-fill the tank at a rate of 2 liters per minute. How many minutes must the pump run to replace the water lost each hour?

## 17.3: Up and Down with the Piccards

1. Challenger Deep is the deepest known point in the ocean, at 35,814 feet below sea level. In 1960, Jacques Piccard and Don Walsh rode down in the Trieste and became the first people to visit the Challenger Deep.

   a. If sea level is represented by 0 feet, explain how you can represent the depth of a submarine descending from sea level to the bottom of Challenger Deep.

   b. Trieste's descent was a change in depth of -3 feet per second. We can use the relationship $y = -3x$ to model this, where $y$ is the depth (in feet) and $x$ is the time (in seconds). Using this model, how much time would the Trieste take to reach the bottom?

   c. It took the Trieste 3 hours to ascend back to sea level. This can be modeled by a different relationship $y = kx$. What is the value of $k$ in this situation?

2. The design of the Trieste was based on the design of a hot air balloon built by Auguste Piccard, Jacques's father. In 1932, Auguste rode in his hot-air balloon up to a record-breaking height.

   a. Auguste's ascent took 7 hours and went up 51,683 feet. Write a relationship $y = kx$ to represent his ascent from his starting location.

   b. Auguste's descent took 3 hours and went down 52,940 feet. Write another relationship to represent his descent.

   c. Did Auguste Piccard end up at a greater or lesser altitude than his starting point? How much higher or lower?

**Are you ready for more?**

During which part of either trip was a Piccard changing vertical position the fastest? Explain your reasoning.

- Jacques's descent
- Jacques's ascent
- Auguste's ascent
- Auguste's descent

## 17.4: Buying and Selling Power

A utility company charges $0.12 per kilowatt-hour for energy a customer uses. They give a credit of $0.025 for every kilowatt-hour of electricity a customer with a solar panel generates that they don't use themselves.

A customer has a charge of $82.04 and a credit of -$4.10 on this month's bill.

1. What is the amount due this month?

2. How many kilowatt-hours did they use?

3. How many kilowatt-hours did they generate that they didn't use themselves?

### Are you ready for more?

1. Find the value of the expression without a calculator.

$$(2)(-30) + (-3)(-20) + (-6)(-10) - (2)(3)(10)$$

2. Write an expression that uses addition, subtraction, multiplication, and division and only negative numbers that has the same value.

### Lesson 17 Summary

We saw earlier that we can represent speed with direction using signed numbers. Speed with direction is called *velocity*. Positive velocities always represent movement in the opposite direction from negative velocities.

We can do this with vertical movement: moving up can be represented with positive numbers, and moving down with negative numbers. The magnitude tells you how fast, and the sign tells you which direction. (We could actually do it the other way around if we wanted to, but usually we make up positive and down negative.)

# Lesson 17 Practice Problems

1. Describe a situation where each of the following quantities might be useful.

    a. -20 gallons per hour

    b. -10 feet per minute

    c. -0.1 kilograms per second

2. A bank charges a service fee of $7.50 per month for a checking account.

    A bank account has $85.00. If no money is deposited or withdrawn except the service charge, how many months until the account balance is negative?

3. A submarine is searching for underwater features. It is accompanied by a small aircraft and an underwater robotic vehicle.

    At one time the aircraft is 200 m above the surface, the submarine is 55 m below the surface, and the underwater robotic vehicle is 227 m below the surface.

    a. What is the difference in height between the submarine and the aircraft?

    b. What is the distance between the underwater robotic vehicle and the submarine?

    (From Unit 7, Lesson 10.)

4. Evaluate each expression. When the answer is not a whole number, write your answer as a fraction.

   a. $-4 \cdot -6$

   b. $-24 \cdot \frac{-7}{6}$

   c. $4 \div -6$

   d. $\frac{4}{3} \div -24$

(From Unit 7, Lesson 16.)

5. a. A restaurant bill is \$21. You leave a 15% tip. How much do you pay including the tip?

   b. Which of the following represents the amount a customer pays including the tip of 15% if the bill was $b$ dollars? Select **all** that apply.

   - $15b$
   - $b + 0.15b$
   - $1.15b$
   - $1.015b$
   - $b + \frac{15}{100}b$
   - $b + 0.15$
   - $0.15b$

(From Unit 6, Lesson 7.)

6. Consider a rectangular prism with length 4 and width and height $d$.

   a. Find an expression for the volume of the prism in terms of $d$.

   b. Compute the volume of the prism when $d = 1$, when $d = 2$, and when $d = \frac{1}{2}$.

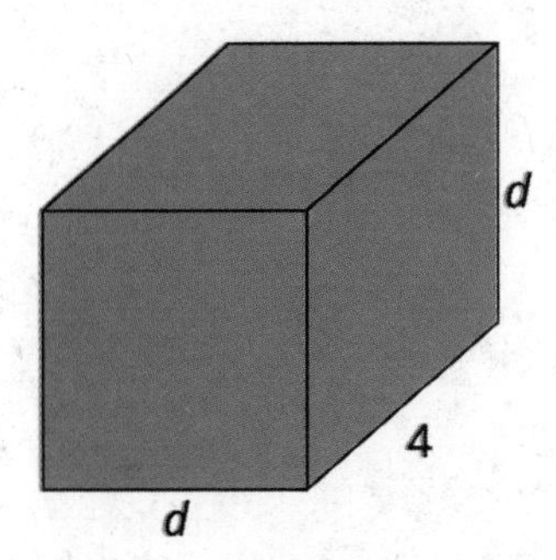


(From Unit 4, Lesson 15.)

# Lesson 18: Expressions with Rational Numbers

Let's develop our signed number sense.

## 18.1: True or False: Rational Numbers

Decide if each statement is true or false. Be prepared to explain your reasoning.

1. $(-38.76)(-15.6)$ is negative

2. $10,000 - 99,999 < 0$

3. $\left(\frac{3}{4}\right)\left(-\frac{4}{3}\right) = 0$

4. $(30)(-80) - 50 = 50 - (30)(-80)$

## 18.2: Card Sort: The Same But Different

Your teacher will give you a set of cards. Group them into pairs of expressions that have the same value.

## 18.3: Near and Far From Zero

| $a$ | $b$ | $-a$ | $-4b$ | $-a+b$ | $a \div -b$ | $a^2$ | $b^3$ |
|---|---|---|---|---|---|---|---|
| $-\frac{1}{2}$ | 6 | | | | | | |
| $\frac{1}{2}$ | -6 | | | | | | |
| -6 | $-\frac{1}{2}$ | | | | | | |

1. For each set of values for $a$ and $b$, evaluate the given expressions and record your answers in the table.

2. When $a = -\frac{1}{2}$ and $b = 6$, which expression:

   has the largest value? has the smallest value? is the closest to zero?

3. When $a = \frac{1}{2}$ and $b = -6$, which expression:

   has the largest value? has the smallest value? is the closest to zero?

4. When $a = -6$ and $b = -\frac{1}{2}$, which expression:

   has the largest value? has the smallest value? is the closest to zero?

### Are you ready for more?

Are there any values could you use for $a$ and $b$ that would make all of these expressions have the same value? Explain your reasoning.

## 18.4: Seagulls and Sharks Again

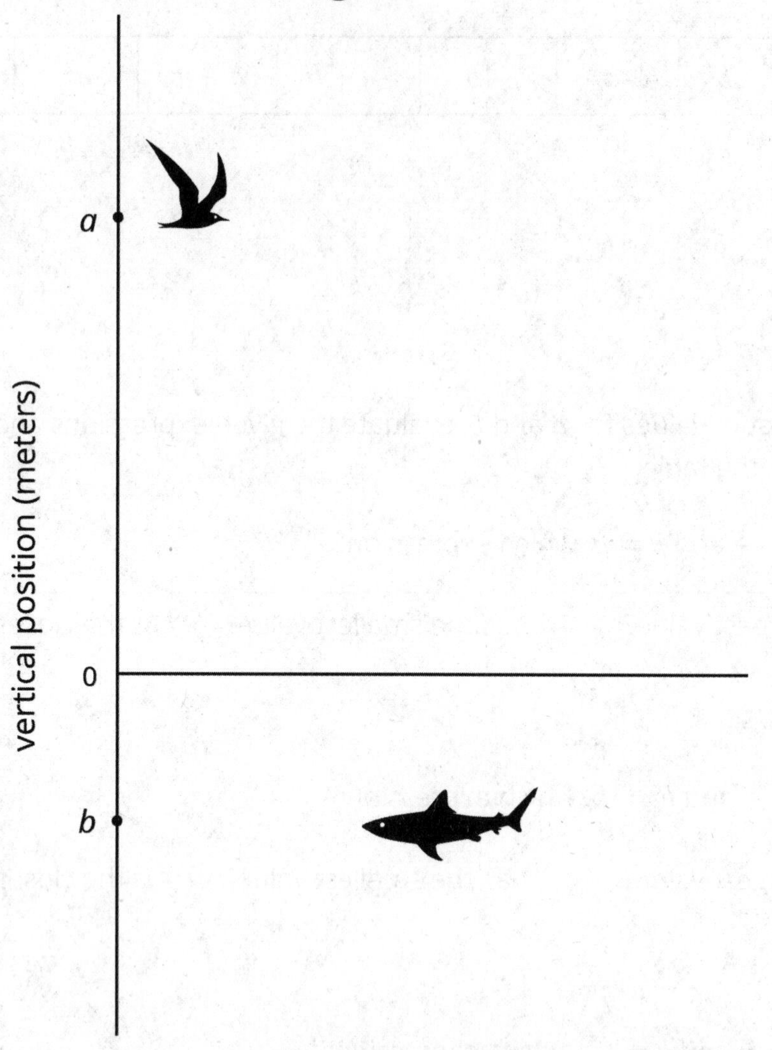


A seagull has a vertical position $a$, and a shark has a vertical position $b$. Draw and label a point on the vertical axis to show the vertical position of each new animal.

1. A dragonfly at $d$, where $d = -b$
2. A jellyfish at $j$, where $j = 2b$
3. An eagle at $e$, where $e = \frac{1}{4}a$.
4. A clownfish at $c$, where $c = \frac{-a}{2}$
5. A vulture at $v$, where $v = a + b$
6. A goose at $g$, where $g = a - b$

## Lesson 18 Summary

We can represent sums, differences, products, and quotients of **rational numbers**, and combinations of these, with numerical and algebraic expressions.

| Sums: | Differences: | Products: | Quotients: |
|---|---|---|---|
| $\frac{1}{2} + -9$ | $\frac{1}{2} - -9$ | $(\frac{1}{2})(-9)$ | $\frac{1}{2} \div -9$ |
| $-8.5 + x$ | $-8.5 - x$ | $-8.5x$ | $\frac{-8.5}{x}$ |

We can write the product of two numbers in different ways.

- By putting a little dot between the factors, like this: $-8.5 \cdot x$.
- By putting the factors next to each other without any symbol between them at all, like this: $-8.5x$.

We can write the quotient of two numbers in different ways as well.

- By writing the division symbol between the numbers, like this: $-8.5 \div x$.
- By writing a fraction bar between the numbers like this: $\frac{-8.5}{x}$.

When we have an algebraic expression like $\frac{-8.5}{x}$ and are given a value for the variable, we can find the value of the expression. For example, if $x$ is 2, then the value of the expression is -4.25, because $-8.5 \div 2 = -4.25$.

## Glossary

- rational number

## Lesson 18 Practice Problems

1. The value of $x$ is $\frac{-1}{4}$. Order these expressions from least to greatest:

   $x$ $\quad$ $1 - x$ $\quad$ $x - 1$ $\quad$ $-1 \div x$

2. Here are four expressions that have the value $\frac{-1}{2}$:

   $\frac{-1}{4} + \left(\frac{-1}{4}\right)$ $\quad$ $\frac{1}{2} - 1$ $\quad$ $-2 \cdot \frac{1}{4}$ $\quad$ $-1 \div 2$

   Write five expressions: a sum, a difference, a product, a quotient, and one that involves at least two operations that have the value $\frac{-3}{4}$.

3. Find the value of each expression.

   a. $-22 + 5$

   b. $-22 - (-5)$

   c. $(-22)(-5)$

   d. $-22 \div 5$

4. The price of an ice cream cone is $3.25, but it costs $3.51 with tax. What is the sales tax rate?

   (From Unit 6, Lesson 7.)

5. Here are 4 points on a coordinate plane.

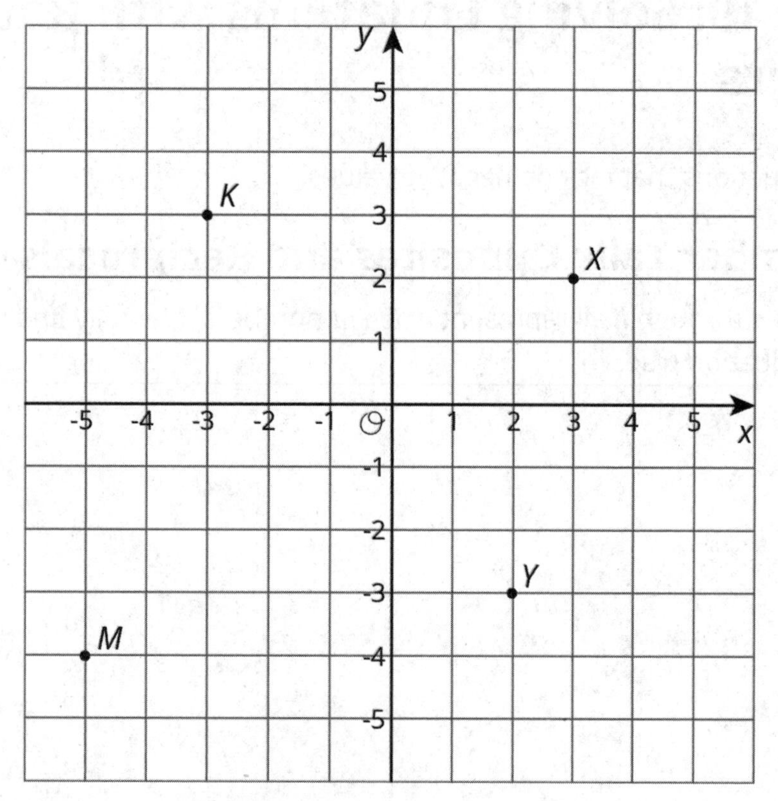


a. Label each point with its coordinates.

b. Plot a point that is 3 units from point $K$. Label it $P$.

c. Plot a point that is 2 units from point $M$. Label it $W$.

(From Unit 7, Lesson 13.)

6. A furniture store pays a wholesale price for a mattress. Then, the store marks up the retail price to 150% of the wholesale price. Later, they put the mattress on sale for 50% off of the retail price. A customer just bought the mattress on sale and paid $1,200.

a. What was the retail price of the mattress, before the discount?

b. What was the wholesale price, before the markup?

(From Unit 6, Lesson 8.)

# Lesson 19: Solving Equations with Rational Numbers

Let's solve equations that include negative values.

## 19.1: Number Talk: Opposites and Reciprocals

The **variables** $a$ through $h$ all represent *different* numbers. Mentally find numbers that make each equation true.

$\frac{3}{5} \cdot \frac{5}{3} = a$

$7 \cdot b = 1$

$c \cdot d = 1$

$-6 + 6 = e$

$11 + f = 0$

$g + h = 0$

## 19.2: Match Solutions

1. Match each equation to its solution.

| | |
|---|---|
| a. $\frac{1}{2}x = -5$ | 1. $x = -4.5$ |
| b. $-2x = -9$ | 2. $x = -\frac{1}{2}$ |
| c. $-\frac{1}{2}x = \frac{1}{4}$ | 3. $x = -10$ |
| d. $-2x = 7$ | 4. $x = 4.5$ |
| e. $x + -2 = -6.5$ | 5. $x = 2\frac{1}{2}$ |
| f. $-2 + x = \frac{1}{2}$ | 6. $x = -3.5$ |

Be prepared to explain your reasoning.

## 19.3: Trip to the Mountains

The Hiking Club is on a trip to hike up a mountain.

1. The members increased their elevation 290 feet during their hike this morning. Now they are at an elevation of 450 feet.

   a. Explain how to find their elevation before the hike.

   b. Han says the equation $e + 290 = 450$ describes the situation. What does the variable $e$ represent?

   c. Han says that he can rewrite his equation as $e = 450 + -290$ to solve for $e$. Compare Han's strategy to your strategy for finding the beginning elevation.

2. The temperature fell 4 degrees in the last hour. Now it is 21 degrees. Write and solve an equation to find the temperature it was 1 hour ago.

3. There are 3 times as many students participating in the hiking trip this year than last year. There are 42 students on the trip this year.

   a. Explain how to find the number of students that came on the hiking trip last year.

b. Mai says the equation $3s = 42$ describes the situation. What does the variable $s$ represent?

c. Mai says that she can rewrite her equation as $s = \frac{1}{3} \cdot 42$ to solve for $s$. Compare Mai's strategy to your strategy for finding the number of students on last year's trip.

4. The cost of the hiking trip this year is $\frac{2}{3}$ of the cost of last year's trip. This year's trip cost \$32. Write and solve an equation to find the cost of last year's trip.

**Are you ready for more?**

A number line is shown below. The numbers 0 and 1 are marked on the line, as are two other rational numbers $a$ and $b$.

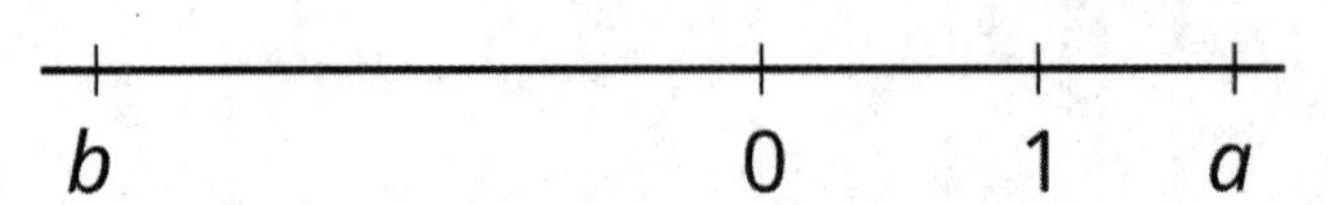


Decide which of the following numbers are positive and which are negative.

$a - 1$ $\quad$ $a - 2$ $\quad$ $-b$ $\quad$ $a + b$ $\quad$ $a - b$ $\quad$ $ab + 1$

## 19.4: Card Sort: Matching Inverses

Your teacher will give you a set of cards with numbers on them.

1. Match numbers with their additive inverses.
2. Next, match numbers with their multiplicative inverses.
3. What do you notice about the numbers and their inverses?

### Lesson 19 Summary

To solve the equation $x + 8 = \text{-}5$, we can add the opposite of 8, or -8, to each side:

$$x + 8 = \text{-}5$$
$$(x + 8) + \text{-}8 = (\text{-}5) + \text{-}8$$
$$x = \text{-}13$$

Because adding the opposite of a number is the same as subtracting that number, we can also think of it as subtracting 8 from each side.

We can use the same approach for this equation:

$$\text{-}12 = t + \text{-}\frac{2}{9}$$
$$(\text{-}12) + \frac{2}{9} = \left(t + \text{-}\frac{2}{9}\right) + \frac{2}{9}$$
$$\text{-}11\frac{7}{9} = t$$

To solve the equation $8x = \text{-}5$, we can multiply each side by the reciprocal of 8, or $\frac{1}{8}$:

$$8x = \text{-}5$$
$$\frac{1}{8}(8x) = \frac{1}{8}(\text{-}5)$$
$$x = \text{-}\frac{5}{8}$$

Because multiplying by the reciprocal of a number is the same as dividing by that number, we can also think of it as dividing by 8.

We can use the same approach for this equation:

$$\text{-}12 = \text{-}\frac{2}{9}t$$
$$\text{-}\frac{9}{2}(\text{-}12) = \text{-}\frac{9}{2}\left(\text{-}\frac{2}{9}t\right)$$
$$54 = t$$

### Glossary

- variable

## Lesson 19 Practice Problems

1. Solve.

   a. $\frac{2}{5}t = 6$

   b. $-4.5 = a - 8$

   c. $\frac{1}{2} + p = -3$

   d. $12 = x \cdot 3$

   e. $-12 = -3y$

2. Match each equation to a step that will help solve the equation.

   A. $5x = 0.4$

   B. $\frac{x}{5} = 8$

   C. $3 = \frac{-x}{5}$

   D. $7 = -5x$

   1. Multiply each side by 5.
   2. Multiply each side by -5.
   3. Multiply each side by $\frac{1}{5}$.
   4. Multiply each side by $\frac{-1}{5}$.

3. a. Write an equation where a number is added to a variable, and a solution is -8.

   b. Write an equation where a number is multiplied by a variable, and a solution is $\frac{-4}{5}$.

4. Evaluate each expression if $x$ is $\frac{2}{5}$, $y$ is -4, and $z$ is -0.2.

a. $x + y$

b. $2x - z$

c. $x + y + z$

d. $y \cdot x$

(From Unit 7, Lesson 18.)

5. The markings on the number line are evenly spaced. Label the other markings on the number line.

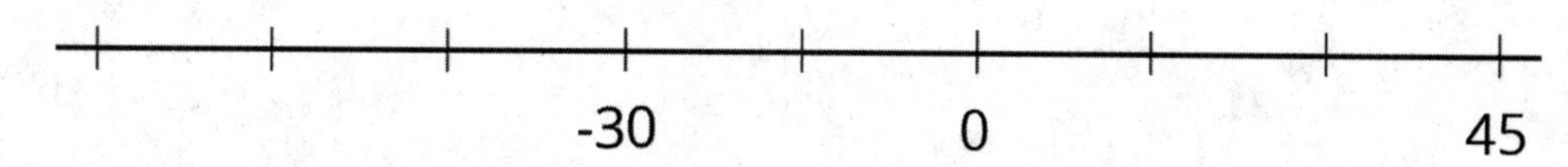


(From Unit 7, Lesson 14.)

6. One night, it is 24°C warmer in Tucson than it was in Minneapolis. If the temperatures in Tucson and Minneapolis are opposites, what is the temperature in Tucson?

A. -24°C

B. -12°C

C. 12°C

D. 24°C

(From Unit 7, Lesson 2.)

# Lesson 20: Representing Contexts with Equations

Let's write equations that represent situations.

## 20.1: Don't Solve It

Is the solution positive or negative?

$(-8.7)(1.4) = a$

$-8.7b = 1.4$

$-8.7 + c = -1.4$

$-8.7 - d = -1.4$

## 20.2: Warmer or Colder than Before?

For each situation,

- Find *two* equations that could represent the situation from the bank of equations. (Some equations will not be used.)
- Explain what the variable $v$ represents in the situation.
- Determine the value of the variable that makes the equation true, and explain your reasoning.

Bank of equations:

| | | | |
|---|---|---|---|
| $-3v = 9$ | $v = -16 + 6$ | $v = \frac{1}{3} \cdot (-6)$ | $v + 12 = 4$ |
| $-4 \cdot 3 = v$ | $v = 4 + (-12)$ | $v = -16 - (6)$ | $v = 9 + 3$ |
| $-4 \cdot -3 = v$ | $-3v = -6$ | $-6 + v = -16$ | $-4 = \frac{1}{3}v$ |
| $v = -\frac{1}{3} \cdot 9$ | $v = -\frac{1}{3} \cdot (-6)$ | $v = 4 + 12$ | $4 = 3v$ |

1. Between 6 a.m. and noon, the temperature rose 12 degrees Fahrenheit to 4 degrees Fahrenheit.

2. At midnight the temperature was -6 degrees. By 4 a.m. the temperature had fallen to -16 degrees.

3. The temperature is 0 degrees at midnight and dropping 3 degrees per hour. The temperature is -6 degrees at a certain time.

4. The temperature is 0 degrees at midnight and dropping 3 degrees per hour. The temperature is 9 degrees at a certain time.

5. The temperature at 9 p.m. is one third the temperature at midnight.

## 20.3: Animals Changing Altitudes

1. Match each situation with a diagram.

   a. A penguin is standing 3 feet above sea level and then dives down 10 feet. What is its depth?

   b. A dolphin is swimming 3 feet below sea level and then jumps up 10 feet. What is its height at the top of the jump?

   c. A sea turtle is swimming 3 feet below sea level and then dives down 10 feet. What is its depth?

   d. An eagle is flying 10 feet above sea level and then dives down to 3 feet above sea level. What was its change in altitude?

   e. A pelican is flying 10 feet above sea level and then dives down reaching 3 feet below sea level. What was its change in altitude?

   f. A shark is swimming 10 feet below sea level and then swims up reaching 3 feet below sea level. What was its change in depth?

2. Next, write an equation to represent each animal's situation and answer the question. Be prepared to explain your reasoning.

Diagrams

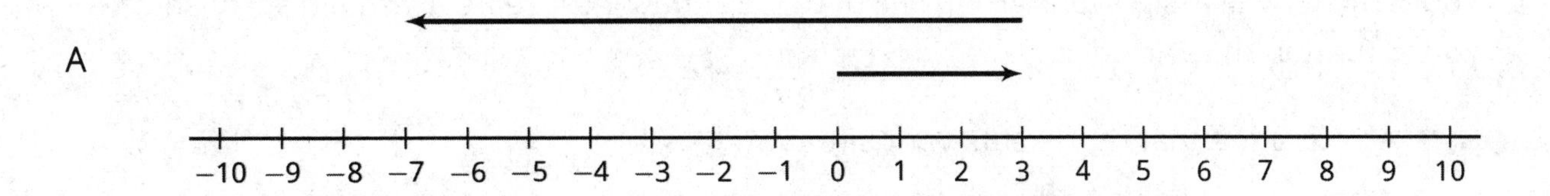
A
−10 −9 −8 −7 −6 −5 −4 −3 −2 −1 0 1 2 3 4 5 6 7 8 9 10

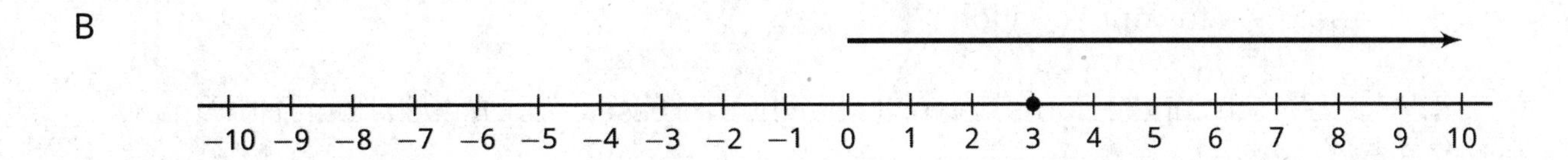
B
−10 −9 −8 −7 −6 −5 −4 −3 −2 −1 0 1 2 3 4 5 6 7 8 9 10

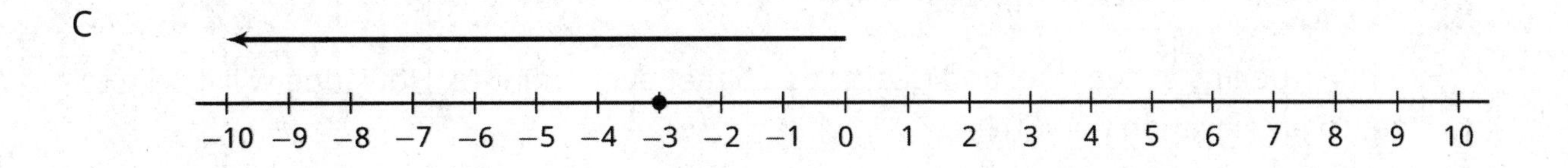
C
−10 −9 −8 −7 −6 −5 −4 −3 −2 −1 0 1 2 3 4 5 6 7 8 9 10

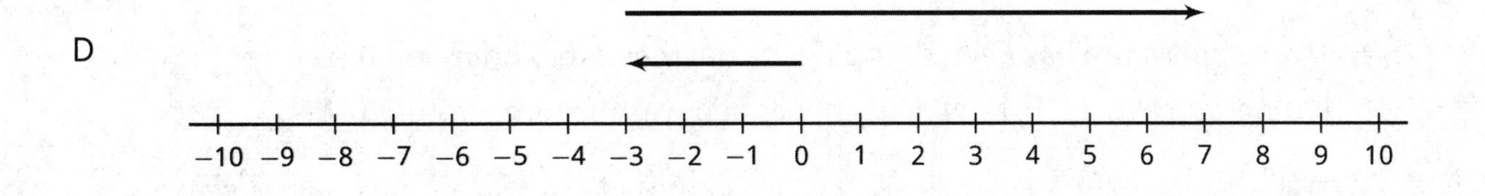
D
−10 −9 −8 −7 −6 −5 −4 −3 −2 −1 0 1 2 3 4 5 6 7 8 9 10

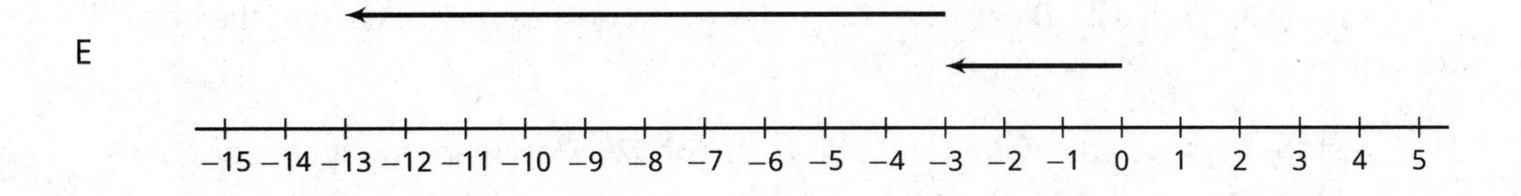
E
−15 −14 −13 −12 −11 −10 −9 −8 −7 −6 −5 −4 −3 −2 −1 0 1 2 3 4 5

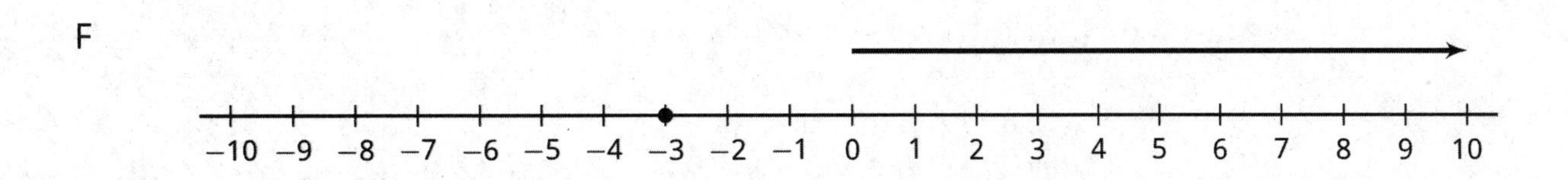
F
−10 −9 −8 −7 −6 −5 −4 −3 −2 −1 0 1 2 3 4 5 6 7 8 9 10

## 20.4: Equations Tell a Story

Your teacher will assign your group *one* of these situations. Create a visual display about your situation that includes:

- An equation that represents your situation
- What your variable and each term in the equation represent
- How the operations in the equation represent the relationships in the story
- How you use inverses to solve for the unknown quantity
- The solution to your equation

1. As a $7\frac{1}{4}$ inch candle burns down, its height decreases $\frac{3}{4}$ inch each hour. How many hours does it take for the candle to burn completely?

2. On Monday $\frac{1}{9}$ of the enrolled students in a school were absent. There were 4,512 students present. How many students are enrolled at the school?

3. A hiker begins at sea level and descends 25 feet every minute. How long will it take to get to an elevation of -750 feet?

4. Jada practices the violin for the same amount of time every day. On Tuesday she practices for 35 minutes. How much does Jada practice in a week?

5. The temperature has been dropping $2\frac{1}{2}$ degrees every hour and the current temperature is $-15^\circ\text{F}$. How many hours ago was the temperature $0^\circ\text{F}$?

6. The population of a school increased by 12%, and now the population is 476. What was the population before the increase?

7. During a 5% off sale, Diego pays $74.10 for a new hockey stick. What was the original price?

8. A store buys sweaters for $8 and sells them for $26. How many sweaters does the store need to sell to make a profit of $990?

### Are you ready for more?

Diego and Elena are 2 miles apart and begin walking towards each other. Diego walks at a rate of 3.7 miles per hour and Elena walks 4.3 miles per hour. While they are walking, Elena's dog runs back and forth between the two of them, at a rate of 6 miles per hour. Assuming the dog does not lose any time in turning around, how far has the dog run by the time Diego and Elena reach each other?

### Lesson 20 Summary

We can use variables and equations involving signed numbers to represent a story or answer questions about a situation.

For example, if the temperature is $-3^\circ\text{C}$ and then falls to $-17^\circ\text{C}$, we can let $x$ represent the temperature change and write the equation:

$$-3 + x = -17$$

We can solve the equation by adding 3 to each side. Since $-17 + 3 = -14$, the change is $-14^\circ\text{C}$.

Here is another example: if a starfish is descending by $\frac{3}{2}$ feet every hour then we can solve

$$-\frac{3}{2}h = -6$$

to find out how many hours $h$ it takes the starfish to go down 6 feet.

We can solve this equation by multiplying each side by $-\frac{2}{3}$. Since $-\frac{2}{3} \cdot -6 = 4$, we know it will take the starfish 4 hours to descend 6 feet.

## Lesson 20 Practice Problems

1. Match each situation to one of the equations.

   A. A whale was diving at a rate of 2 meters per second. How long will it take for the whale to get from the surface of the ocean to an elevation of -12 meters at that rate?

   B. A swimmer dove below the surface of the ocean. After 2 minutes, she was 12 meters below the surface. At what rate was she diving?

   C. The temperature was -12 degrees Celsius and rose to 2 degrees Celsius. What was the change in temperature?

   D. The temperature was 2 degrees Celsius and fell to -12 degrees Celsius. What was the change in temperature?

   1. $-12 + x = 2$
   2. $2 + x = -12$
   3. $-2x = -12$
   4. $2x = -12$

2. Starting at noon, the temperature dropped steadily at a rate of 0.8 degrees Celsius every hour.

   For each of these situations, write and solve an equation and describe what your variable represents.

   a. How many hours did it take for the temperature to decrease by 4.4 degrees Celsius?

   b. If the temperature after the 4.4 degree drop was -2.5 degrees Celsius, what was the temperature at noon?

3. Find the value of each expression.

   a. $12 + -10$

   b. $-5 - 6$

   c. $-42 + 17$

   d. $35 - -8$

   e. $-4\frac{1}{2} + 3$

(From Unit 7, Lesson 10.)

4. A shopper bought a watermelon, a pack of napkins, and some paper plates. In his state, there is no tax on food. The tax rate on non-food items is 5%. The total for the three items he bought was \$8.25 before tax, and he paid \$0.19 in tax. How much did the watermelon cost?

(From Unit 6, Lesson 7.)

5. A 50-centimeter piece of wire is bent into a circle. What is the area of this circle?

(From Unit 5, Lesson 15.)

# Lesson 21: Drawing on the Coordinate Plane

- Let's draw on the coordinate plane.

## 21.1: Cat Pictures

Use graphing technology to recreate this image. If graphing technology is not available, list the ordered pairs that make up this image. Then compare your list with a partner.

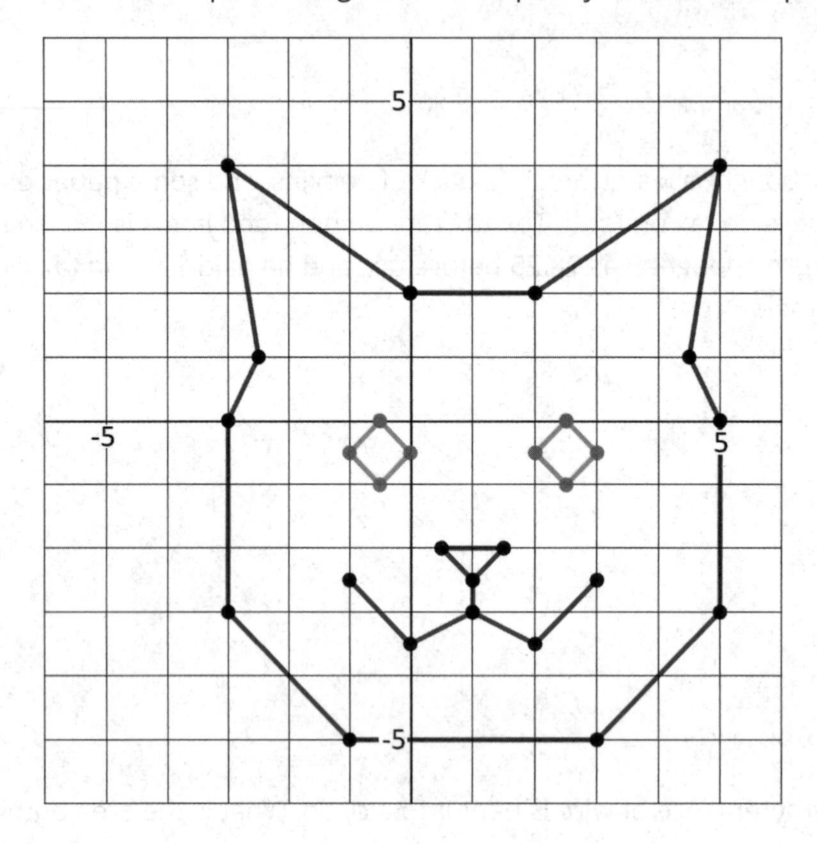


If you have time, consider adding more details to your image such as whiskers, the inside of the ears, a bow, or a body.

**Are you ready for more?**

If you are using graphing technology, add these statements to the list of things being graphed:

$$x > 6$$
$$y > 5$$
$$x < -4$$
$$y < -6$$

Describe the result. Why do you think that happened?

## 21.2: Design Your Own Image

Use graphing technology to create an image of your own design. You could draw a different animal, a vehicle, a building, or something else. Make sure your image includes at least 4 points in each quadrant of the coordinate plane.

If graphing technology is not available, create your image on graph paper, and then list the ordered pairs that make up your image. Trade lists with a partner but do not show them your image. Graph your partner's ordered pairs and see if your images match.

# Lesson 22: The Stock Market

Let's learn about the Stock Market.

## 22.1: Revisiting Interest and Depreciation

1. Lin deposited $300 in a savings account that has a 2% interest rate per year. How much is in her account after 1 year? After 2 years?

2. Diego wants to sell his bicycle. It cost $150 when he bought it but has depreciated by 15%. How much should he sell it for?

## 22.2: Gains and Losses

1. Here is some information from the stock market in September 2016. Complete the table.

| company | value on day 1 (dollars) | value on day 2 (dollars) | change in value (dollars) | change in value as a percentage of day 1 value |
|---|---|---|---|---|
| Mobile Tech Company | 107.95 | 111.77 | 3.82 | 3.54% |
| Electrical Appliance Company | | 114.03 | 2.43 | 2.18% |
| Oil Corporation | 26.14 | 25.14 | | -3.83% |
| Department Store Company | 7.38 | 7.17 | | |
| Jewelry Company | | 70.30 | | 2.27% |

2. Which company's change in dollars had the largest magnitude?

3. Which company's change in percentage had the largest magnitude?

## 22.3: What is a Stock Portfolio?

A person who wants to make money by investing in the stock market usually buys a portfolio, or a collection of different stocks. That way, if one of the stocks decreases in value, they won't lose all of their money at once.

1. Here is an example of someone's stock portfolio. Complete the table to show the total value of each investment.

| name | price (dollars) | number of shares | total value (dollars) |
|---|---|---|---|
| Technology Company | 107.75 | 98 | |
| Airline Company | 133.54 | 27 | |
| Film Company | 95.95 | 135 | |
| Sports Clothing Company | 58.96 | 100 | |

2. Here is the same portfolio the next year. Complete the table to show the new total value of each investment.

| company | old price (dollars) | price change | new price (dollars) | number of shares | total value (dollars) |
|---|---|---|---|---|---|
| Technology Company | 107.75 | +2.43% | | 98 | |
| Airline Company | 133.54 | -7.67% | | 27 | |
| Film Company | 95.95 | | 87.58 | 135 | |
| Sports Clothing Company | 58.96 | -5.56% | | 100 | |

3. Did the entire portfolio increase or decrease in value over the year?

## 22.4: Your Own Stock Portfolio

Your teacher will give you a list of stocks.

1. Select a combination of stocks with a total value close to, but no more than, $100.

2. Using the new list, how did the total value of your selected stocks change?

# Learning Targets

### Lesson 1: Positive and Negative Numbers

- I can explain what 0, positive numbers, and negative numbers mean in the context of temperature and elevation.
- I can use positive and negative numbers to describe temperature and elevation.
- I know what positive and negative numbers are.

### Lesson 2: Comparing Positive and Negative Numbers

- I can explain how to use the positions of numbers on a number line to compare them.
- I can explain what a rational number is.
- I can use inequalities to compare positive and negative numbers.
- I understand what it means for numbers to be opposites.

### Lesson 3: Ordering Rational Numbers

- I can compare and order rational numbers.
- I can use phrases like "greater than," "less than," and "opposite" to compare rational numbers.

### Lesson 4: Absolute Value of Numbers

- I can explain what the absolute value of a number is.
- I can find the absolute values of rational numbers.
- I can recognize and use the notation for absolute value.

### Lesson 5: Comparing Numbers and Distance from Zero

- I can explain what absolute value means in situations involving elevation.
- I can use absolute values to describe elevations.
- I can use inequalities to compare rational numbers and the absolute values of rational numbers.

### Lesson 6: Changing Temperatures

- I can use a number line to add positive and negative numbers.

### Lesson 7: Changing Elevation

- I understand how to add positive and negative numbers in general.

### Lesson 8: Money and Debts

- I understand what positive and negative numbers mean in a situation involving money.

### Lesson 9: Representing Subtraction

- I can explain the relationship between addition and subtraction of rational numbers.
- I can use a number line to subtract positive and negative numbers.

### Lesson 10: Subtracting Rational Numbers

- I can find the difference between two rational numbers.
- I can solve problems that involve adding and subtracting rational numbers.
- I understand how to subtract positive and negative numbers in general.

### Lesson 11: Constructing the Coordinate Plane

- I can plot points with negative coordinates in the coordinate plane.
- I know what negative numbers in coordinates tell us.
- When given points to plot, I can construct a coordinate plane with an appropriate scale and pair of axes.

### Lesson 12: Interpreting Points on a Coordinate Plane

- I can explain how rational numbers represent balances in a money context.
- I can explain what points in a four-quadrant coordinate plane represent in a situation.
- I can plot points in a four-quadrant coordinate plane to represent situations and solve problems.

### Lesson 13: Distances and Shapes on the Coordinate Plane

- I can find horizontal and vertical distances between points on the coordinate plane.
- I can plot polygons on the coordinate plane when I have the coordinates for the vertices.

### Lesson 14: Position, Speed, Direction

- I can explain what it means when time is represented with a negative number in a situation about speed and direction.
- I can multiply two negative numbers.
- I can use rational numbers to represent speed and direction.

### Lesson 15: Multiplying Rational Numbers

- I can solve problems that involve multiplying rational numbers.

### Lesson 16: Dividing Rational Numbers

- I can divide rational numbers.

### Lesson 17: Negative Rates

- I can solve problems that involve multiplying and dividing rational numbers.
- I can solve problems that involve negative rates.

### Lesson 18: Expressions with Rational Numbers

- I can add, subtract, multiply, and divide rational numbers.
- I can evaluate expressions that involve rational numbers.

### Lesson 19: Solving Equations with Rational Numbers

- I can solve equations that include rational numbers and have rational solutions.

### Lesson 20: Representing Contexts with Equations

- I can explain what the solution to an equation means for the situation.
- I can write and solve equations to represent situations that involve rational numbers.

### Lesson 21: Drawing on the Coordinate Plane

- I can use ordered pairs to draw a picture.

## Lesson 22: The Stock Market

- I can solve problems about the stock market using rational numbers and percentages.

# Illustrative Mathematics

STUDENT EDITION

Book 2

# Lesson 1: Representing Data Graphically

Let's represent data with dot plots and bar graphs.

## 1.1: Curious about Caps

Clare collects bottle caps and keeps them in plastic containers.

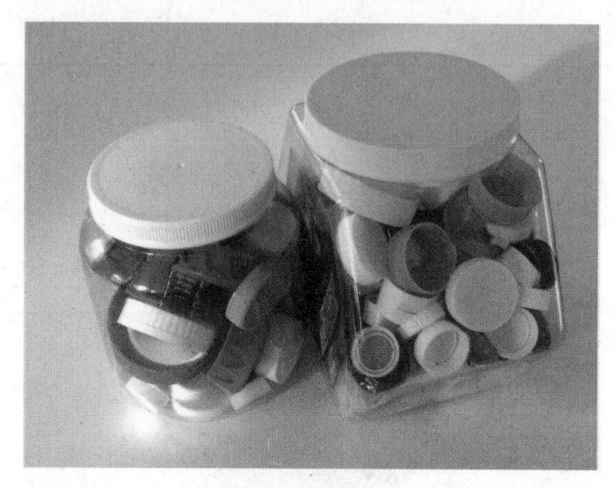

Write one statistical question that someone could ask Clare about her collection. Be prepared to explain your reasoning.

## 1.2: Estimating Caps

1. Write down the statistical question your class is trying to answer.

2. Look at the dot plot that shows the data from your class. Write down one thing you notice and one thing you wonder about the dot plot.

3. Use the dot plot to answer the statistical question. Be prepared to explain your reasoning.

## 1.3: Been There, Done That!

Priya wants to know if basketball players on a men's team and a women's team have had prior experience in international competitions. She gathered data on the number of times the players were on a team before 2016.

men's team

3 0 0 0 0 1 0 0 0 0 0 0

women's team

2 3 3 1 0 2 0 1 1 0 3 1

1. Did Priya collect categorical or numerical data?

2. Organize the information on the two basketball teams into these tables.

Men's Basketball Team Players

| number of prior competitions | frequency (number) |
|---|---|
| 0 | |
| 1 | |
| 2 | |
| 3 | |
| 4 | |

Women's Basketball Team Players

| number of prior competitions | frequency (number) |
|---|---|
| 0 | |
| 1 | |
| 2 | |
| 3 | |
| 4 | |

3. Make a dot plot for each table.

Men's Basketball Team Players

Women's Basketball Team Players

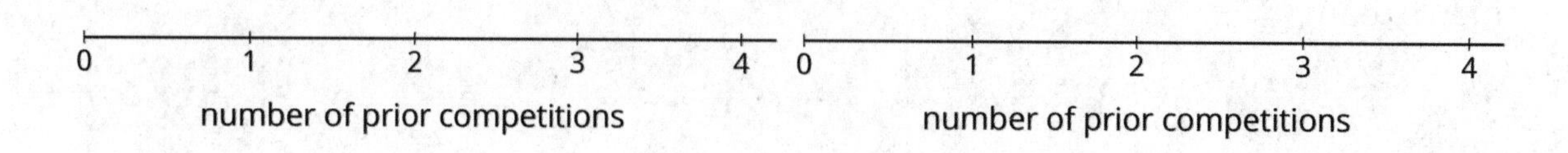


4. Study your dot plots. What do they tell you about the competition participation of:

   a. the players on the men's basketball team?

   b. the players on the women's basketball team?

5. Explain why a dot plot is an appropriate representation for Priya's data.

**Are you ready for more?**

Combine the data for the players on the men's and women's teams and represent it as a single dot plot. What can you say about the repeat participation of the basketball players?

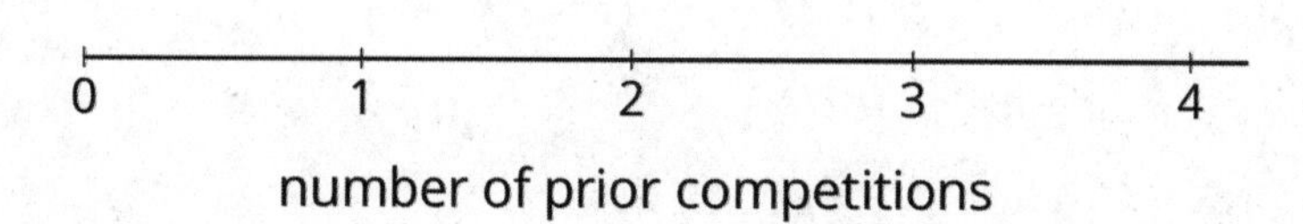

## 1.4: Favorite Summer Sports

Kiran wants to know which three summer sports are most popular in his class. He surveyed his classmates on their favorite summer sport. Here are their responses.

| | | | |
|---|---|---|---|
| swimming | gymnastics | track and field | volleyball |
| swimming | swimming | diving | track and field |
| gymnastics | basketball | basketball | volleyball |
| track and field | track and field | volleyball | gymnastics |
| diving | gymnastics | volleyball | rowing |
| track and field | track and field | soccer | swimming |
| gymnastics | track and field | swimming | rowing |
| diving | soccer | | |

1. Did Kiran collect categorical or numerical data?
2. Organize the responses in a table to help him find which summer sports are most popular in his class.

| sport | frequency |
|---|---|
| | |
| | |
| | |
| | |
| | |
| | |
| | |
| | |

3. Represent the information in the table as a bar graph.

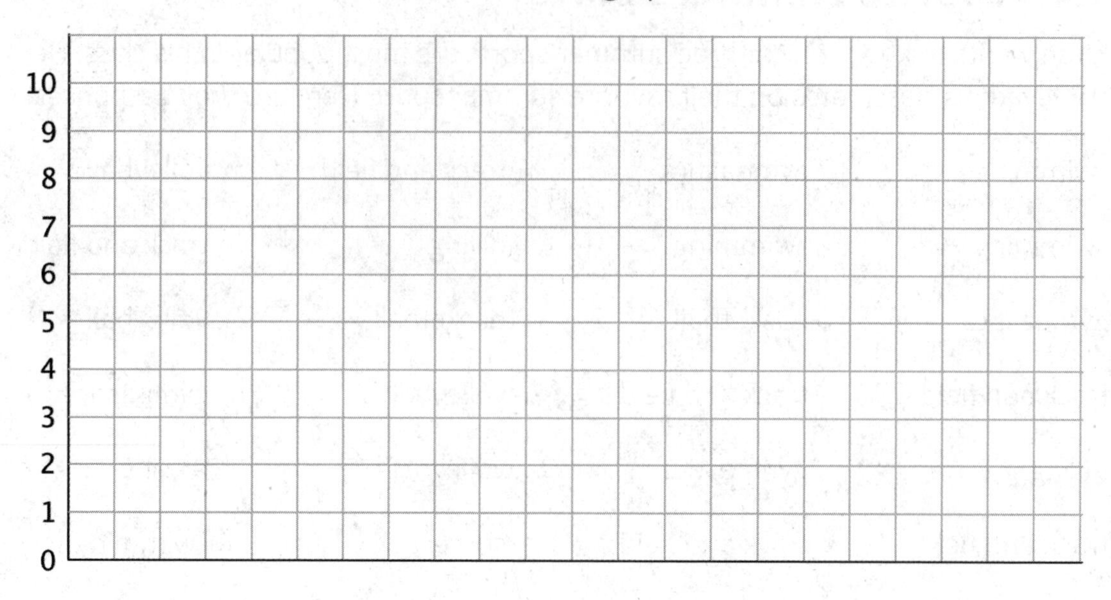


4. a. How can you use the bar graph to find how many classmates Kiran surveyed?

b. Which three summer sports are most popular in Kiran's class?

c. Use your bar graph to describe at least one observation about Kiran's classmates' preferred summer sports.

5. Could a dot plot be used to represent Kiran's data? Explain your reasoning.

## Lesson 1 Summary

When we analyze data, we are often interested in the **distribution**, which is information that shows all the data values and how often they occur.

In a previous lesson, we saw data about 10 dogs. We can see the distribution of the dog weights in a table such as this one.

| weight in kilograms | frequency |
|---|---|
| 6 | 1 |
| 7 | 3 |
| 10 | 2 |
| 32 | 1 |
| 35 | 2 |
| 36 | 1 |

The term **frequency** refers to the number of times a data value occurs. In this case, we see that there are three dogs that weigh 7 kilograms, so "3" is the frequency for the value "7 kilograms."

Recall that dot plots are often used to to represent numerical data. Like a frequency table, a dot plot also shows the distribution of a data set. This dot plot, which you saw in an earlier lesson, shows the distribution of dog weights.

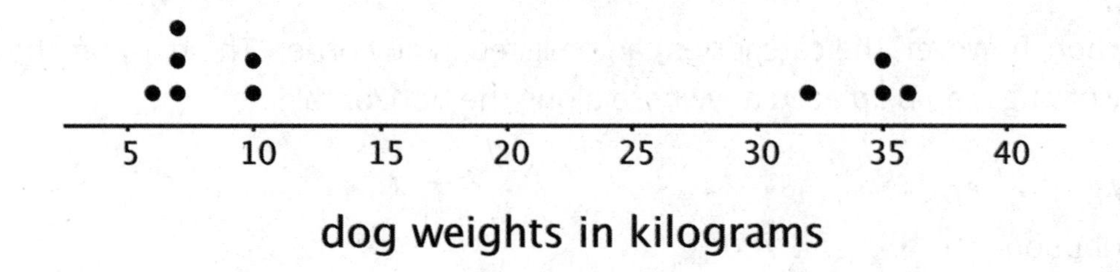


A dot plot uses a horizontal number line. We show the frequency of a value by the number of dots drawn above that value. Here, the two dots above the number 35 tell us that there are two dogs weighing 35 kilograms.
The distribution of categorical data can also be shown in a table. This table shows the distribution of dog breeds.

| breed | frequency |
|---|---|
| pug | 9 |
| beagle | 9 |
| German shepherd | 12 |

We often represent the distribution of categorical data using a bar graph.

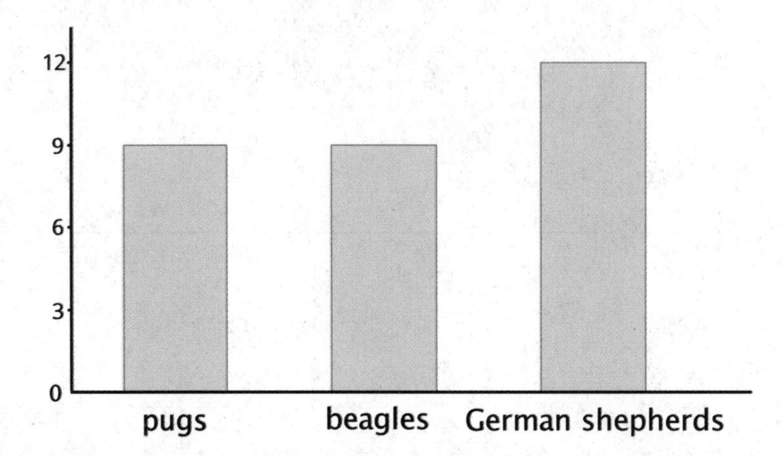


A bar graph also uses a horizontal line. Above it we draw a rectangle (or "bar") to represent each category in the data set. The height of a bar tells us the frequency of the category. There are 12 German shepherds in the data set, so the bar for this category is 12 units tall. Below the line we write the labels for the categories.

In a dot plot, a data value is placed according to its position on the number line. A weight of 10 kilograms must be shown as a dot above 10 on the number line.

In a bar graph, however, the categories can be listed in any order. The bar that shows the frequency of pugs can be placed anywhere along the horizontal line.

## Glossary

- distribution
- frequency

## Lesson 1 Practice Problems

1. A teacher drew a line segment that was 20 inches long on the blackboard. She asked each of her students to estimate the length of the segment and used their estimates to draw this dot plot.

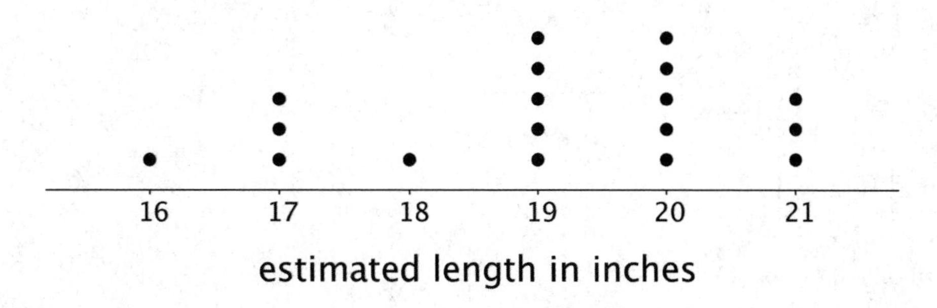


   a. How many students were in the class?

   b. Were students generally accurate in their estimates of the length of the line? Explain your reasoning.

2. Here are descriptions of data sets. Select **all** descriptions of data sets that could be graphed as dot plots.

   A. Class size for the classes at an elementary school

   B. Colors of cars in a parking lot

   C. Favorite sport of each student in a sixth-grade class

   D. Birth weights for the babies born during October at a hospital

   E. Number of goals scored in each of 20 games played by a school soccer team

3. Priya recorded the number of attempts it took each of 12 of her classmates to successfully throw a ball into a basket. Make a dot plot of Priya's data.

   1 2 1 3 1 4 4 3 1 2 5 2

4. Solve each equation.

a. $9v = 1$

b. $1.37w = 0$

c. $1 = \frac{7}{10}x$

d. $12.1 = 12.1 + y$

e. $\frac{3}{5} + z = 1$

(From Unit 4, Lesson 4.)

5. Find the quotients.

a. $\frac{2}{5} \div 2$

b. $\frac{2}{5} \div 5$

c. $2 \div \frac{2}{5}$

d. $5 \div \frac{2}{5}$

(From Unit 3, Lesson 7.)

6. Find the area of each triangle.

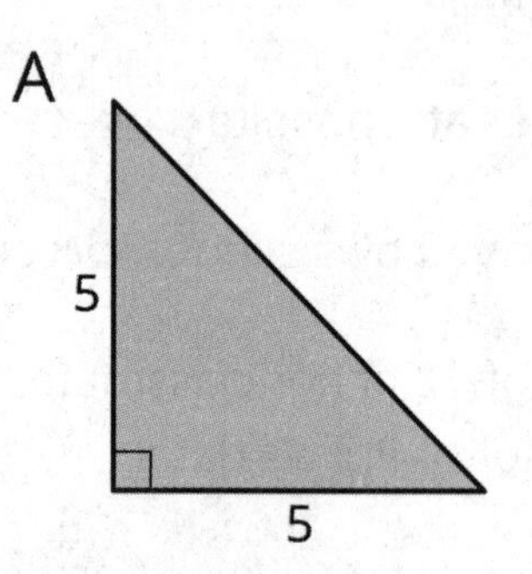


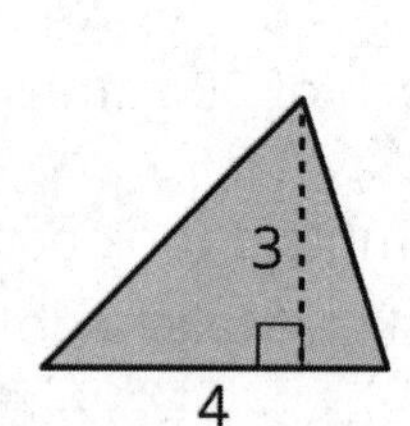


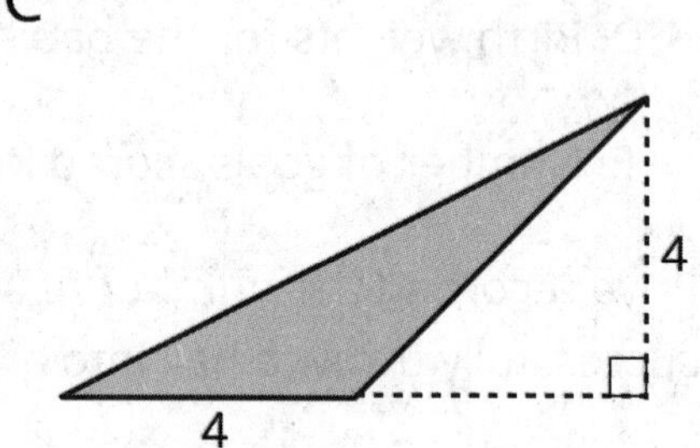


(From Unit 1, Lesson 8.)

# Lesson 2: Using Dot Plots to Answer Statistical Questions

Let's use dot plots to describe distributions and answer questions.

## 2.1: Packs on Backs

This dot plot shows the weights of backpacks, in kilograms, of 50 sixth-grade students at a school in New Zealand.

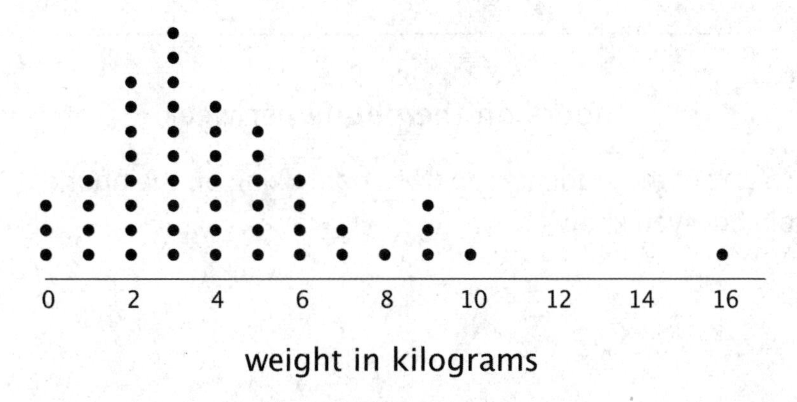


1. The dot plot shows several dots at 0 kilograms. What could a value of 0 mean in this context?

2. Clare and Tyler studied the dot plot.

    - Clare said, "I think we can use 3 kilograms to describe a typical backpack weight of the group because it represents 20%—or the largest portion—of the data."
    - Tyler disagreed and said, "I think 3 kilograms is too low to describe a typical weight. Half of the dots are for backpacks that are heavier than 3 kilograms, so I would use a larger value."

    Do you agree with either of them? Explain your reasoning.

## 2.2: On the Phone

Twenty-five sixth-grade students were asked to estimate how many hours a week they spend talking on the phone. This dot plot represents their reported number of hours of phone usage per week.

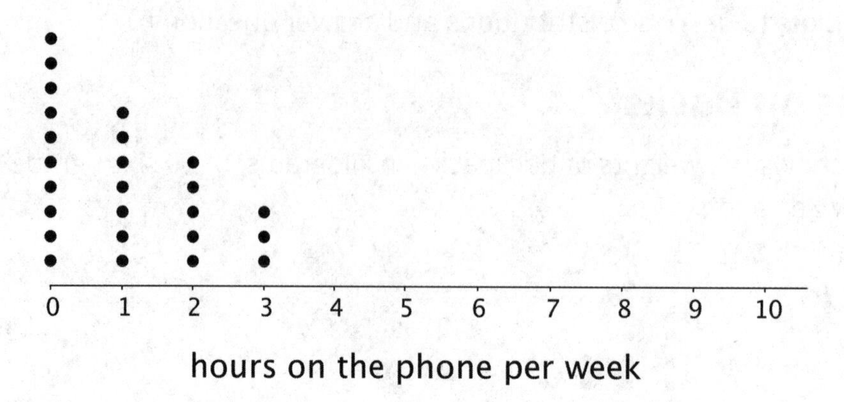


1. a. How many of the students reported not talking on the phone during the week? Explain how you know.

   b. What percentage of the students reported not talking on the phone?

2. a. What is the largest number of hours a student spent talking on the phone per week?

   b. What percentage of the group reported talking on the phone for this amount of time?

3. a. How many hours would you say that these students typically spend talking on the phone?

   b. How many minutes per day would that be?

4. a. How would you describe the **spread** of the data? Would you consider these students' amounts of time on the phone to be alike or different? Explain your reasoning.

   b. Here is the dot plot from an earlier activity. It shows the number of hours per week the same group of 25 sixth-grade students reported spending on homework.

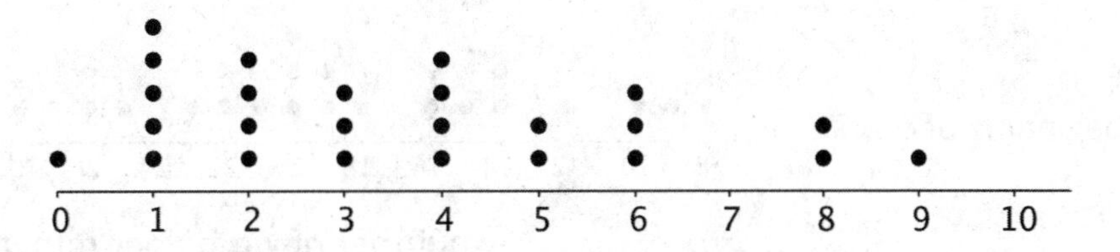


   Overall, are these students more alike in the amount of time they spend talking on the phone or in the amount of time they spend on homework? Explain your reasoning.

5. Suppose someone claimed that these sixth-grade students spend too much time on the phone. Do you agree? Use your analysis of the dot plot to support your answer.

## 2.3: Click-Clack

1. A keyboarding teacher wondered: "Do typing speeds of students improve after taking a keyboarding course?" Explain why her question is a statistical question.

2. The teacher recorded the number of words that her students could type per minute at the beginning of a course and again at the end. The two dot plots show the two data sets.

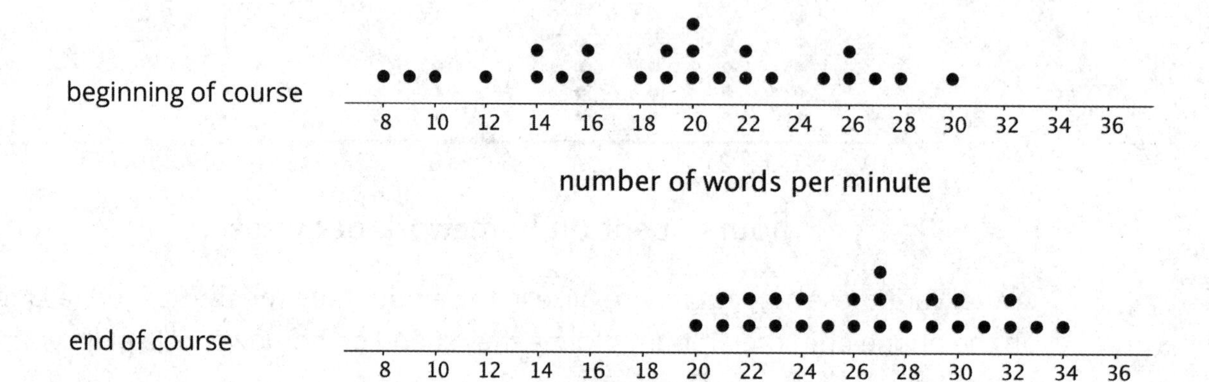


Based on the dot plots, do you agree with each of the following statements about this group of students? Be prepared to explain your reasoning.

a. Overall, the students' typing speed did not improve. They typed at the same speed at the end of the course as they did at the beginning.

b. 20 words per minute is a good estimate for how fast, in general, the students typed at the beginning of the course.

c. 20 words per minute is a good description of the **center** of the data set at the end of the course.

d. There was more variability in the typing speeds at the beginning of the course than at the end, so the students' typing speeds were more alike at the end.

3. Overall, how fast would you say that the students typed after completing the course? What would you consider the center of the end-of-course data?

## Are you ready for more?

Use one of these suggestions (or make up your own). Research to create a dot plot with at least 10 values. Then, describe the center and spread of the distribution.

- Points scored by your favorite sports team in its last 10 games
- Length of your 10 favorite movies (in minutes)
- Ages of your favorite 10 celebrities

## Lesson 2 Summary

One way to describe what is typical or characteristic for a data set is by looking at the **center** and **spread** of its distribution.

Let's compare the distribution of cat weights and dog weights shown on these dot plots.

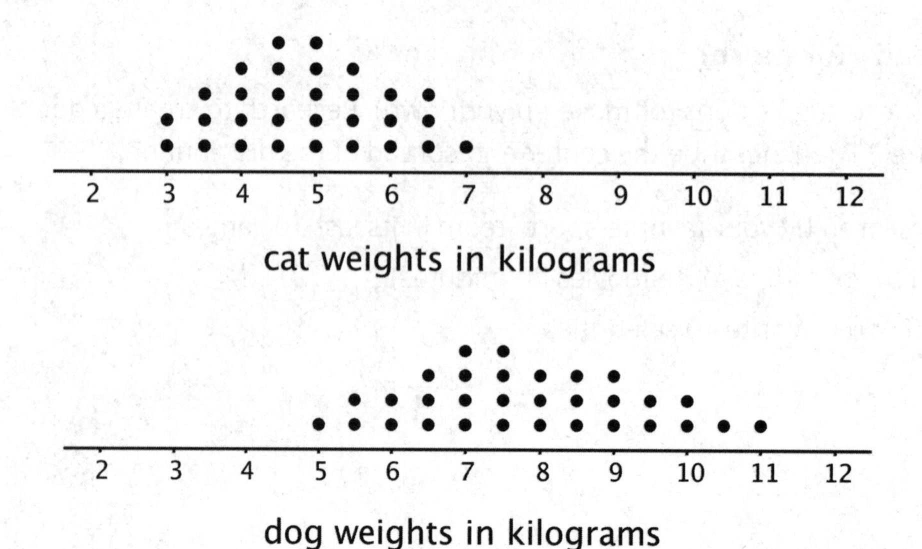


The collection of points for the cat data is further to the left on the number line than the dog data. Based on the dot plots, we may describe the center of the distribution for cat weights to be between 4 and 5 kilograms and the center for dog weights to be between 7 and 8 kilograms.

We often say that values at or near the center of a distribution are typical for that group. This means that a weight of 4–5 kilograms is typical for a cat in the data set, and weight of 7–8 kilograms is typical for a dog.

We also see that the dog weights are more spread out than the cat weights. The difference between the heaviest and lightest cats is only 4 kilograms, but the difference between the heaviest and lightest dogs is 6 kilograms.

A distribution with greater spread tells us that the data have greater variability. In this case, we could say that the cats are more similar in their weights than the dogs.

In future lessons, we will discuss how to measure the center and spread of a distribution.

### Glossary

- center
- spread

## Lesson 2 Practice Problems

1. Three sets of data about ten sixth-grade students were used to make three dot plots. The person who made these dot plots forgot to label them. Match each dot plot with the appropriate label.

A

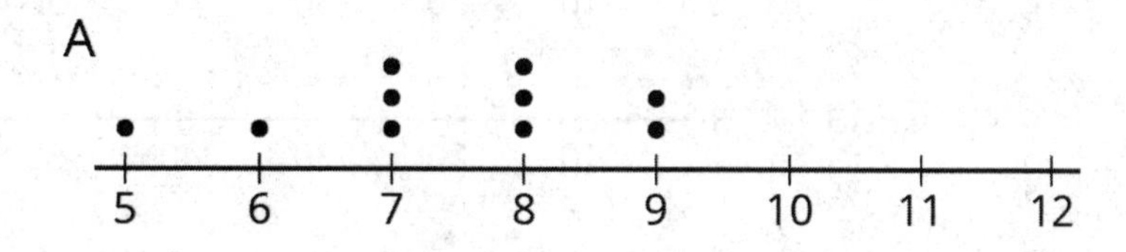


B

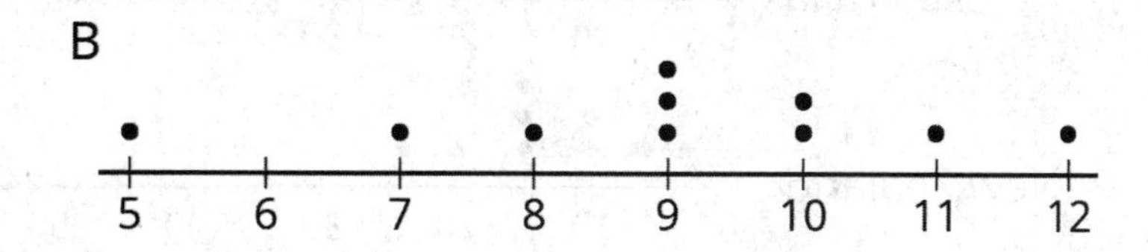


C

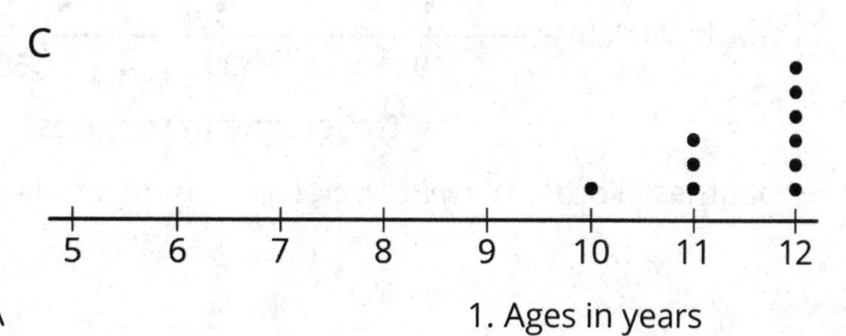


A. Dot Plot A

B. Dot Plot B

C. Dot Plot C

1. Ages in years

2. Numbers of hours of sleep on nights before school days

3. Numbers of hours of sleep on nights before non-school days

2. The dot plots show the time it takes to get to school for ten sixth-grade students from the United States, Canada, Australia, New Zealand, and South Africa.

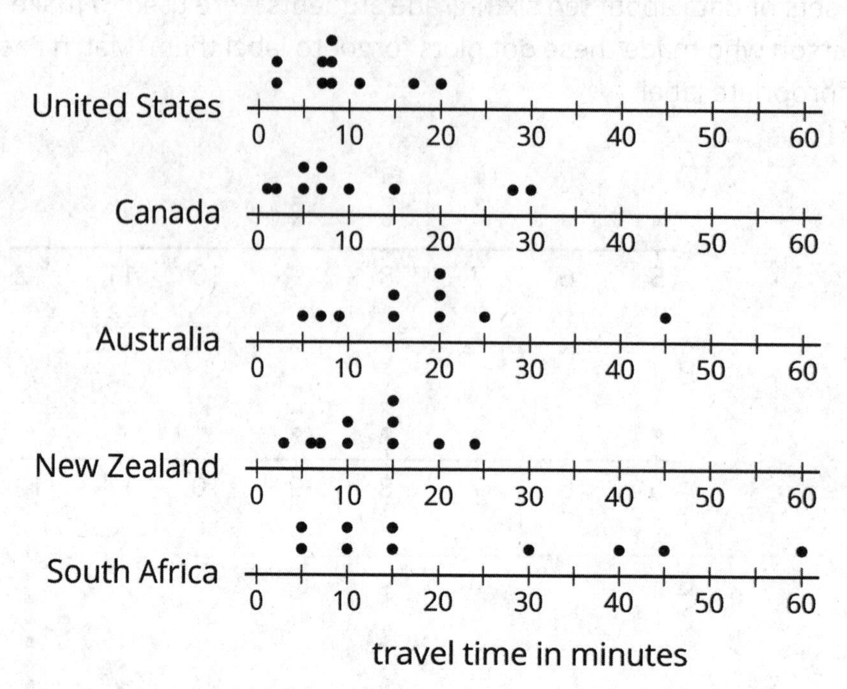


a. List the countries in order of *typical travel times*, from shortest to longest.

b. List the countries in order of *variability in travel times*, from the least variability to the greatest.

3. Twenty-five students were asked to rate—on a scale of 0 to 10—how important it is to reduce pollution. A rating of 0 means "not at all important" and a rating of 10 means "very important." Here is a dot plot of their responses.

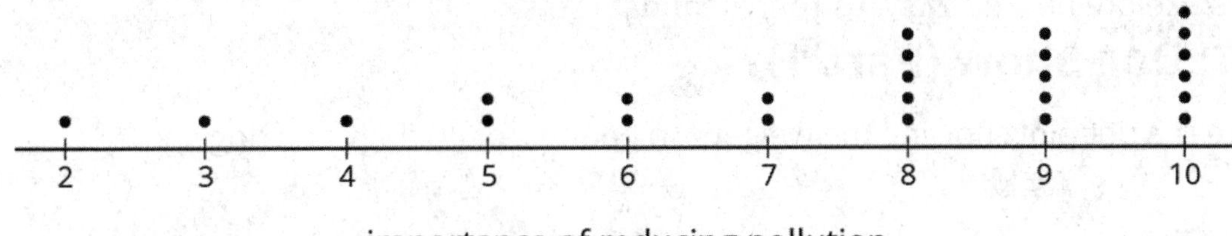


Explain why a rating of 6 is *not* a good description of the center of this data set.

# Lesson 3: Interpreting Histograms

Let's explore how histograms represent data sets.

## 3.1: Dog Show (Part 1)

Here is a dot plot showing the weights, in pounds, of 40 dogs at a dog show.

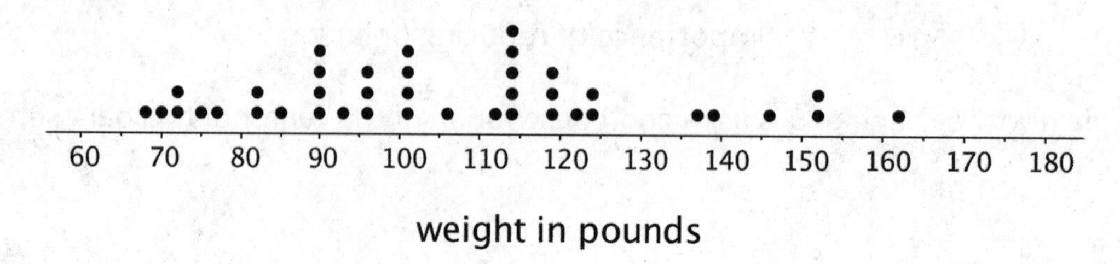


1. Write two statistical questions that can be answered using the dot plot.

2. What would you consider a typical weight for a dog at this dog show? Explain your reasoning.

## 3.2: Dog Show (Part 2)

Here is a **histogram** that shows some dog weights in pounds.

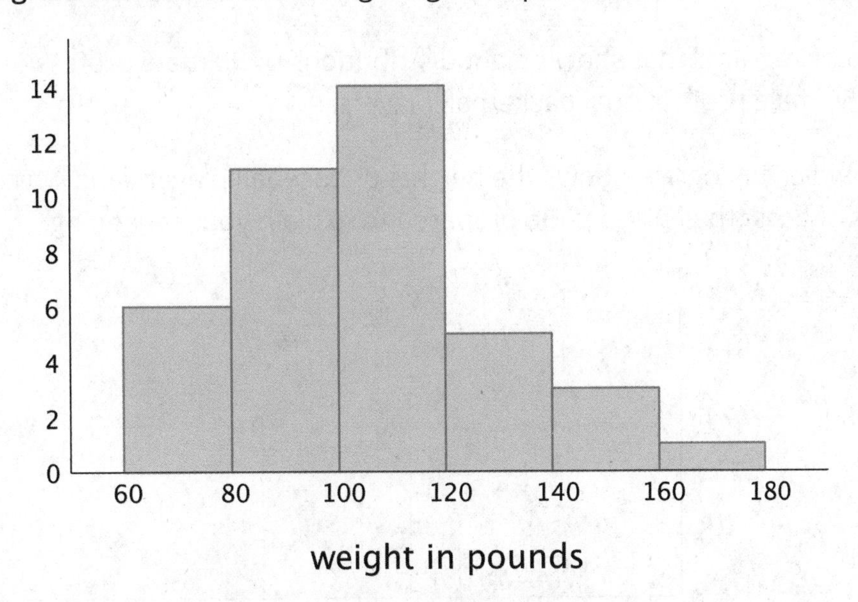


Each bar includes the left-end value but not the right-end value. For example, the first bar includes dogs that weigh 60 pounds and 68 pounds but not 80 pounds.

1. Use the histogram to answer the following questions.

    a. How many dogs weigh at least 100 pounds?

    b. How many dogs weigh exactly 70 pounds?

    c. How many dogs weigh at least 120 and less than 160 pounds?

    d. How much does the heaviest dog at the show weigh?

    e. What would you consider a typical weight for a dog at this dog show? Explain your reasoning.

2. Discuss with a partner:

    - If you used the dot plot to answer the same five questions you just answered, how would your answers be different?

    - How are the histogram and the dot plot alike? How are they different?

## 3.3: Tall and Taller Players

Professional basketball players tend to be taller than professional baseball players.

Here are two histograms that show height distributions of 50 male professional baseball players and 50 male professional basketball players.

1. Decide which histogram shows the heights of baseball players and which shows the heights of basketball players. Be prepared to explain your reasoning.

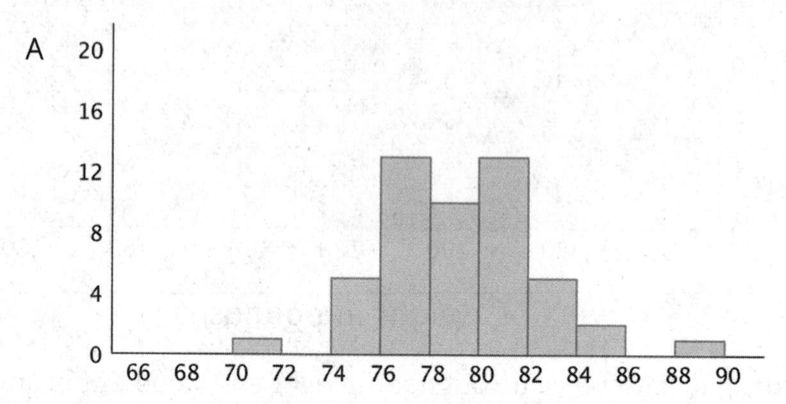


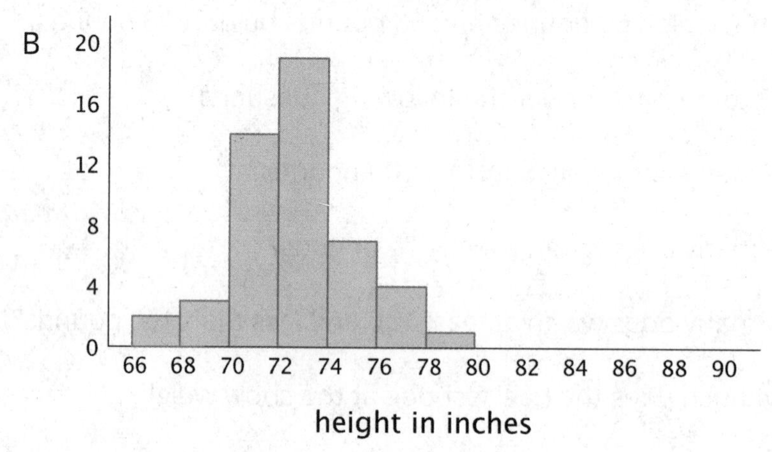


2. Write 2–3 sentences that describe the distribution of the heights of the basketball players. Comment on the center and spread of the data.

3. Write 2–3 sentences that describe the distribution of the heights of the baseball players. Comment on the center and spread of the data.

## Lesson 3 Summary

In addition to using dot plots, we can also represent distributions of numerical data using **histograms.**

Here is a dot plot that shows the weights, in kilograms, of 30 dogs, followed by a histogram that shows the same distribution.

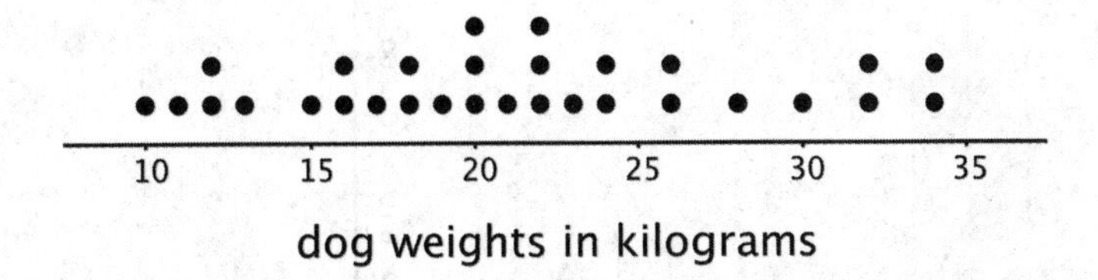


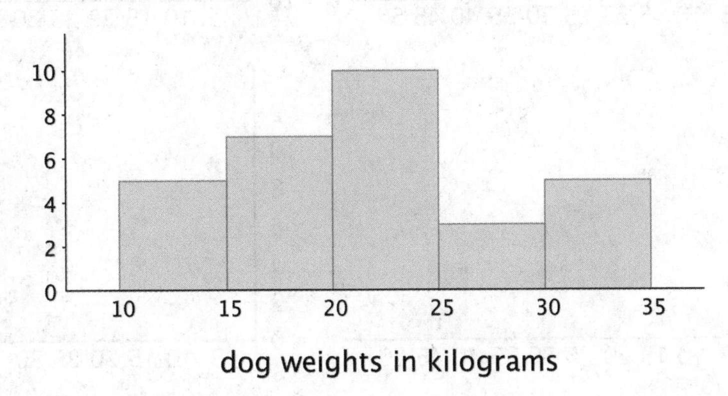


In a histogram, data values are placed in groups or "bins" of a certain size, and each group is represented with a bar. The height of the bar tells us the frequency for that group.

For example, the height of the tallest bar is 10, and the bar represents weights from 20 to less than 25 kilograms, so there are 10 dogs whose weights fall in that group. Similarly, there are 3 dogs that weigh anywhere from 25 to less than 30 kilograms.

Notice that the histogram and the dot plot have a similar shape. The dot plot has the advantage of showing all of the data values, but the histogram is easier to draw and to interpret when there are a lot of values or when the values are all different.

## Glossary

- histogram

# Lesson 3 Practice Problems

1. Match histograms A through E to dot plots 1 through 5 so that each match represents the same data set.

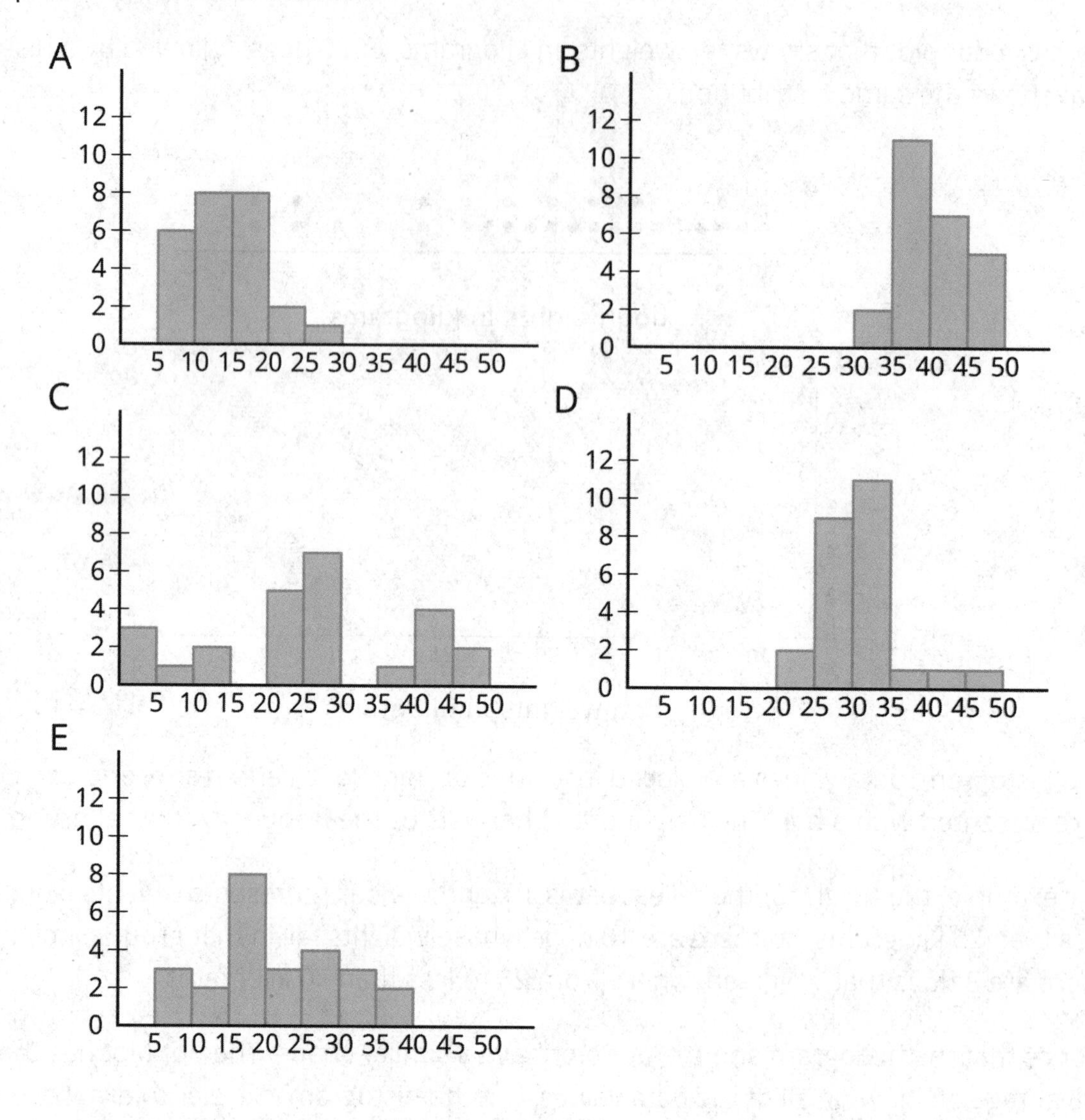


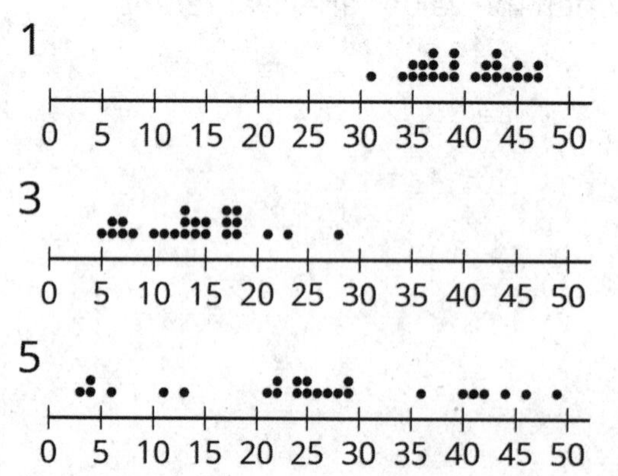


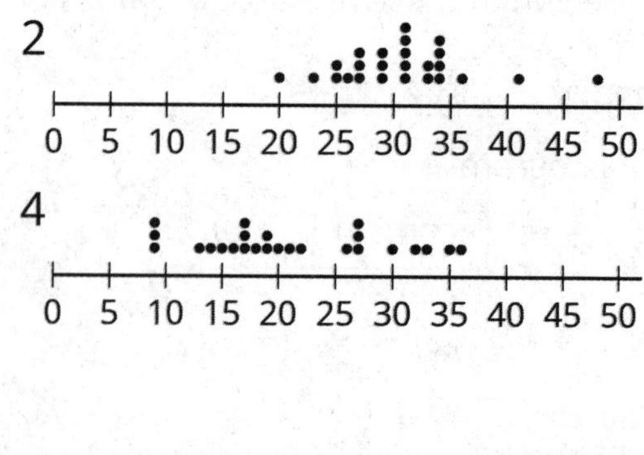

2. Here is a histogram that summarizes the lengths, in feet, of a group of adult female sharks. Select **all** the statements that are true, according to the histogram.

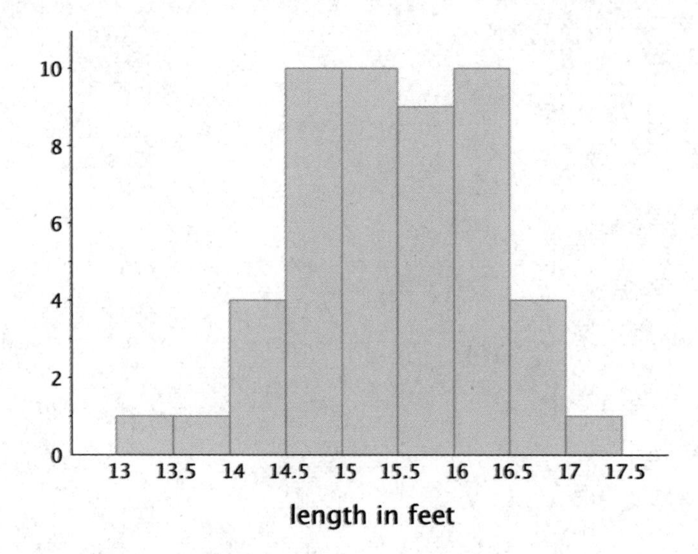


A. A total of 9 sharks were measured.

B. A total of 50 sharks were measured.

C. The longest shark that was measured was 10 feet long.

D. Most of the sharks that were measured were over 16 feet long.

E. Two of the sharks that were measured were less than 14 feet long.

3. This table shows the times, in minutes, it took 40 sixth-grade students to run 1 mile.

| time (minutes) | frequency |
|---|---|
| 4 to less than 6 | 1 |
| 6 to less than 8 | 5 |
| 8 to less than 10 | 13 |
| 10 to less than 12 | 12 |
| 12 to less than 14 | 7 |
| 14 to less than 16 | 2 |

Draw a histogram for the information in the table.

4. Order these numbers from greatest to least: -4, $\frac{1}{4}$, 0, 4, $-3\frac{1}{2}$, $\frac{7}{4}$, $-\frac{5}{4}$

(From Unit 7, Lesson 3.)

# Lesson 4: The Mean

Let's explore the mean of a data set and what it tells us.

## 4.1: Which One Doesn't Belong: Division

Which expression does not belong? Be prepared to explain your reasoning.

$\frac{8+8+4+4}{4}$ $\frac{10+10+4}{4}$ $\frac{9+9+5+5}{4}$ $\frac{6+6+6+6+6}{5}$

## 4.2: Spread Out and Share

1. The kittens in a room at an animal shelter are placed in 5 crates.

   a. The manager of the shelter wants the kittens distributed equally among the crates. How might that be done? How many kittens will end up in each crate?

   b. The number of kittens in each crate after they are equally distributed is called the **mean** number of kittens per crate, or the **average** number of kittens per crate. Explain how the expression $10 \div 5$ is related to the average.

c. Another room in the shelter has 6 crates. No two crates has the same number of kittens, and there is an average of 3 kittens per crate. Draw or describe at least two different arrangements of kittens that match this description.

2. Five servers were scheduled to work the number of hours shown. They decided to share the workload, so each one would work equal hours.

server A: 3 server B: 6 server C: 11 server D: 7 server E: 4

a. On the grid on the left, draw 5 bars whose heights represent the hours worked by servers A, B, C, D, and E.

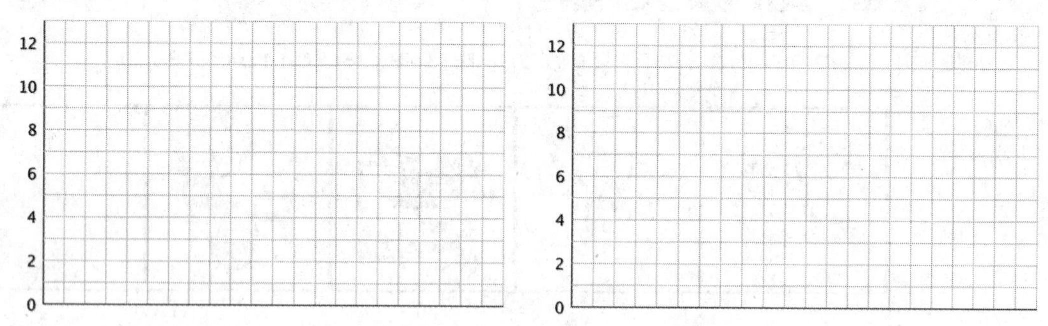


b. Think about how you would rearrange the hours so that each server gets a fair share. Then, on the grid on the right, draw a new graph to represent the rearranged hours. Be prepared to explain your reasoning.

c. Based on your second drawing, what is the average or mean number of hours that the servers will work?

d. Explain why we can also find the mean by finding the value of the expression $31 \div 5$.

e. Which server will see the biggest change to work hours? Which server will see the least change?

## Are you ready for more?

Server F, working 7 hours, offers to join the group of five servers, sharing their workload. If server F joins, will the mean number of hours worked increase or decrease? Explain how you know.

# 4.3: Travel Times (Part 2)

1. Here are dot plots showing how long Diego's trips to school took in minutes—which you studied earlier—and how long Andre's trips to school took in minutes. The dot plots include the means for each data set, marked by triangles.

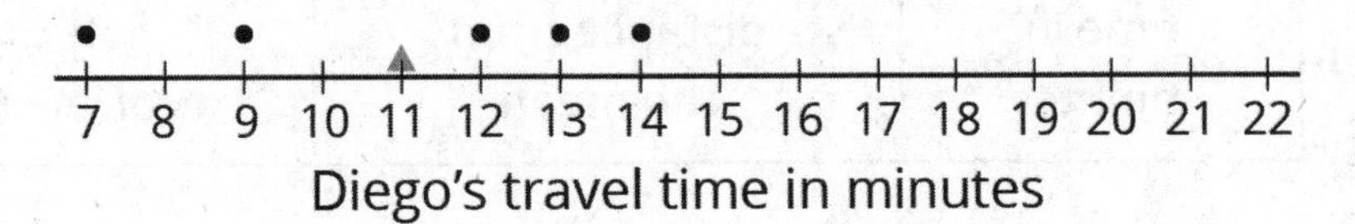


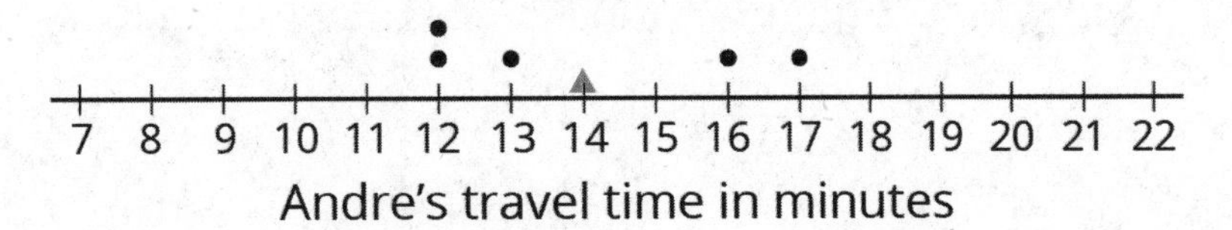


   a. Which of the two data sets has a larger mean? In this context, what does a larger mean tell us?

   b. Which of the two data sets has larger sums of distances to the left and right of the mean? What do these sums tell us about the variation in Diego's and Andre's travel times?

2. Here is a dot plot showing lengths of Lin's trips to school.

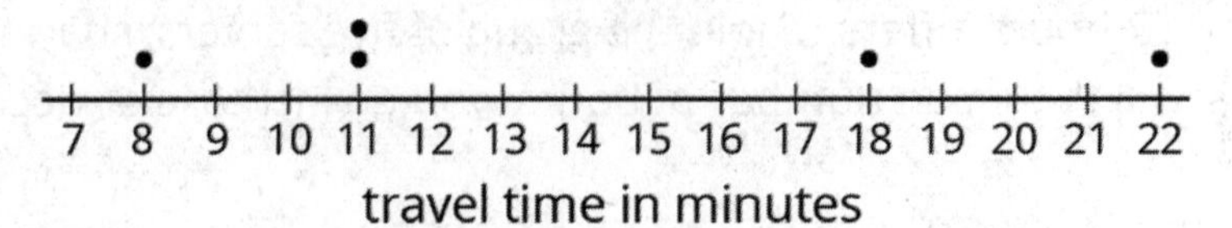


a. Calculate the mean of Lin's travel times.

b. Complete the table with the distance between each point and the mean as well whether the point is to the left or right of the mean.

| time in minutes | distance from the mean | left or right of the mean? |
|---|---|---|
| 22 | | |
| 18 | | |
| 11 | | |
| 8 | | |
| 11 | | |

c. Find the sum of distances to the left of the mean and the sum of distances to the right of the mean.

d. Use your work to compare Lin's travel times to Andre's. What can you say about their average travel times? What about the variability in their travel times?

## Lesson 4 Summary

Sometimes a general description of a distribution does not give enough information, and a more precise way to talk about center or spread would be more useful. The **mean**, or **average**, is a number we can use to summarize a distribution.

We can think about the mean in terms of "fair share" or "leveling out." That is, a mean can be thought of as a number that each member of a group would have if all the data values were combined and distributed equally among the members.

For example, suppose there are 5 bottles which have the following amounts of water: 1 liter, 4 liters, 2 liters, 3 liters, and 0 liters.

To find the mean, first we add up all of the values. We can think of as putting all of the water together: $1 + 4 + 2 + 3 + 0 = 10$.

To find the "fair share," we divide the 10 liters equally into the 5 containers: $10 \div 5 = 2$.

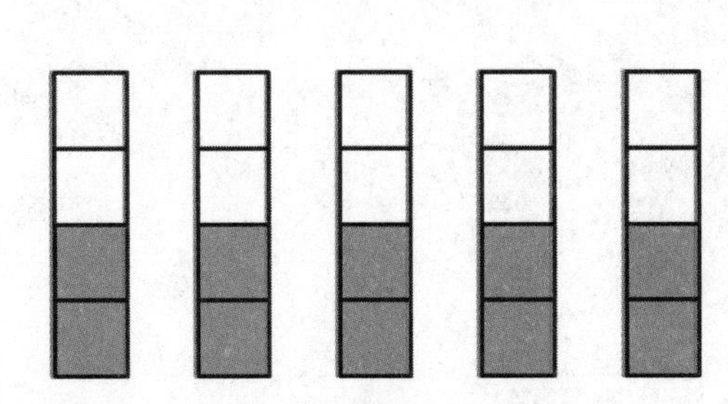

Suppose the quiz scores of a student are 70, 90, 86, and 94. We can find the mean (or average) score by finding the sum of the scores ($70 + 90 + 86 + 94 = 340$) and dividing the sum by four ($340 \div 4 = 85$). We can then say that the student scored, on average, 85 points on the quizzes.

In general, to find the mean of a data set with $n$ values, we add all of the values and divide the sum by $n$.

The mean is often used as a **measure of center** of a distribution. This is because the mean of a distribution can be seen as the "balance point" for the distribution.

The sum of the distances for the data points to the left of the mean is equal to the sum of the distances for the data points to the right of the mean. So, the mean is often near the middle of the distribution, especially when the data is symmetric.

## Glossary

- average
- mean
- measure of center

## Lesson 4 Practice Problems

1. A preschool teacher is rearranging four boxes of playing blocks so that each box contains an equal number of blocks. Currently Box 1 has 32 blocks, Box 2 has 18, Box 3 has 41, and Box 4 has 9.

   Select **all** the ways he could make each box have the same number of blocks.

   A. Remove all the blocks and make four equal piles of 25, then put each pile in one of the boxes.

   B. Remove 7 blocks from Box 1 and place them in Box 2.

   C. Remove 21 blocks from Box 3 and place them in Box 4.

   D. Remove 7 blocks from Box 1 and place them in Box 2, and remove 21 blocks from Box 3 and place them in Box 4.

   E. Remove 7 blocks from Box 1 and place them in Box 2, and remove 16 blocks from Box 3 and place them in Box 4.

2. Three sixth-grade classes raised $25.50, $49.75, and $37.25 for their classroom libraries. They agreed to share the money raised equally. What is each class's equal share? Explain or show your reasoning.

3. An earthworm farmer examined two containers of a certain species of earthworms so that he could learn about their lengths. He measured 25 earthworms in each container and recorded their lengths in millimeters.

   Here are histograms of the lengths for each container.

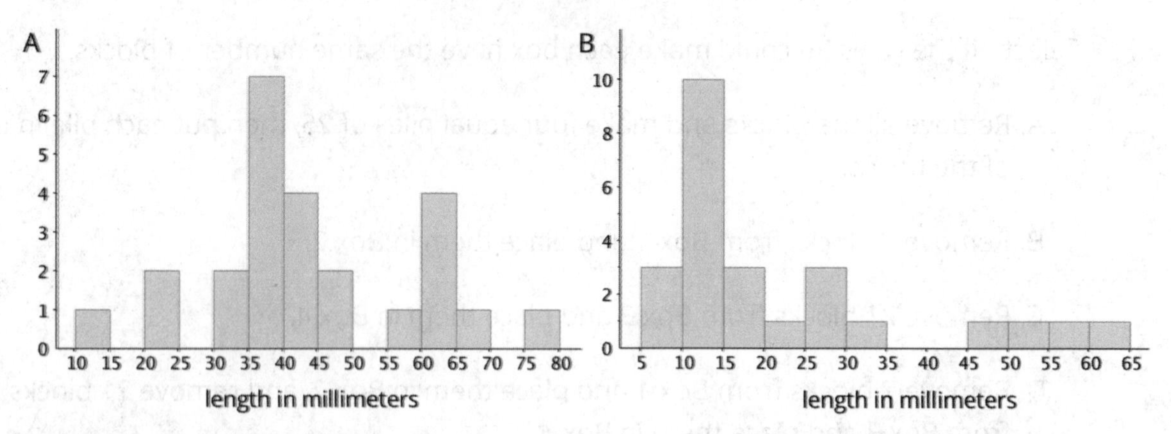


   a. Which container tends to have longer worms than the other container?

   b. For which container would 15 millimeters be a reasonable description of a typical length of the worms in the container?

   c. If length is related to age, which container had the most young worms?

   (From Unit 8, Lesson 3.)

4. Noah scored 20 points in a game. Mai's score was 30 points. The mean score for Noah, Mai, and Clare was 40 points. What was Clare's score? Explain or show your reasoning.

5. a. Plot $\frac{2}{3}$ and $\frac{3}{4}$ on a number line.

   b. Is $\frac{2}{3} < \frac{3}{4}$, or is $\frac{3}{4} < \frac{2}{3}$? Explain how you know.

   (From Unit 7, Lesson 2.)

6. Select **all** the expressions that represent the total area of the large rectangle.

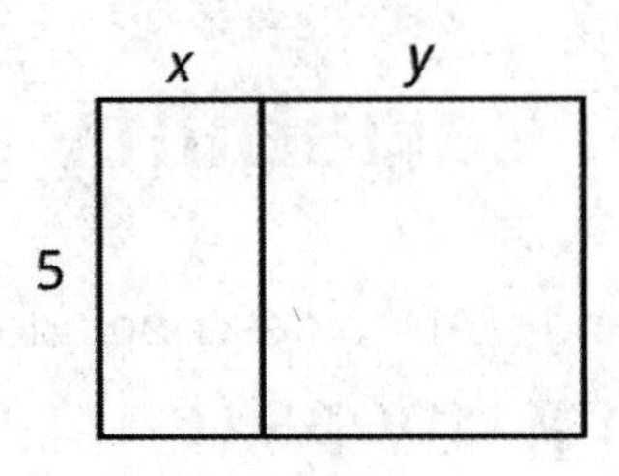


A. $5(x + y)$

B. $5 + xy$

C. $5x + 5y$

D. $2(5 + x + y)$

E. $5xy$

(From Unit 4, Lesson 10.)

# Lesson 5: Variability and MAD

Let's use mean and MAD to describe and compare distributions.

## 5.1: Shooting Hoops (Part 1)

Elena, Jada, and Lin enjoy playing basketball during recess. Lately, they have been practicing free throws. They record the number of baskets they make out of 10 attempts. Here are their data sets for 12 school days.

Elena

| 4 | 5 | 1 | 6 | 9 | 7 | 2 | 8 | 3 | 3 | 5 | 7 |
|---|---|---|---|---|---|---|---|---|---|---|---|

Jada

| 2 | 4 | 5 | 4 | 6 | 6 | 4 | 7 | 3 | 4 | 8 | 7 |
|---|---|---|---|---|---|---|---|---|---|---|---|

Lin

| 3 | 6 | 6 | 4 | 5 | 5 | 3 | 5 | 4 | 6 | 6 | 7 |
|---|---|---|---|---|---|---|---|---|---|---|---|

1. Calculate the mean number of baskets each player made, and compare the means. What do you notice?

2. What do the means tell us in this context?

## 5.2: Shooting Hoops (Part 3)

The tables show Elena, Jada, and Lin's basketball data from an earlier activity. Recall that the mean of Elena's data, as well as that of Jada and Lin's data, was 5.

1. Record the distance between each of Elena's scores and the mean.

| Elena | 4 | 5 | 1 | 6 | 9 | 7 | 2 | 8 | 3 | 3 | 5 | 7 |
|---|---|---|---|---|---|---|---|---|---|---|---|---|
| distance from 5 | 1 | | | 1 | | | | | | | | |

Now find *the average of the distances* in the table. Show your reasoning and round your answer to the nearest tenth.

This value is the **mean absolute deviation (MAD)** of Elena's data.

Elena's MAD: ________

2. Find the mean absolute deviation of Jada's data. Round it to the nearest tenth.

| Jada | 2 | 4 | 5 | 4 | 6 | 6 | 4 | 7 | 3 | 4 | 8 | 7 |
|---|---|---|---|---|---|---|---|---|---|---|---|---|
| distance from 5 | | | | | | | | | | | | |

Jada's MAD: ________

3. Find the mean absolute deviation of Lin's data. Round it to the nearest tenth.

| Lin | 3 | 6 | 6 | 4 | 5 | 5 | 3 | 5 | 4 | 6 | 6 | 7 |
|---|---|---|---|---|---|---|---|---|---|---|---|---|
| distance from 5 | | | | | | | | | | | | |

Lin's MAD: ________

4. Compare the MADs and dot plots of the three students' data. Do you see a relationship between each student's MAD and the distribution on her dot plot? Explain your reasoning.

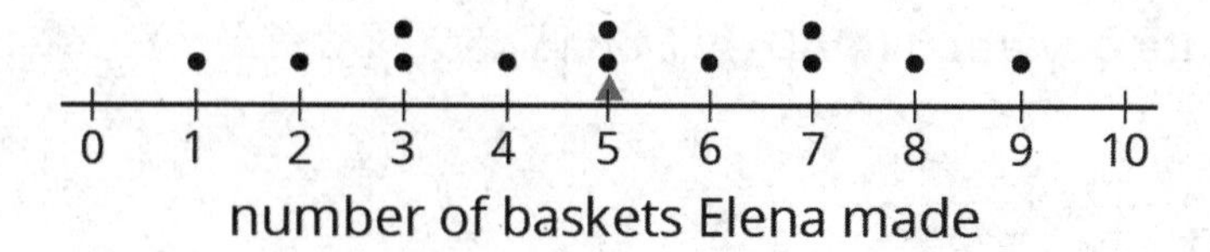


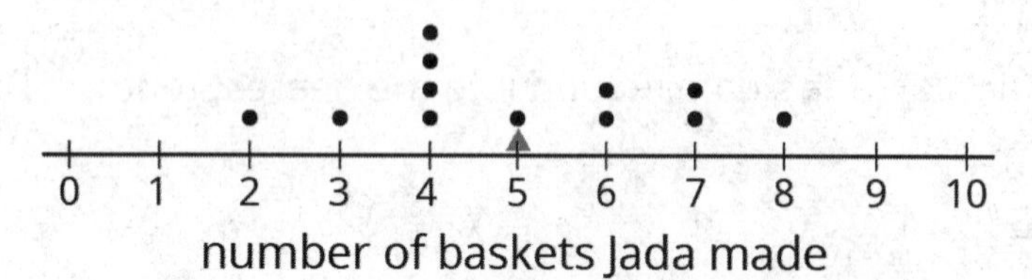


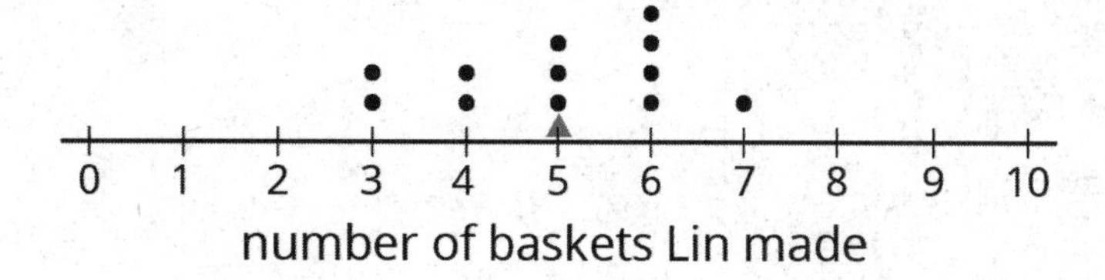


### Are you ready for more?

Invent another data set that also has a mean of 5 but has a MAD greater than 2. Remember, the values in the data set must be whole numbers from 0 to 10.

## 5.3: Which Player Would You Choose?

1. Andre and Noah joined Elena, Jada, and Lin in recording their basketball scores. They all recorded their scores in the same way: the number of baskets made out of 10 attempts. Each collected 12 data points.

    - Andre's mean number of baskets was 5.25, and his MAD was 2.6.
    - Noah's mean number of baskets was also 5.25, but his MAD was 1.

Here are two dot plots that represent the two data sets. The triangle indicates the location of the mean.

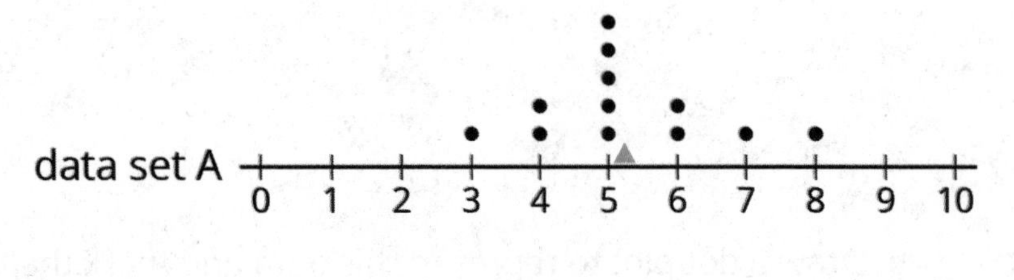


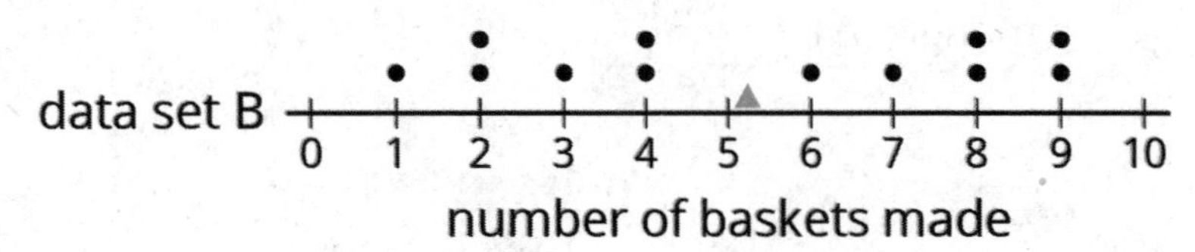


a. Without calculating, decide which dot plot represents Andre's data and which represents Noah's. Explain how you know.

b. If you were the captain of a basketball team and could use one more player on your team, would you choose Andre or Noah? Explain your reasoning.

2. An eighth-grade student decided to join Andre and Noah and kept track of his scores. His data set is shown here. The mean number of baskets he made is 6.

| eighth-grade student | 6 | 5 | 4 | 7 | 6 | 5 | 7 | 8 | 5 | 6 | 5 | 8 |
|---|---|---|---|---|---|---|---|---|---|---|---|---|
| distance from 6 | | | | | | | | | | | | |

a. Calculate the MAD. Show your reasoning.

b. Draw a dot plot to represent his data and mark the location of the mean with a triangle (Δ).

c. Compare the eighth-grade student's mean and MAD to Noah's mean and MAD. What do you notice?

d. Compare their dot plots. What do you notice about the distributions?

e. What can you say about the two players' shooting accuracy and consistency?

## Are you ready for more?

Invent a data set with a mean of 7 and a MAD of 1.

## 5.4: Swimmers Over the Years

In 1984, the mean age of swimmers on the U.S. women's swimming team was 18.2 years and the MAD was 2.2 years. In 2016, the mean age of the swimmers was 22.8 years, and the MAD was 3 years.

1. How has the average age of the women on the U.S. swimming team changed from 1984 to 2016? Explain your reasoning.

2. Are the swimmers on the 1984 team closer in age to one another than the swimmers on the 2016 team are to one another? Explain your reasoning.

3. Here are dot plots showing the ages of the women on the U.S. swimming team in 1984 and in 2016. Use them to make two other comments about how the women's swimming team has changed over the years.

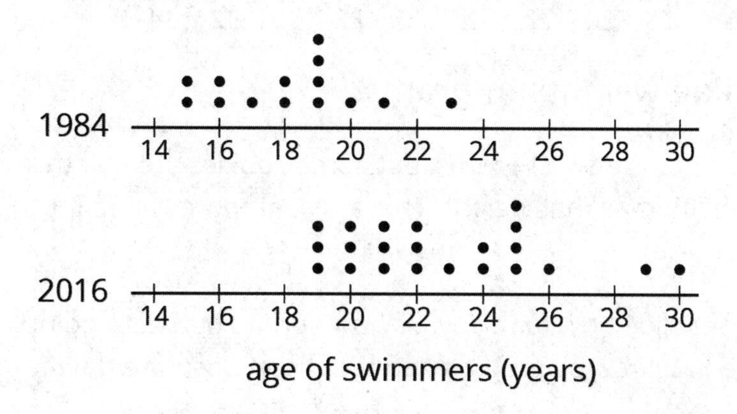

## Lesson 5 Summary

We use the mean of a data set as a measure of center of its distribution, but two data sets with the same mean could have very different distributions.

This dot plot shows the weights, in grams, of 22 cookies.

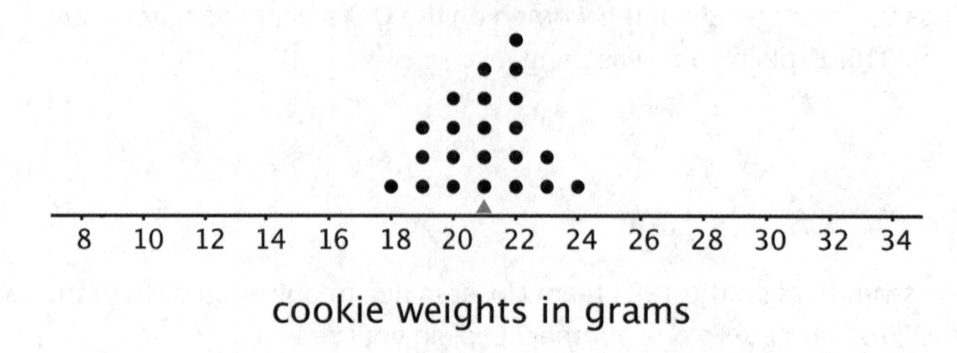


The mean weight is 21 grams. All the weights are within 3 grams of the mean, and most of them are even closer. These cookies are all fairly close in weight.

This dot plot shows the weights, in grams, of a different set of 30 cookies.

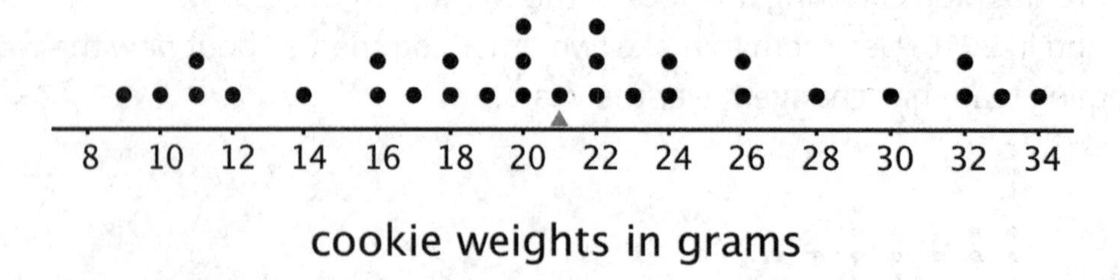


The mean weight for this set of cookies is also 21 grams, but some cookies are half that weight and others are one-and-a-half times that weight. There is a lot more variability in the weight.

There is a number we can use to describe how far away, or how spread out, data points generally are from the mean. This *measure of spread* is called the **mean absolute deviation (MAD)**.

Here the MAD tells us how far cookie weights typically are from 21 grams. To find the MAD, we find the distance between each data value and the mean, and then calculate the mean of those distances.

For instance, the point that represents 18 grams is 3 units away from the mean of 21 grams. We can find the distance between each point and the mean of 21 grams and organize the distances into a table, as shown.

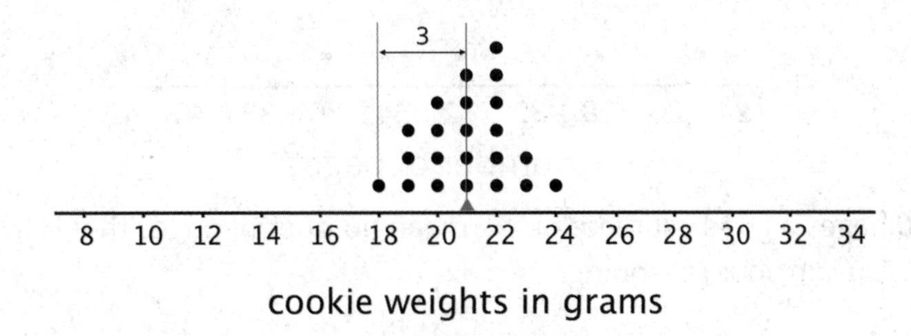


| weight in grams | 18 | 19 | 19 | 19 | 20 | 20 | 20 | 20 | 21 | 21 | 21 | 21 | 21 | 22 | 22 | 22 | 22 | 22 | 22 | 23 | 23 | 24 |
|---|---|---|---|---|---|---|---|---|---|---|---|---|---|---|---|---|---|---|---|---|---|---|
| distance from mean | 3 | 2 | 2 | 2 | 1 | 1 | 1 | 1 | 0 | 0 | 0 | 0 | 0 | 1 | 1 | 1 | 1 | 1 | 1 | 2 | 2 | 3 |

The values in the first row of the table are the cookie weights for the first set of cookies. Their mean, 21 grams, is the *mean of the cookie weights*.

The values in the second row of the table are the *distances* between the values in the first row and 21. The mean of these distances is the *MAD of the cookie weights*.

What can we learn from the averages of these distances once they are calculated?

- In the first set of cookies, the distances are all between 0 and 3. The MAD is 1.2 grams, which tells us that the cookie weights are typically within 1.2 grams of 21 grams. We could say that a typical cookie weighs between 19.8 and 22.2 grams.

- In the second set of cookies, the distances are all between 0 and 13. The MAD is 5.6 grams, which tells us that the cookie weights are typically within 5.6 grams of 21 grams. We could say a typical cookie weighs between 15.4 and 26.6 grams.

The MAD is also called a *measure of the variability* of the distribution. In these examples, it is easy to see that a higher MAD suggests a distribution that is more spread out, showing more variability.

## Glossary

- mean absolute deviation (MAD)

## Lesson 5 Practice Problems

1. Han recorded the number of pages that he read each day for five days. The dot plot shows his data.

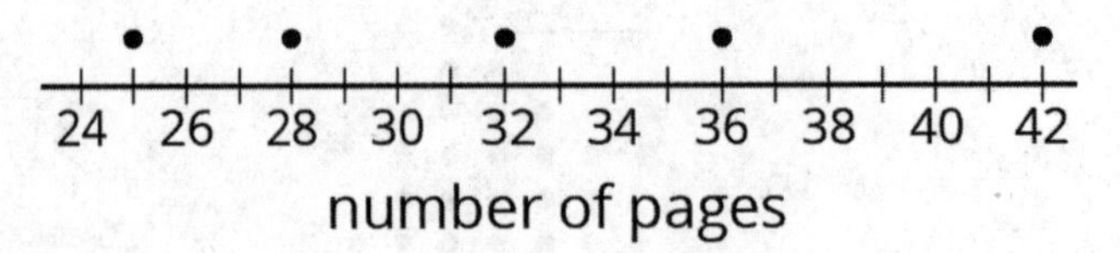


   a. Is 30 pages a good estimate of the mean number of pages that Han read each day? Explain your reasoning.

   b. Find the mean number of pages that Han read during the five days. Draw a triangle to mark the mean on the dot plot.

   c. Use the dot plot and the mean to complete the table.

| number of pages | distance from mean | left or right of mean |
|---|---|---|
| 25 | | left |
| 28 | | |
| 32 | | |
| 36 | | |
| 42 | | |

   d. Calculate the mean absolute deviation (MAD) of the data. Explain or show your reasoning.

2. Ten sixth-grade students recorded the amounts of time each took to travel to school. The dot plot shows their travel times.

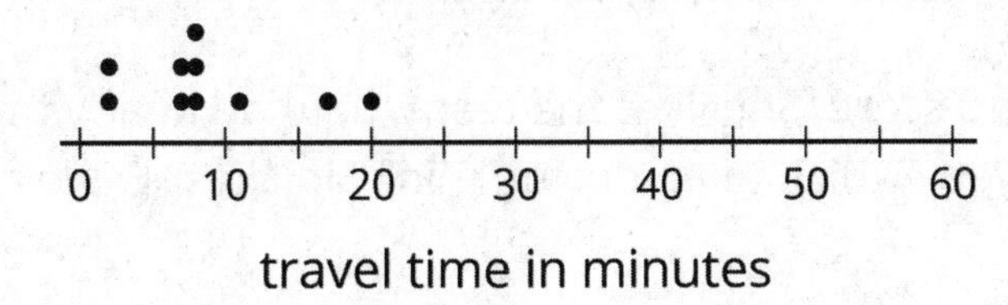


The mean travel time for these students is approximately 9 minutes. The MAD is approximately 4.2 minutes.

a. Which number of minutes—9 or 4.2—is a typical amount of time for the ten sixth-grade students to travel to school? Explain your reasoning.

b. Based on the mean and MAD, Jada believes that travel times between 5 and 13 minutes are common for this group. Do you agree? Explain your reasoning.

c. A different group of ten sixth-grade students also recorded their travel times to school. Their mean travel time was also 9 minutes, but the MAD was about 7 minutes. What could the dot plot of this second data set be? Describe or draw how it might look.

3. In an archery competition, scores for each round are calculated by averaging the distance of 3 arrows from the center of the target.

An archer has a mean distance of 1.6 inches and a MAD distance of 1.3 inches in the first round. In the second round, the archer's arrows are farther from the center but are more consistent. What values for the mean and MAD would fit this description for the second round? Explain your reasoning.

4. Two high school basketball teams have identical records of 15 wins and 2 losses. Sunnyside High School's mean score is 50 points and its MAD is 4 points. Shadyside High School's mean score is 60 points and its MAD is 15 points.

   Lin read the records of each team's score. She likes the team that had nearly the same score for every game it played. Which team do you think Lin likes? Explain your reasoning.

5. Jada thinks the perimeter of this rectangle can be represented with the expression $a + a + b + b$. Andre thinks it can be represented with $2a + 2b$. Do you agree with either of them? Explain your reasoning.

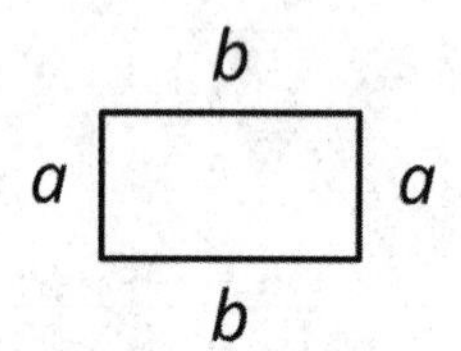


(From Unit 4, Lesson 8.)

# Lesson 6: The Median

Let's explore the median of a data set and what it tells us.

## 6.1: The Plot of the Story

1. Here are two dot plots and two stories. Match each story with a dot plot that could represent it. Be prepared to explain your reasoning.

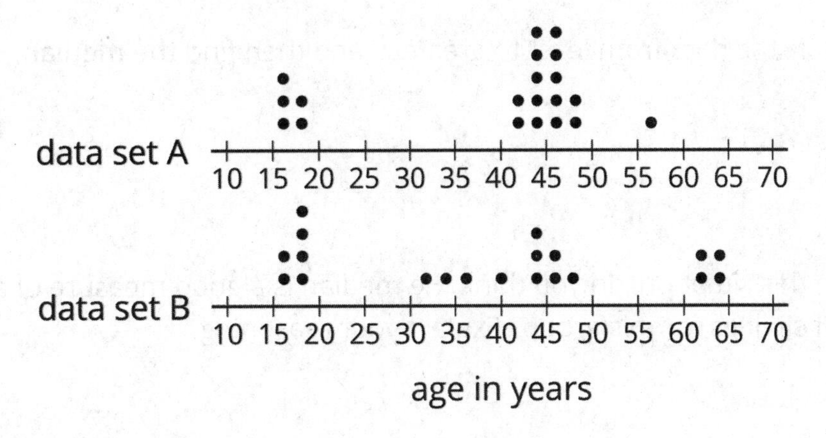


   - Twenty people—high school students, teachers, and invited guests—attended a rehearsal for a high school musical. The mean age was 38.5 years and the MAD was 16.5 years.
   - High school soccer team practice is usually watched by supporters of the players. One evening, twenty people watched the team practice. The mean age was 38.5 years and the MAD was 12.7 years.

2. Another evening, twenty people watched the soccer team practice. The mean age was similar to that from the first evening, but the MAD was greater (about 20 years). Make a dot plot that could illustrate the distribution of ages in this story.

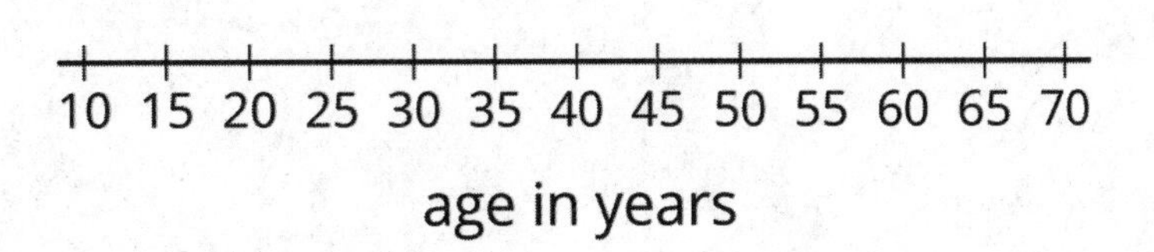

## 6.2: Finding the Middle

1. Your teacher will give you an index card. Write your first and last names on the card. Then record the total number of letters in your name. After that, pause for additional instructions from your teacher.

2. Here is the data set on numbers of siblings from an earlier activity.

   1 0 2 1 7 0 2 0 1 10

   a. Sort the data from least to greatest, and then find the **median**.

   b. In this situation, do you think the median is a good measure of a typical number of siblings for this group? Explain your reasoning.

3. Here is the dot plot showing the travel time, in minutes, of Elena's bus rides to school.

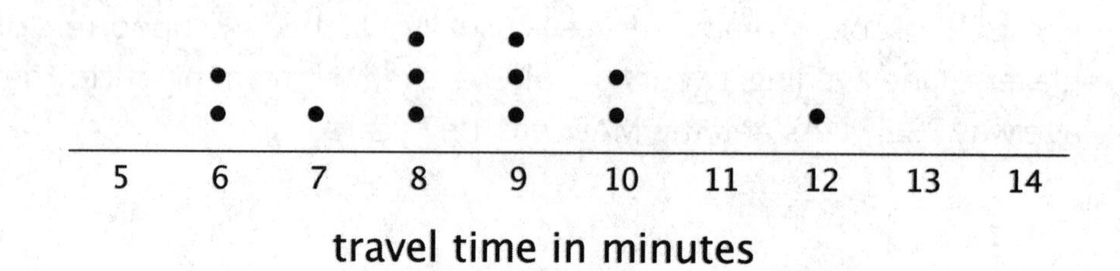


   a. Find the median travel time. Be prepared to explain your reasoning.

   b. What does the median tell us in this context?

## 6.3: Mean or Median?

1. Your teacher will give you six cards. Each has either a dot plot or a histogram. Sort the cards into *two* piles based on the distributions shown. Be prepared to explain your reasoning.

2. Discuss your sorting decisions with another group. Did you have the same cards in each pile? If so, did you use the same sorting categories? If not, how are your categories different?

   Pause here for a class discussion.

3. Use the information on the cards to answer the following questions.

   a. Card A: What is a typical age of the dogs being treated at the animal clinic?

   b. Card B: What is a typical number of people in the Irish households?

   c. Card C: What is a typical travel time for the New Zealand students?

   d. Card D: Would 15 years old be a good description of a typical age of the people who attended the birthday party?

   e. Card E: Is 15 minutes or 24 minutes a better description of a typical time it takes the students in South Africa to get to school?

   f. Card F: Would 21.3 years old be a good description of a typical age of the people who went on a field trip to Washington, D.C.?

4. How did you decide which measure of center to use for the dot plots on Cards A–C? What about for those on Cards D–F?

**Are you ready for more?**

Most teachers use the mean to calculate a student's final grade, based on that student's scores on tests, quizzes, homework, projects, and other graded assignments.

Diego thinks that the median might be a better way to measure how well a student did in a course. Do you agree with Diego? Explain your reasoning.

## Lesson 6 Summary

The **median** is another measure of center of a distribution. It is the middle value in a data set when values are listed in order. Half of the values in a data set are less than or equal to the median,and half of the values are greater than or equal to the median.

To find the median, we order the data values from least to greatest and find the number in the middle.

Suppose we have 5 dogs whose weights, in pounds, are shown in the table. The median weight for this group of dogs is 32 pounds because three dogs weigh less than or equal to 32 pounds and three dogs weigh greater than or equal to 32 pounds.

| 20 | 25 | 32 | 40 | 55 |
|---|---|---|---|---|

Now suppose we have 6 cats whose weights, in pounds, are as shown in the table. Notice that there are *two* values in the middle: 7 and 8.

| 4 | 6 | 7 | 8 | 10 | 10 |
|---|---|---|---|---|---|

The median weight must be between 7 and 8 pounds, because half of the cats weigh less or equal to 7 pounds and half of the cats weigh greater than or equal to 8 pounds.

In general, when we have an even number of values, we take the number exactly in between the two middle values. In this case, the median cat weight is 7.5 pounds because $(7 + 8) \div 2 = 7.5$.

Here is a set of 30 cookies. It has a mean weight of 21 grams, but the median weight is 23 grams.

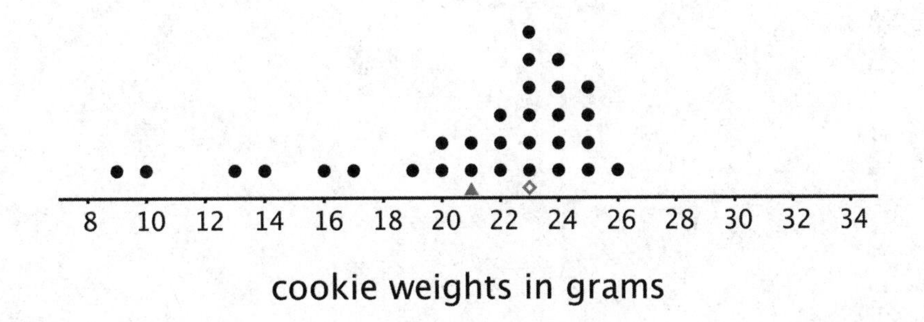


In this case, the median is closer to where most of the data points are clustered and is therefore a better measure of center for this distribution. That is, it is a better description of a typical cookie weight. The mean weight is influenced (in this case, pulled down) by a handful of much smaller cookies, so it is farther away from most data points.

In general, when a distribution is symmetrical or approximately symmetrical, the mean and median values are close. But when a distribution is not roughly symmetrical, the two values tend to be farther apart. Because the mean is fairly influenced by each value in the data set, it is generally preferred for distributions where it makes sense to use it. In cases when the distribution is less symmetric, the median is often reported as the typical value.

## Glossary

- median

## Lesson 6 Practice Problems

1. Here is data that shows a student's scores for 10 rounds of a video game.

   130 150 120 170 130 120 160 160 190 140

   What is the median score?

   A. 125

   B. 145

   C. 147

   D. 150

2. When he sorts the class's scores on the last test, the teacher notices that exactly 12 students scored better than Clare and exactly 12 students scored worse than Clare. Does this mean that Clare's score on the test is the median? Explain your reasoning.

3. The medians of the following dot plots are 6, 12, 13, and 15, but not in that order. Match each dot plot with its median.

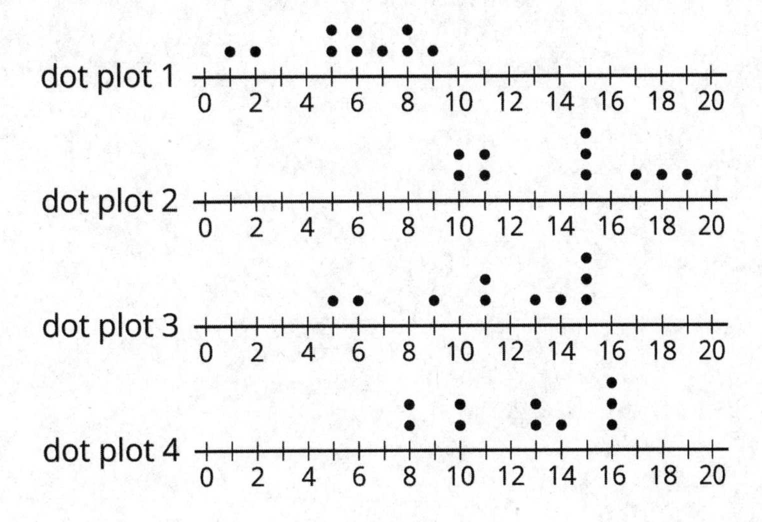

4. Ten sixth-grade students reported the hours of sleep they get on nights before a school day. Their responses are recorded in the dot plot.

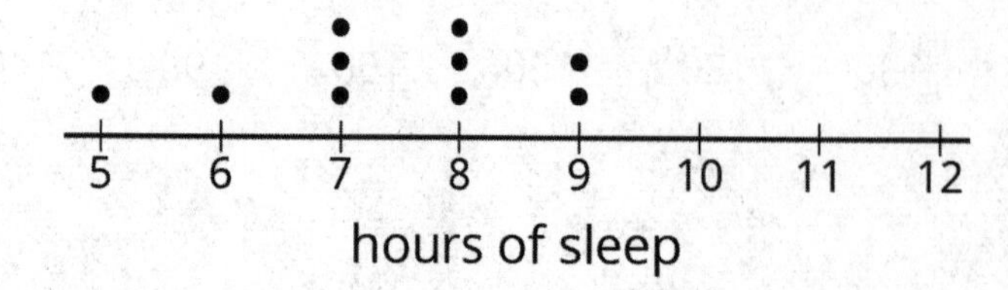


Looking at the dot plot, Lin estimated the mean number of hours of sleep to be 8.5 hours. Noah's estimate was 7.5 hours. Diego's estimate was 6.5 hours.

Which estimate do you think is best? Explain how you know.

(From Unit 8, Lesson 4.)

5. In his history class, Han's homework scores are:

100 100 100 100 95 100 90 100 0

The history teacher uses the mean to calculate the grade for homework. Write an argument for Han to explain why median would be a better measure to use for his homework grades.

6. The dot plots show how much time, in minutes, students in a class took to complete each of five different tasks. Select **all** the dot plots of tasks for which the mean time is approximately equal to the median time.

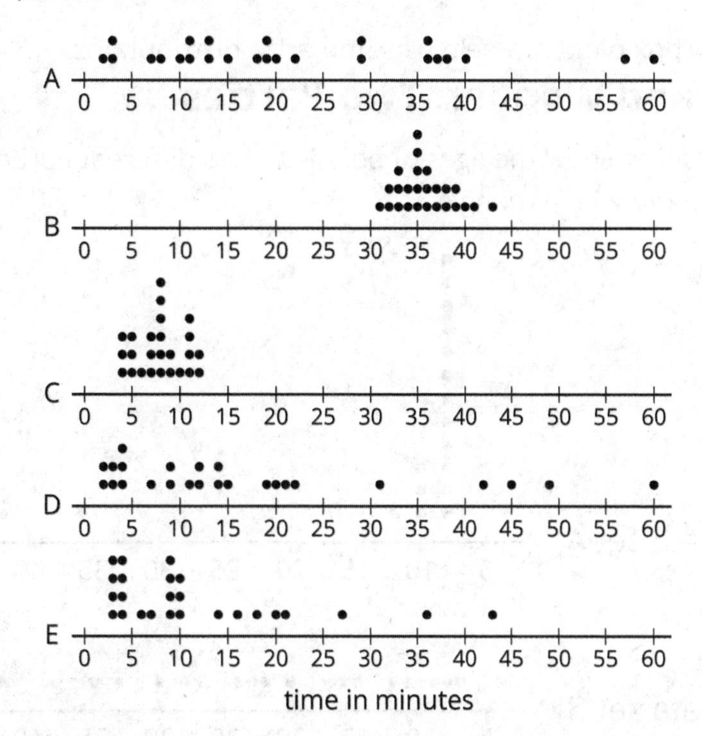


7. Here is a set of coordinates. Draw and label an appropriate pair of axes and plot the points. $A = (1, 0)$, $B = (0, 0.5)$, $C = (4, 3.5)$, $D = (1.5, 0.5)$

(From Unit 7, Lesson 11.)

# Lesson 7: Box Plots and Interquartile Range

Let's explore how box plots can help us summarize distributions.

## 7.1: Notice and Wonder: Two Parties

Here are dot plots that show the ages of people at two different parties. The mean of each distribution is marked with a triangle.

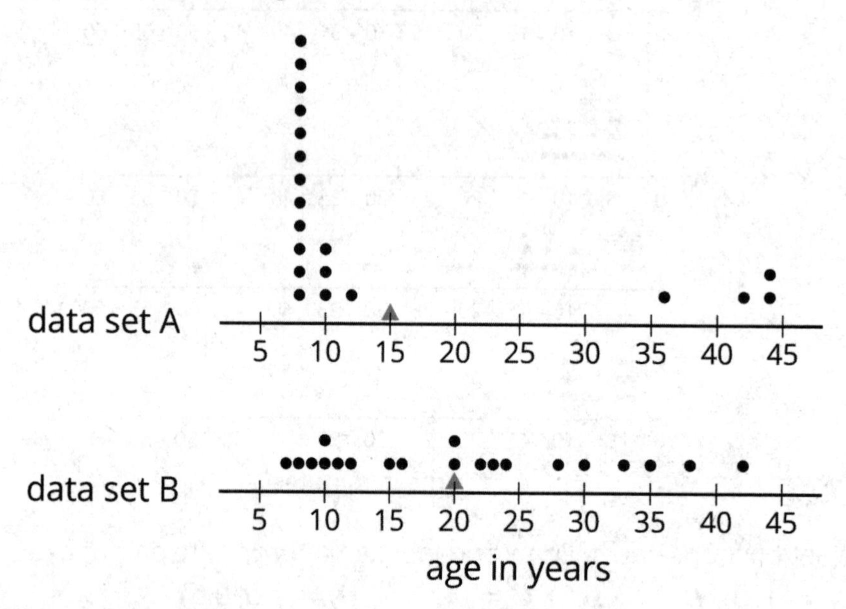


What do you notice and what do you wonder about the distributions in the two dot plots?

## 7.2: The Five-Number Summary

Here are the ages of the people at one party, listed from least to greatest.

| 7 | 8 | 9 | 10 | 10 | 11 | 12 | 15 |
|---|---|---|---|---|---|---|---|
| 16 | 20 | 20 | 22 | 23 | 24 | 28 | 30 |
| 33 | 35 | 38 | 42 | | | | |

1. a. Find the median of the data set and label it "50th percentile." This splits the data into an upper half and a lower half.

   b. Find the middle value of the *lower* half of the data, without including the median. Label this value "25th percentile."

   c. Find the middle value of the *upper* half of the data, without including the median. Label this value "75th percentile."

2. You have split the data set into four pieces. Each of the three values that split the data is called a **quartile**.

   - We call the 25th percentile the *first quartile*. Write "Q1" next to that number.
   - The median can be called the *second quartile*. Write "Q2" next to that number.
   - We call the 75th percentile the *third quartile*. Write "Q3" next to that number.

3. Label the lowest value in the set "minimum" and the greatest value "maximum."

4. The values you have identified make up the *five-number summary* for the data set. Record them here.

   minimum: ____ Q1: ____ Q2: ____ Q3: ____ maximum: ____

5. The median of this data set is 20. This tells us that half of the people at the party were 20 years old or younger, and the other half were 20 or older. What do each of these other values tell us about the ages of the people at the party?

   a. the third quartile

   b. the minimum

   c. the maximum

**Are you ready for more?**

There was another party where 21 people attended. Here is the five-number summary of their ages.

minimum: 5 Q1: 6 Q2: 27 Q3: 32 maximum: 60

1. Do you think this party had more children or fewer children than the earlier one? Explain your reasoning.

2. Were there more children or adults at this party? Explain your reasoning.

## 7.3: Human Box Plot

Your teacher will give you the data on the lengths of names of students in your class. Write the five-number summary by finding the data set's minimum, Q1, Q2, Q3, and the maximum.

Pause for additional instructions from your teacher.

## 7.4: Studying Blinks

Twenty people participated in a study about blinking. The number of times each person blinked while watching a video for one minute was recorded. The data values are shown here, in order from smallest to largest.

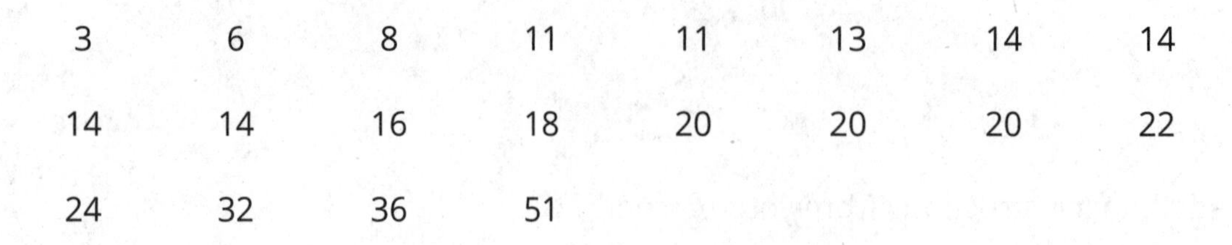

| | | | | | | | |
|---|---|---|---|---|---|---|---|
| 3 | 6 | 8 | 11 | 11 | 13 | 14 | 14 |
| 14 | 14 | 16 | 18 | 20 | 20 | 20 | 22 |
| 24 | 32 | 36 | 51 | | | | |

1. a. Use the grid and axis to make a dot plot of this data set.

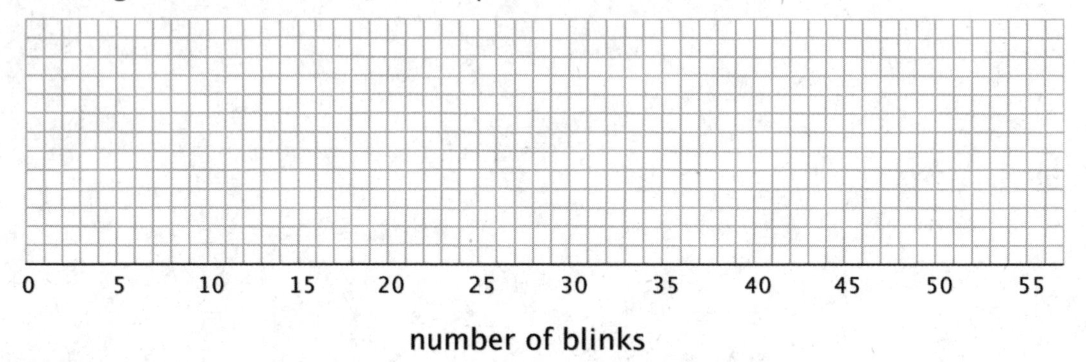


b. Find the median (Q2) and mark its location on the dot plot.

c. Find the first quartile (Q1) and the third quartile (Q3). Mark their locations on the dot plot.

d. What are the minimum and maximum values?

2. A **box plot** can be used to represent the five-number summary graphically. Let's draw a box plot for the number-of-blinks data. On the grid, *above* the dot plot:

   a. Draw a box that extends from the first quartile (Q1) to the third quartile (Q3). Label the quartiles.

   b. At the median (Q2), draw a vertical line from the top of the box to the bottom of the box. Label the median.

   c. From the left side of the box (Q1), draw a horizontal line (a whisker) that extends to the minimum of the data set. On the right side of the box (Q3), draw a similar line that extends to the maximum of the data set.

3. You have now created a box plot to represent the number of blinks data. What fraction of the data values are represented by each of these elements of the box plot?

    a. The left whisker

    b. The box

    c. The right whisker

**Are you ready for more?**

Suppose there were some errors in the data set: the smallest value should have been 6 instead of 3, and the largest value should have been 41 instead of 51. Determine if any part of the five-number summary would change. If you think so, describe how it would change. If not, explain how you know.

## Lesson 7 Summary

Earlier we learned that the mean is a measure of the center of a distribution and the MAD is a measure of the variability (or spread) that goes with the mean. There is also a measure of spread that goes with the median. It is called the interquartile range (IQR).

Finding the IQR involves splitting a data set into fourths. Each of the three values that splits the data into fourths is called a **quartile**.

- The median, or second quartile (Q2), splits the data into two halves.
- The first quartile (Q1) is the middle value of the lower half of the data.
- The third quartile (Q3) is the middle value of the upper half of the data.

For example, here is a data set with 11 values.

| 12 | 19 | 20 | 21 | 22 | 33 | 34 | 35 | 40 | 40 | 49 |
|---|---|---|---|---|---|---|---|---|---|---|
| | | Q1 | | | Q2 | | | Q3 | | |

- The median is 33.
- The first quartile is 20. It is the median of the numbers less than 33.
- The third quartile 40. It is the median of the numbers greater than 33.

The difference between the maximum and minimum values of a data set is the **range**. The difference between Q3 and Q1 is the **interquartile range (IQR)**. Because the distance between Q1 and Q3 includes the middle two-fourths of the distribution, the values between those two quartiles are sometimes called the *middle half of the data*.

The bigger the IQR, the more spread out the middle half of the data values are. The smaller the IQR, the closer together the middle half of the data values are. This is why we can use the IQR as a measure of spread.

A *five-number summary* can be used to summarize a distribution. It includes the minimum, first quartile, median, third quartile, and maximum of the data set. For the previous example, the five-number summary is 12, 20, 33, 40, and 49. These numbers are marked with diamonds on the dot plot.

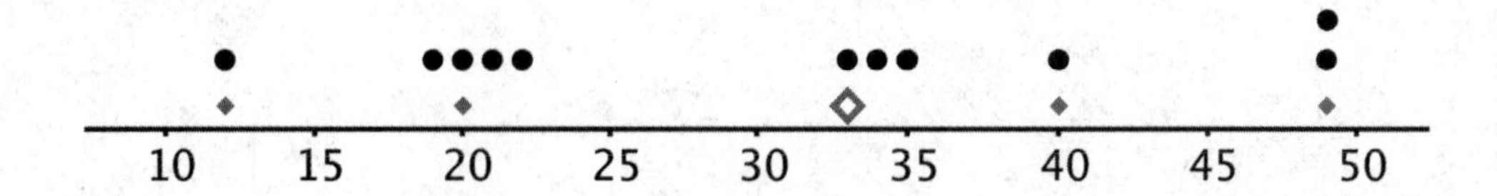

Different data sets can have the same five-number summary. For instance, here is another data set with the same minimum, maximum, and quartiles as the previous example.

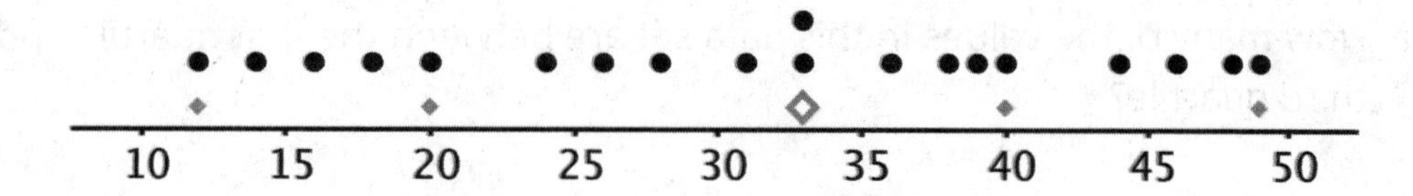


A **box plot** represents the five-number summary of a data set.

It shows the first quartile (Q1) and the third quartile (Q3) as the left and right sides of a rectangle or a box. The median (Q2) is shown as a vertical segment inside the box. On the left side, a horizontal line segment—a "whisker"—extends from Q1 to the minimum value. On the right, a whisker extends from Q3 to the maximum value.

The rectangle in the middle represents the middle half of the data. Its width is the IQR. The whiskers represent the bottom quarter and top quarter of the data set.

The box plots for these data sets are shown above the corresponding dot plots.

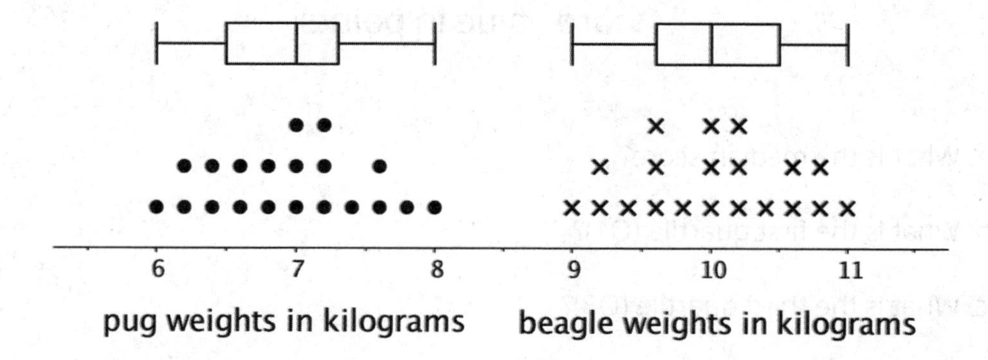


We can tell from the box plots that, in general, the pugs in the group are lighter than the beagles: the median weight of pugs is 7 kilograms and the median weight of beagles is 10 kilograms. Because the two box plots are on the same scale and the rectangles have similar widths, we can also tell that the IQRs for the two breeds are very similar. This suggests that the variability in the beagle weights is very similar to the variability in the pug weights.

## Glossary

- box plot
- interquartile range (IQR)
- quartile
- range

## Lesson 7 Practice Problems

1. Suppose that there are 20 numbers in a data set and that they are all different.

   a. How many of the values in this data set are between the first quartile and the third quartile?

   b. How many of the values in this data set are between the first quartile and the median?

2. In a word game, 1 letter is worth 1 point. This dot plot shows the scores for 20 common words.

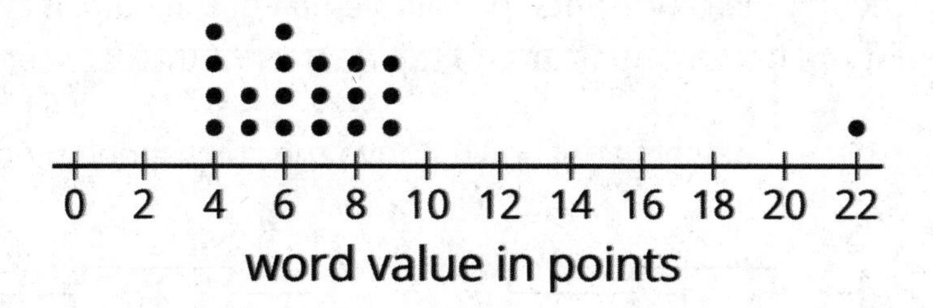


   a. What is the median score?

   b. What is the first quartile (Q1)?

   c. What is the third quartile (Q3)?

   d. What is the interquartile range (IQR)?

3. Mai and Priya each played 10 games of bowling and recorded the scores. Mai's median score was 120, and her IQR was 5. Priya's median score was 118, and her IQR was 15. Whose scores probably had less variability? Explain how you know.

4. Here are five dot plots that show the amounts of time that ten sixth-grade students in five countries took to get to school. Match each dot plot with the appropriate median and IQR.

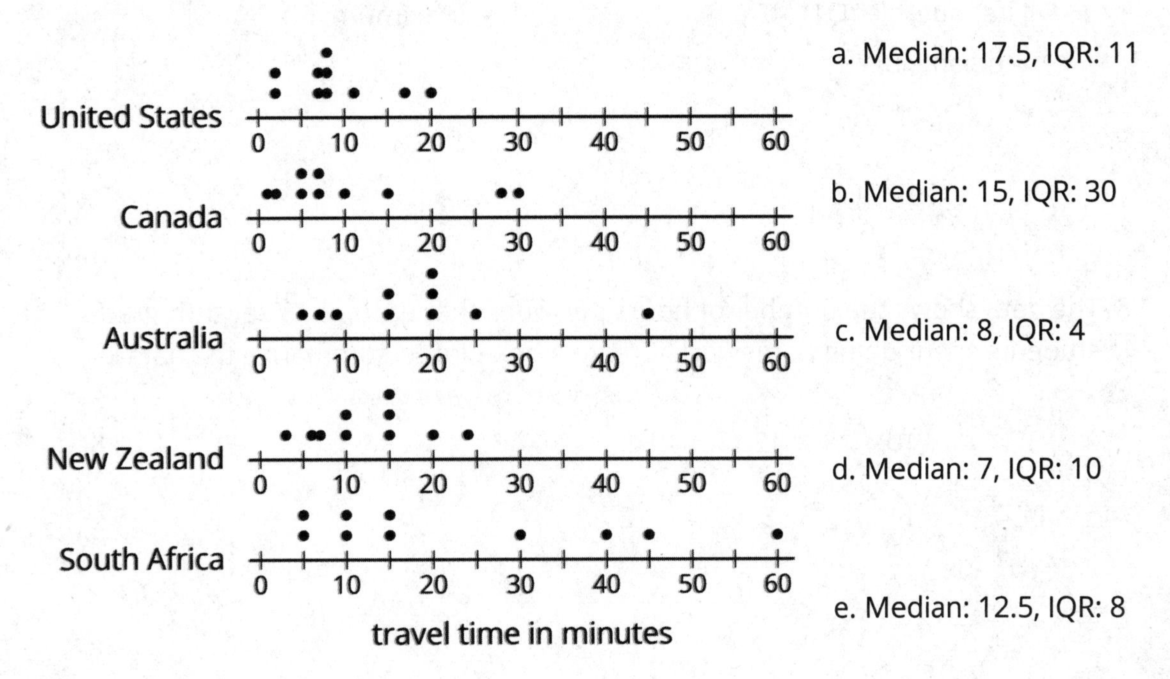


5. There are 20 pennies in a jar. If 16% of the coins in the jar are pennies, how many coins are there in the jar?

(From Unit 4, Lesson 7.)

6. Each student in a class recorded how many books they read during the summer. Here is a box plot that summarizes their data.

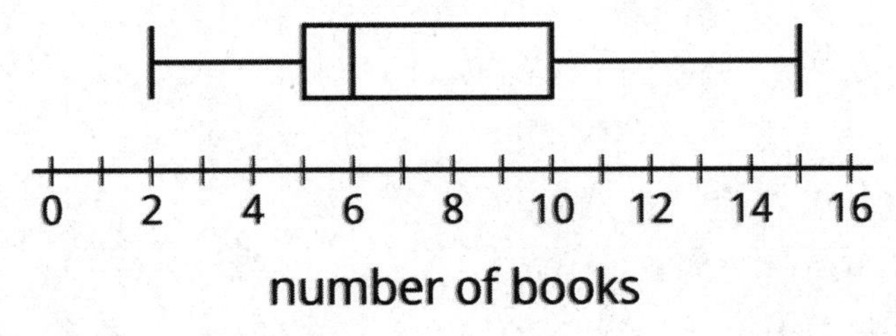


a. What is the greatest number of books read by a student in this group?

b. What is the median number of books read by the students?

c. What is the interquartile range (IQR)?

7. Use this five-number summary to draw a box plot. All values are in seconds.

- Minimum: 40
- First quartile (Q1): 45
- Median: 48
- Third quartile (Q3): 50
- Maximum: 60

8. The data shows the number of hours per week that each of 13 seventh-grade students spent doing homework. Create a box plot to summarize the data.

| 3 | 10 | 12 | 4 | 7 | 9 | 5 | 5 |
|---|---|---|---|---|---|---|---|
| 11 | 11 | 5 | 12 | 11 | | | |

# Lesson 8: Larger Populations

Let's compare larger groups.

## 8.1: First Name versus Last Name

Consider the question: In general, do the students at this school have more letters in their first name or last name? How many more letters?

1. What are some ways you might get some data to answer the question?

2. The other day, we compared the heights of people on different teams and the lengths of songs on different albums. What makes this question about first and last names harder to answer than those questions?

## 8.2: John Jacobjingleheimerschmidt

Continue to consider the question from the warm-up: In general, do the students at this school have more letters in their first name or last name? How many more letters?

1. How many letters are in your first name? In your last name?

2. Do the number of letters in your own first and last names give you enough information to make conclusions about students' names in your entire school? Explain your reasoning.

3. Your teacher will provide you with data from the class. Record the mean number of letters as well as the mean absolute deviation for each data set.

    a. The first names of the students in your class.

    b. The last names of the students in your class.

4. Which mean is larger? By how much? What does this difference tell you about the situation?

5. Do the mean numbers of letters in the first and last names for everyone in your class give you enough information to make conclusions about students' names in your entire school? Explain your reasoning.

## 8.3: Siblings and Pets

Consider the question: Do people who are the only child *have more pets*?

1. Earlier, we used information about the people in your class to answer a question about the entire school. Would surveying only the people in your class give you enough information to answer this new question? Explain your reasoning.

2. If you had to have an answer to this question by the end of class today, how would you gather data to answer the question?

3. If you could come back tomorrow with your answer to this question, how would you gather data to answer the question?

4. If someone else in the class came back tomorrow with an answer that was different than yours, what would that mean? How would you determine which answer was better?

## 8.4: Sampling the Population

For each question, identify the **population** and a possible **sample**.

1. What is the mean number of pages for novels that were on the best seller list in the 1990s?

2. What fraction of new cars sold between August 2010 and October 2016 were built in the United States?

3. What is the median income for teachers in North America?

4. What is the average lifespan of Tasmanian devils?

## Are you ready for more?

Political parties often use samples to poll people about important issues. One common method is to call people and ask their opinions. In most places, though, they are not allowed to call cell phones. Explain how this restriction might lead to inaccurate samples of the population.

## Lesson 8 Summary

A **population** is a set of people or things that we want to study. Here are some examples of populations:

- All people in the world
- All seventh graders at a school
- All apples grown in the U.S.

A **sample** is a subset of a population. Here are some examples of samples from the listed populations:

- The leaders of each country
- The seventh graders who are in band
- The apples in the school cafeteria

When we want to know more about a population but it is not feasible to collect data from everyone in the population, we often collect data from a sample. In the lessons that follow, we will learn more about how to pick a sample that can help answer questions about the entire population.

## Glossary

- population
- sample

## Lesson 8 Practice Problems

1. Suppose you are interested in learning about how much time seventh grade students at your school spend outdoors on a typical school day.

   Select **all** the samples that are a part of the population you are interested in.

   A. The 20 students in a seventh grade math class.

   B. The first 20 students to arrive at school on a particular day.

   C. The seventh grade students participating in a science fair put on by the four middle schools in a school district.

   D. The 10 seventh graders on the school soccer team.

   E. The students on the school debate team.

2. For each sample given, list two possible populations they could belong to.

   a. Sample: The prices for apples at two stores near your house.

   b. Sample: The days of the week the students in your math class ordered food during the past week.

   c. Sample: The daily high temperatures for the capital cities of all 50 U.S. states over the past year.

3. If 6 coins are flipped, find the probability that there is at least 1 heads.

   (From Unit 8, Lesson 16.)

4. A school's art club holds a bake sale on Fridays to raise money for art supplies. Here are the number of cookies they sold each week in the fall and in the spring:

| | | | | | | | | | | |
|---|---|---|---|---|---|---|---|---|---|---|
| fall | 20 | 26 | 25 | 24 | 29 | 20 | 19 | 19 | 24 | 24 |
| spring | 19 | 27 | 29 | 21 | 25 | 22 | 26 | 21 | 25 | 25 |

a. Find the mean number of cookies sold in the fall and in the spring.

b. The MAD for the fall data is 2.8 cookies. The MAD for the spring data is 2.6 cookies. Express the difference in means as a multiple of the larger MAD.

c. Based on this data, do you think that sales were generally higher in the spring than in the fall?

This problem is from an earlier lesson

5. A school is selling candles for a fundraiser. They keep 40% of the total sales as their commission, and they pay the rest to the candle company.

| price of candle | number of candles sold |
|---|---|
| small candle: $11 | 68 |
| medium candle: $18 | 45 |
| large candle: $25 | 21 |

How much money must the school pay to the candle company?

(From Unit 6, Lesson 8.)

# Lesson 9: What Makes a Good Sample?

Let's see what makes a good sample.

## 9.1: Number Talk: Division by Powers of 10

Find the value of each quotient mentally.

$34,000 \div 10$

$340 \div 100$

$34 \div 10$

$3.4 \div 100$

## 9.2: Selling Paintings

Your teacher will assign you to work with either means or medians.

1. A young artist has sold 10 paintings. Calculate the measure of center you were assigned for each of these samples:

    a. The first two paintings she sold were for \$50 and \$350.

    b. At a gallery show, she sold three paintings for \$250, \$400, and \$1,200.

    c. Her oil paintings have sold for \$410, \$400, and \$375.

2. Here are the selling prices for all 10 of her paintings:

    \$50 \$200 \$250 \$275 \$280 \$350 \$375 \$400 \$410 \$1,200

    Calculate the measure of center you were assigned for all of the selling prices.

3. Compare your answers with your partner. Were the measures of center for any of the samples close to the same measure of center for the population?

## 9.3: Sampling the Fish Market

The price per pound of catfish at a fish market was recorded for 100 weeks.

1. Here are dot plots showing the population and three different samples from that population. What do you notice? What do you wonder?

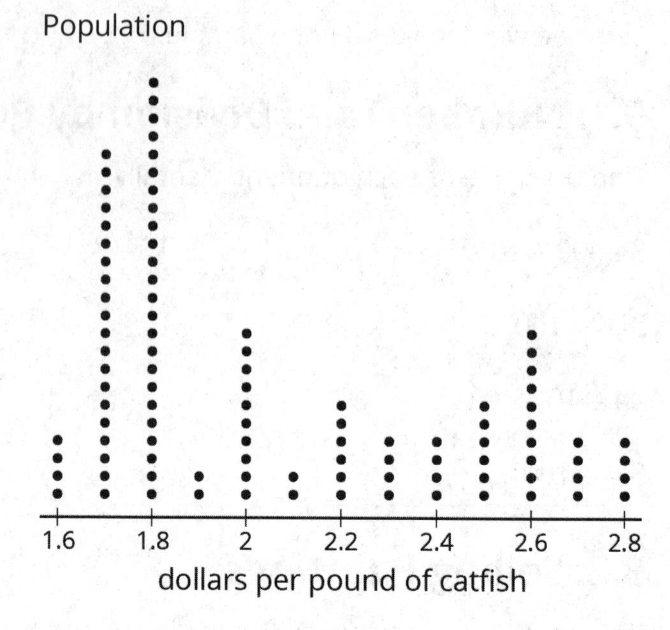


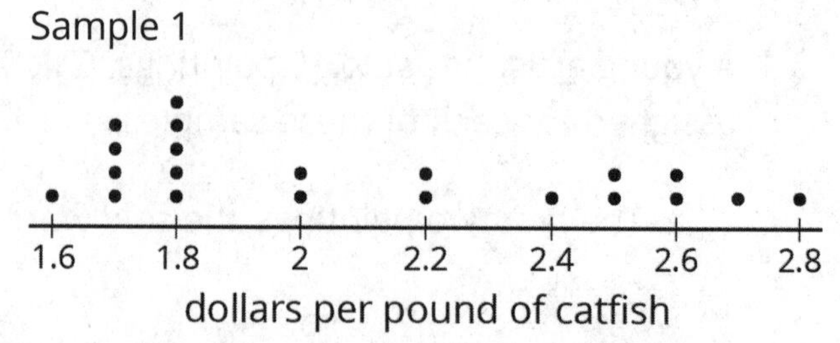


2. If the goal is to have the sample represent the population, which of the samples would work best? Which wouldn't work so well? Explain your reasoning.

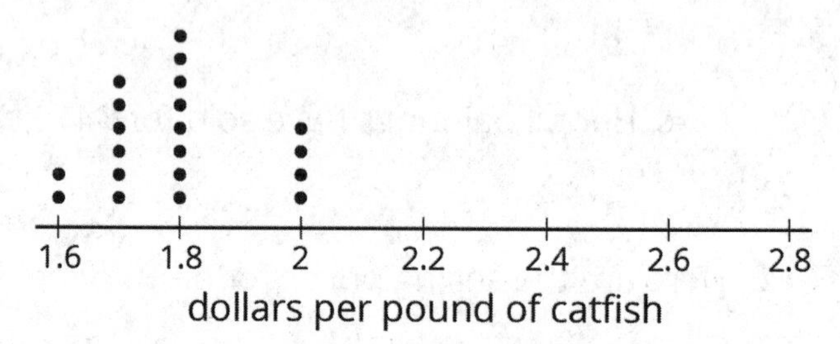


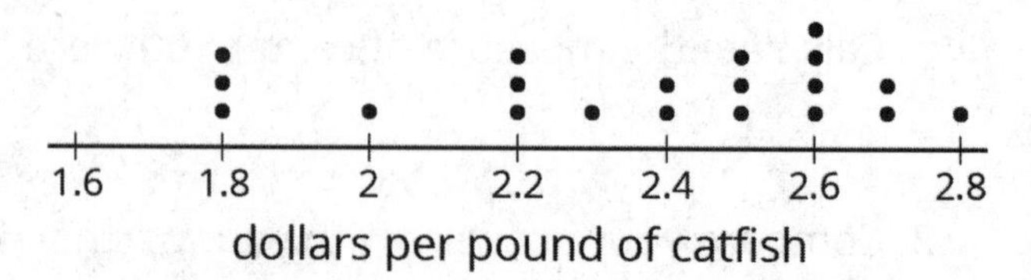

### Are you ready for more?

When doing a statistical study, it is important to keep the goal of the study in mind. Representative samples give us the best information about the distribution of the population as a whole, but sometimes a representative sample won't work for the goal of a study!

For example, suppose you want to study how discrimination affects people in your town. Surveying a representative sample of people in your town would give information about how the population generally feels, but might miss some smaller groups. Describe a way you might choose a sample of people to address this question.

## 9.4: Auditing Sales

An online shopping company tracks how many items they sell in different categories during each month for a year. Three different auditors each take samples from that data. Use the samples to draw dot plots of what the population data might look like for the furniture and electronics categories.

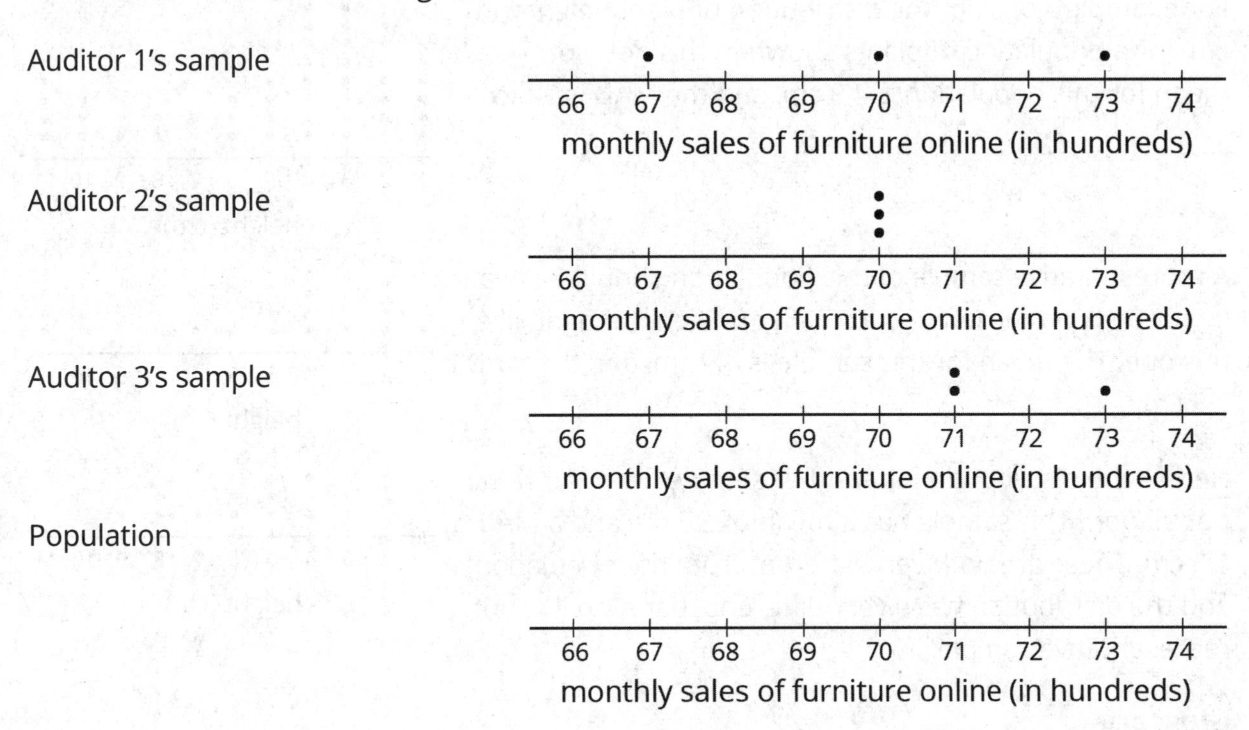

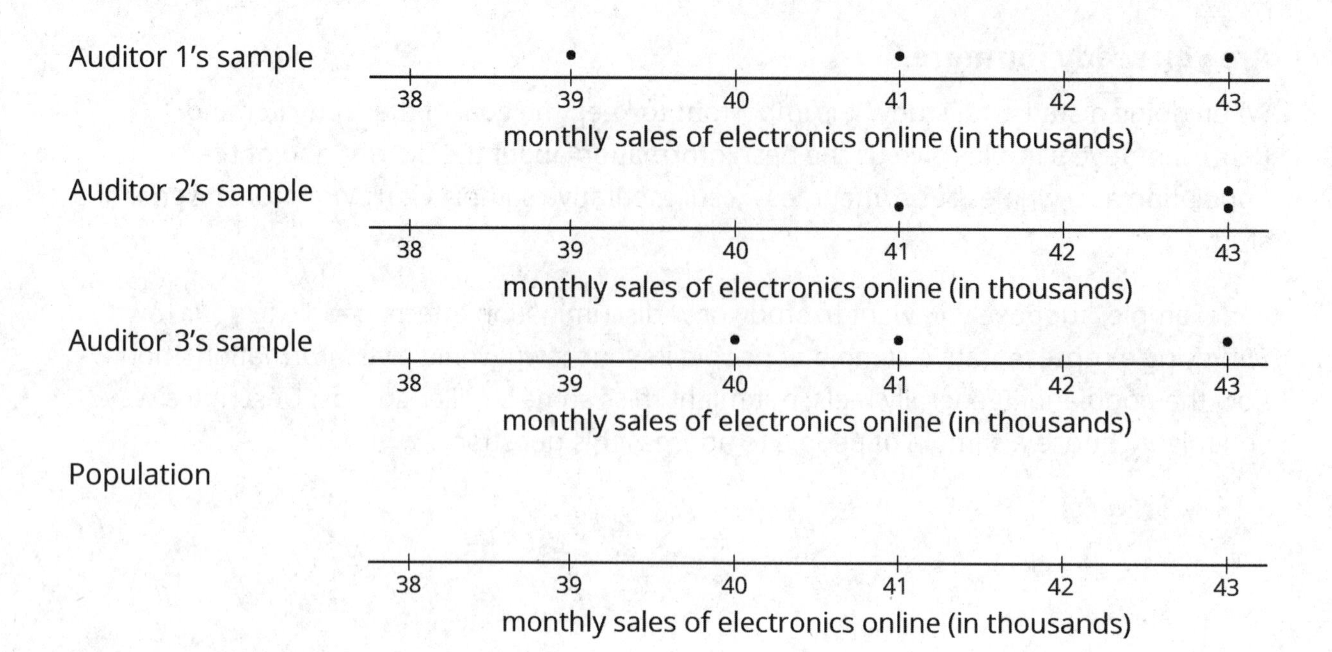


## Lesson 9 Summary

A sample that is **representative** of a population has a distribution that closely resembles the distribution of the population in shape, center, and spread.

For example, consider the distribution of plant heights, in cm, for a population of plants shown in this dot plot. The mean for this population is 4.9 cm, and the MAD is 2.6 cm.

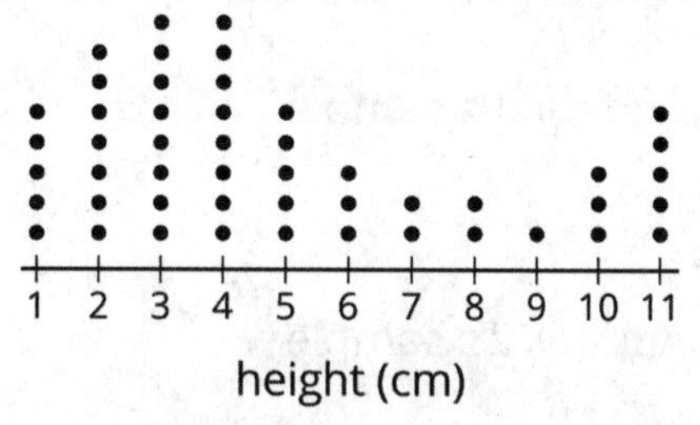


A representative sample of this population should have a larger peak on the left and a smaller one on the right, like this one. The mean for this sample is 4.9 cm, and the MAD is 2.3 cm.

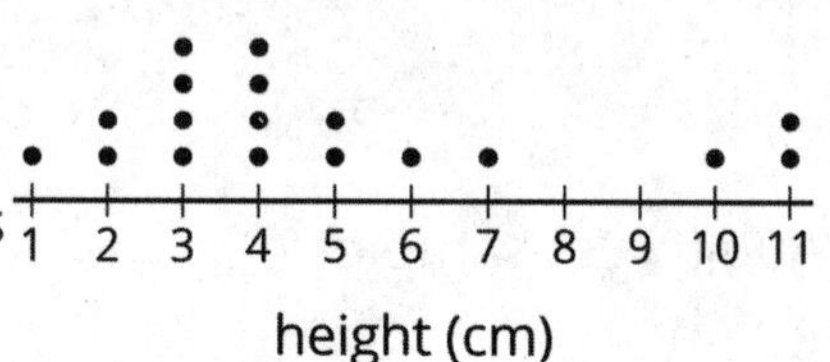


Here is the distribution for another sample from the same population. This sample has a mean of 5.7 cm and a MAD of 1.5 cm. These are both very different from the population, and the distribution has a very different shape, so it is not a representative sample.

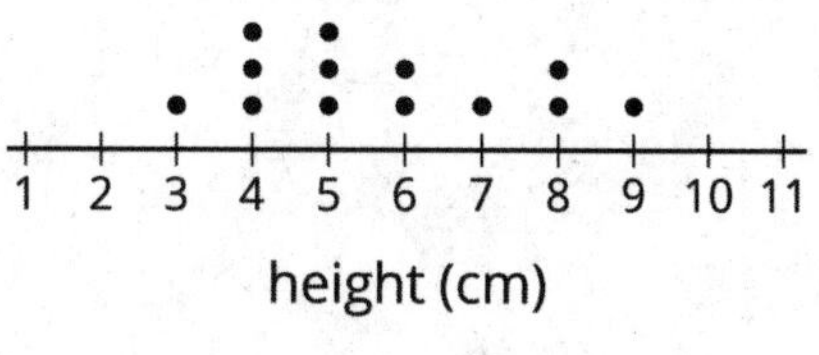


## Glossary

- representative

## Lesson 9 Practice Problems

1. Suppose 45% of all the students at Andre's school brought in a can of food to contribute to a canned food drive. Andre picks a representative sample of 25 students from the school and determines the sample's percentage.

   He expects the percentage for this sample will be 45%. Do you agree? Explain your reasoning.

2. This is a dot plot of the scores on a video game for a population of 50 teenagers.

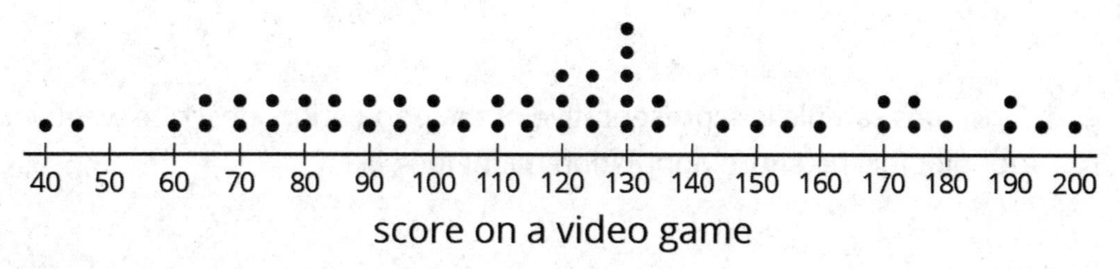


   The three dot plots together are the scores of teenagers in three samples from this population. Which of the three samples is most representative of the population? Explain how you know.

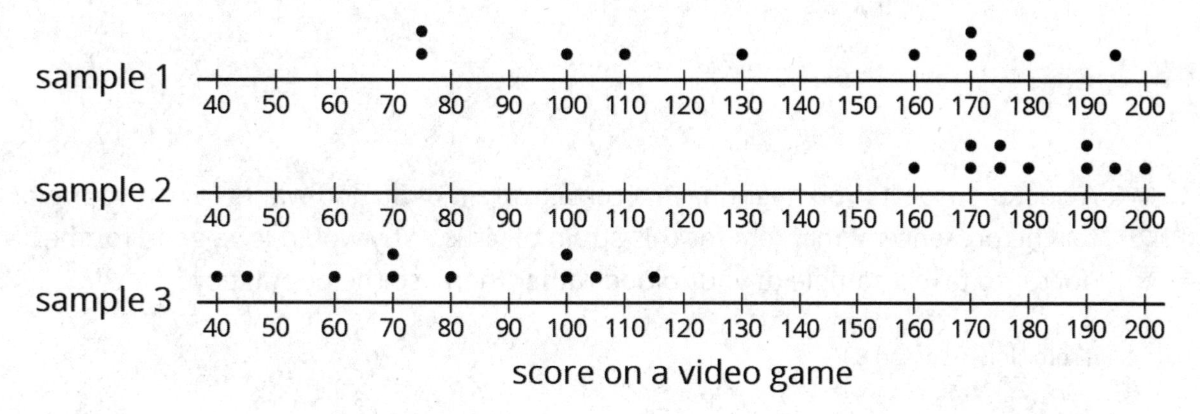

3. This is a dot plot of the number of text messages sent one day for a sample of the students at a local high school. The sample consisted of 30 students and was selected to be representative of the population.

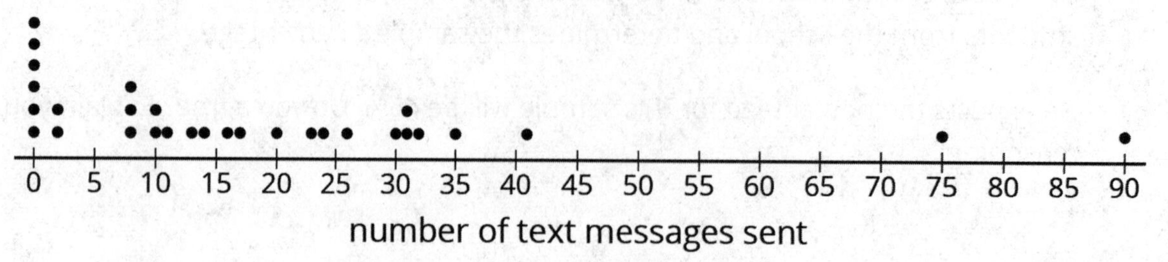


a. What do the six values of 0 in the dot plot represent?

b. Since this sample is representative of the population, describe what you think a dot plot for the entire population might look like.

4. A doctor suspects you might have a certain strain of flu and wants to test your blood for the presence of markers for this strain of virus. Why would it be good for the doctor to take a sample of your blood rather than use the population?

(From Unit 8, Lesson 8.)

# Lesson 10: Sampling in a Fair Way

Let's explore ways to get representative samples.

## 10.1: Ages of Moviegoers

A survey was taken at a movie theater to estimate the average age of moviegoers.

Here is a dot plot showing the ages of the first 20 people surveyed.

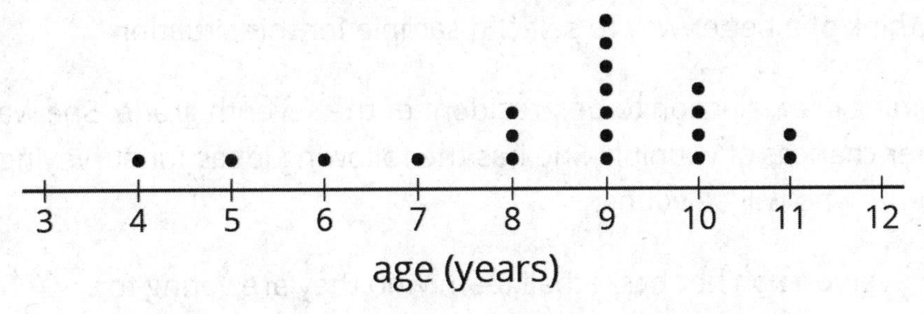


1. What questions do you have about the data from survey?

2. What assumptions would you make based on these results?

## 10.2: Comparing Methods for Selecting Samples

Take turns with your partner reading each option aloud. For each situation, discuss:

- Would the different methods for selecting a sample lead to different conclusions about the population?
- What are the benefits of each method?
- What might each method overlook?
- Which of the methods listed would be the most likely to produce samples that are representative of the population being studied?
- Can you think of a better way to select a sample for this situation?

1. Lin is running in an election to be president of the seventh grade. She wants to predict her chances of winning. She has the following ideas for surveying a sample of the students who will be voting:

    a. Ask everyone on her basketball team who they are voting for.

    b. Ask every third girl waiting in the lunch line who they are voting for.

    c. Ask the first 15 students to arrive at school one morning who they are voting for.

2. A nutritionist wants to collect data on how much caffeine the average American drinks per day. She has the following ideas for how she could obtain a sample:

    a. Ask the first 20 adults who arrive at a grocery store after 10:00 a.m. about the average amount of caffeine they consume each day.

    b. Every 30 minutes, ask the first adult who comes into a coffee shop about the average amount of caffeine they consume each day.

## 10.3: That's the First Straw

Your teacher will have some students draw straws from a bag.

1. As each straw is taken out and measured, record its length (in inches) in the table.

| | straw 1 | straw 2 | straw 3 | straw 4 | straw 5 |
|---|---|---|---|---|---|
| sample 1 | | | | | |
| sample 2 | | | | | |

2. Estimate the mean length of all the straws in the bag based on:

    a. the mean of the first sample.

    b. the mean of the second sample.

3. Were your two estimates the same? Did the mean length of all the straws in the bag change in between selecting the two samples? Explain your reasoning.

4. The actual mean length of all of the straws in the bag is about 2.37 inches. How do your estimates compare to this mean length?

5. If you repeated the same process again but you selected a larger sample (such as 10 or 20 straws, instead of just 5), would your estimate be more accurate? Explain your reasoning.

## 10.4: That's the Last Straw

There were a total of 35 straws in the bag. Suppose we put the straws in order from shortest to longest and then assigned each straw a number from 1 to 35. For each of these methods, decide whether it would be fair way to select a sample of 5 straws. Explain your reasoning.

1. Select the straws numbered 1 through 5.

2. Write the numbers 1 through 35 on pieces of paper that are all the same size. Put the papers into a bag. Without looking, select five papers from the bag. Use the straws with those numbers for your sample.

3. Using the same bag as the previous question, select one paper from the bag. Use the number on that paper to select the first straw for your sample. Then use the next 4 numbers in order to complete your sample. (For example, if you select number 17, then you also use straws 18, 19, 20, and 21 for your sample.)

4. Create a spinner with 35 sections that are all the same size, and number them 1 through 35. Spin the spinner 5 times and use the straws with those numbers for your sample.

## Are you ready for more?

Computers accept inputs, follow instructions, and produce outputs, so they cannot produce truly random numbers. If you knew the input, you could predict the output by following the same instructions the computer is following. When truly random numbers are needed, scientists measure natural phenomena such as radioactive decay or temperature variations. Before such measurements were possible, statisticians used random number tables, like this:

85 67 95 02 42 61 21 35 15 34 41
85 94 61 72 53 24 15 67 85 94 12
67 88 15 32 42 65 75 98 46 25 13
07 53 60 75 82 34 67 44 20 42 33
99 37 40 33 40 88 90 50 75 22 90
00 03 84 57 91 15 70 08 90 03 02
78 07 16 51 13 89 67 64 54 05 26
62 06 61 43 02 60 73 58 38 53 88
02 50 88 44 37 05 13 54 78 97 30

Use this table to select a sample of 5 straws. Pick a starting point at random in the table. If the number is between 01 and 35, include that number straw in your sample. If the number has already been selected, or is not between 01 and 35, ignore it, and move on to the next number.

## Lesson 10 Summary

A sample is *selected at random* from a population if it has an equal chance of being selected as every other sample of the same size. For example, if there are 25 students in a class, then we can write each of the students' names on a slip of paper and select 5 papers from a bag to get a sample of 5 students selected at random from the class.

Other methods of selecting a sample from a population are likely to be *biased*. This means that it is less likely that the sample will be representative of the population as a whole. For example, if we select the first 5 students who walk in the door, that will not give us a random sample because students who typically come late are not likely to be selected. A sample that is selected at random may not always be a representative sample, but it is more likely to be representative than using other methods.

It is not always possible to select a sample at random. For example, if we want to know the average length of wild salmon, it is not possible to identify each one individually, select a few at random from the list, and then capture and measure those exact fish. When a sample cannot be selected at random, it is important to try to reduce bias as much as possible when selecting the sample.

## Lesson 10 Practice Problems

1. The meat department manager at a grocery store is worried some of the packages of ground beef labeled as having one pound of meat may be under-filled. He decides to take a sample of 5 packages from a shipment containing 100 packages of ground beef. The packages were numbered as they were put in the box, so each one has a different number between 1 and 100.

   Describe how the manager can select a fair sample of 5 packages.

2. Select **all** the reasons why random samples are preferred over other methods of getting a sample.

   A. If you select a random sample, you can determine how many people you want in the sample.

   B. A random sample is always the easiest way to select a sample from a population.

   C. A random sample is likely to give you a sample that is representative of the population.

   D. A random sample is a fair way to select a sample, because each person in the population has an equal chance of being selected.

   E. If you use a random sample, the sample mean will always be the same as the population mean.

3. Jada is using a computer's random number generator to produce 6 random whole numbers between 1 and 100 so she can use a random sample. The computer produces the numbers: 1, 2, 3, 4, 5, and 6. Should she use these numbers or have the computer generate a new set of random numbers? Explain your reasoning.

4. Data collected from a survey of American teenagers aged 13 to 17 was used to estimate that 29% of teens believe in ghosts. This estimate was based on data from 510 American teenagers. What is the population that people carrying out the survey were interested in?

   A. All people in the United States.

   B. The 510 teens that were surveyed.

   C. All American teens who are between the ages of 13 and 17.

   D. The 29% of the teens surveyed who said they believe in ghosts.

   (From Unit 8, Lesson 8.)

# Lesson 11: Estimating Population Measures of Center

Let's use samples to estimate measures of center for the population.

## 11.1: Describing the Center

Would you use the median or mean to describe the center of each data set? Explain your reasoning.

Heights of 50 basketball players

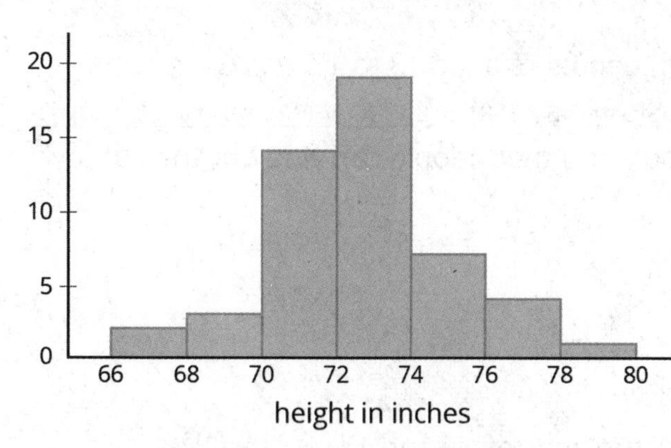


Ages of 30 people at a family dinner party

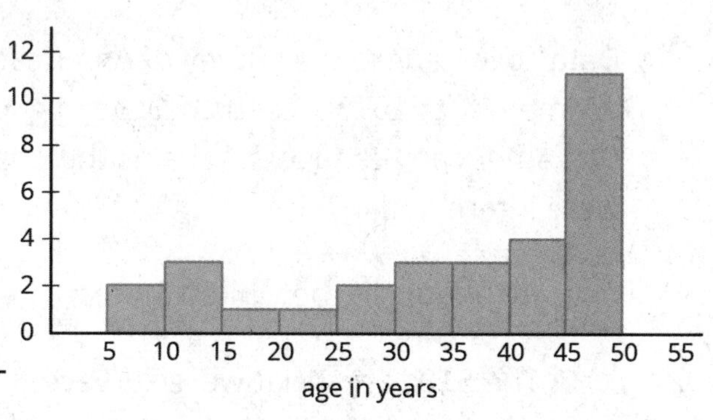


Backpack weights of sixth-grade students

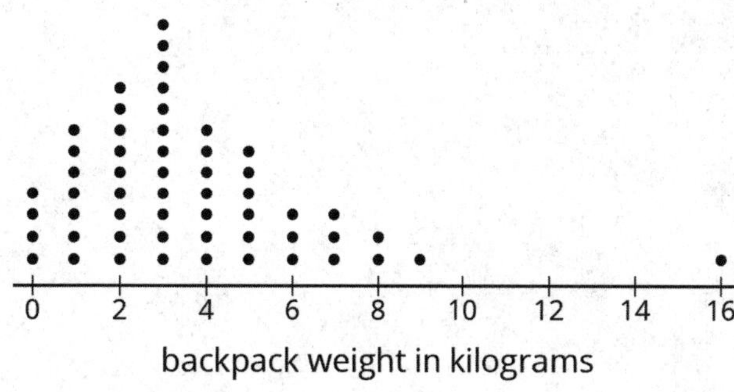


How many books students read over summer break

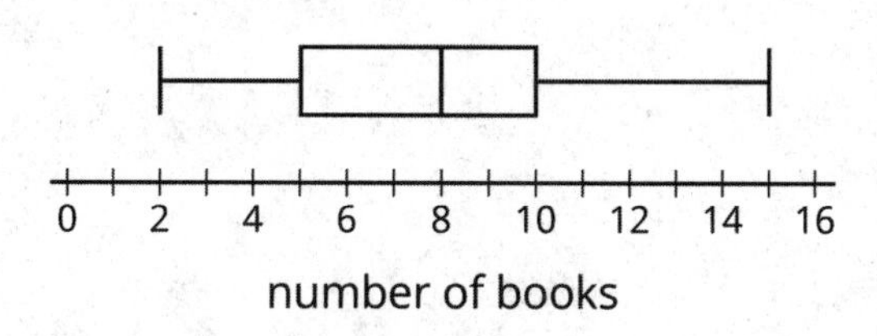

## 11.2: Three Different TV Shows

Here are the ages (in years) of a random sample of 10 viewers for 3 different television shows. The shows are titled, "Science Experiments YOU Can Do," "Learning to Read," and "Trivia the Game Show."

| | | | | | | | | | | |
|---|---|---|---|---|---|---|---|---|---|---|
| sample 1 | 6 | 6 | 5 | 4 | 8 | 5 | 7 | 8 | 6 | 6 |
| sample 2 | 15 | 14 | 12 | 13 | 12 | 10 | 12 | 11 | 10 | 8 |
| sample 3 | 43 | 60 | 50 | 36 | 58 | 50 | 73 | 59 | 69 | 51 |

1. Calculate the mean for *one* of the samples. Make sure each person in your group works with a different sample. Record the answers for all three samples.

2. Which show do you think each sample represents? Explain your reasoning
.

## 11.3: Who's Watching What?

Here are three more samples of viewer ages collected for these same 3 television shows.

| | | | | | | | | | | |
|---|---|---|---|---|---|---|---|---|---|---|
| sample 4 | 57 | 71 | 5 | 54 | 52 | 13 | 59 | 65 | 10 | 71 |
| sample 5 | 15 | 5 | 4 | 5 | 4 | 3 | 25 | 2 | 8 | 3 |
| sample 6 | 6 | 11 | 9 | 56 | 1 | 3 | 11 | 10 | 11 | 2 |

1. Calculate the mean for *one* of these samples. Record all three answers.

2. Which show do you think each of these samples represents? Explain your reasoning.

3. For each show, estimate the mean age for all the show's viewers.

4. Calculate the mean absolute deviation for *one* of the shows' samples. Make sure each person in your group works with a different sample. Record all three answers.

| | Learning to Read | Science Experiments YOU Can Do | Trivia the Game Show |
|---|---|---|---|
| **Which sample?** | | | |
| **MAD** | | | |

5. What do the different values for the MAD tell you about each group?

6. An advertiser has a commercial that appeals to 15- to 16-year-olds. Based on these samples, are any of these shows a good fit for this commercial? Explain or show your reasoning.

## 11.4: Movie Reviews

A movie rating website has many people rate a new movie on a scale of 0 to 100. Here is a dot plot showing a random sample of 20 of these reviews.

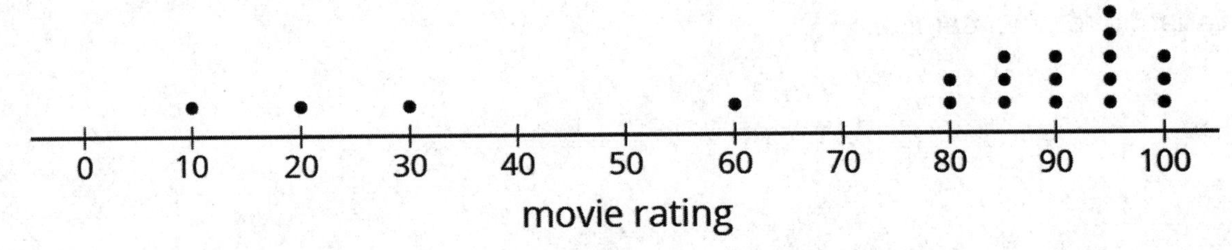


1. Would the mean or median be a better measure for the center of this data? Explain your reasoning.

2. Use the sample to estimate the measure of center that you chose for *all* the reviews.

3. For this sample, the mean absolute deviation is 19.6, and the interquartile range is 15. Which of these values is associated with the measure of center that you chose?

4. Movies must have an average rating of 75 or more from all the reviews on the website to be considered for an award. Do you think this movie will be considered for the award? Use the measure of center and measure of variability that you chose to justify your answer.

## Are you ready for more?

Estimate typical temperatures in the United States today by looking up current temperatures in several places across the country. Use the data you collect to decide on the appropriate measure of center for the country, and calculate the related measure of variation for your sample.

## Lesson 11 Summary

Some populations have greater variability than others. For example, we would expect greater variability in the weights of dogs at a dog park than at a beagle meetup.

Dog park:

Mean weight: 12.8 kg MAD: 2.3 kg

Beagle meetup:

Mean weight: 10.1 kg MAD: 0.8 kg

The lower MAD indicates there is less variability in the weights of the beagles. We would expect that the mean weight from a sample that is randomly selected from a group of beagles will provide a more accurate estimate of the mean weight of all the beagles than a sample of the same size from the dogs at the dog park.

In general, a sample of a similar size from a population with *less* variability is *more likely* to have a mean that is close to the population mean.

## Glossary

- interquartile range (IQR)

## Lesson 11 Practice Problems

1. A random sample of 15 items were selected.

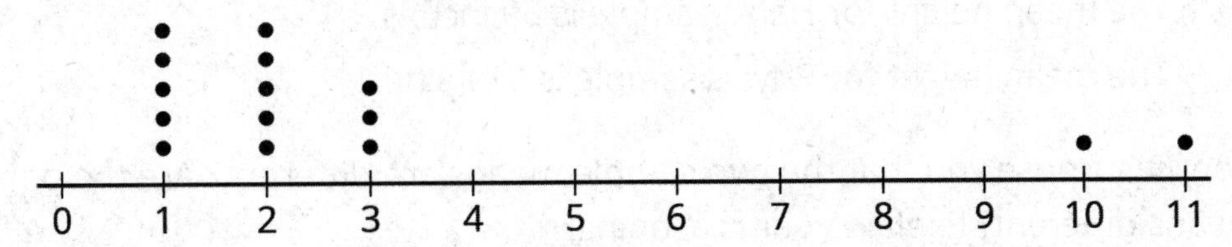


For this data set, is the mean or median a better measure of center? Explain your reasoning.

2. A video game developer wants to know how long it takes people to finish playing their new game. They surveyed a random sample of 13 players and asked how long it took them (in minutes).

| 1,235 | 952 | 457 | 1,486 | 1,759 | 1,148 | 548 | 1,037 |
|---|---|---|---|---|---|---|---|
| 1,864 | 1,245 | 976 | 866 | 1,431 | | | |

   a. Estimate the median time it will take *all* players to finish this game.

   b. Find the interquartile range for this sample.

3. Han and Priya want to know the mean height of the 30 students in their dance class. They each select a random sample of 5 students.

    - The mean height for Han's sample is 59 inches.
    - The mean height for Priya's sample is 61 inches.

    Does it surprise you that the two sample means are different? Are the population means different? Explain your reasoning.

4. Clare and Priya each took a random sample of 25 students at their school.

    - Clare asked each student in her sample how much time they spend doing homework each night. The sample mean was 1.2 hours and the MAD was 0.6 hours.
    - Priya asked each student in her sample how much time they spend watching TV each night. The sample mean was 2 hours and the MAD was 1.3 hours.

    a. At their school, do you think there is more variability in how much time students spend doing homework or watching TV? Explain your reasoning.

    b. Clare estimates the students at her school spend an average of 1.2 hours each night doing homework. Priya estimates the students at her school spend an average of 2 hours each night watching TV. Which of these two estimates is likely to be closer to the actual mean value for all the students at their school? Explain your reasoning.

# Lesson 12: More about Sampling Variability

Let's compare samples from the same population.

## 12.1: Average Reactions

The other day, you worked with the reaction times of twelfth graders to see if they were fast enough to help out at the track meet. Look back at the sample you collected.

1. Calculate the mean reaction time for your sample.

2. Did you and your partner get the same sample mean? Explain why or why not.

## 12.2: Reaction Population

Your teacher will display a blank dot plot.

1. Plot your sample mean from the previous activity on your teacher's dot plot.

2. What do you notice about the distribution of the sample means from the class?

    a. Where is the center?

    b. Is there a lot of variability?

    c. Is it approximately symmetric?

3. The population mean is 0.442 seconds. How does this value compare to the sample means from the class?

Pause here so your teacher can display a dot plot of the population of reaction times.

4. What do you notice about the distribution of the population?

   a. Where is the center?

   b. Is there a lot of variability?

   c. Is it approximately symmetric?

5. Compare the two displayed dot plots.

6. Based on the distribution of sample means from the class, do you think the mean of a random sample of 20 items is likely to be:

   a. within 0.01 seconds of the actual population mean?

   b. within 0.1 seconds of the actual population mean?

   Explain or show your reasoning.

## 12.3: How Much Do You Trust the Answer?

The other day you worked with 2 different samples of viewers from each of 3 different television shows. Each sample included 10 viewers. Here are the mean ages for 100 different samples of viewers from each show.

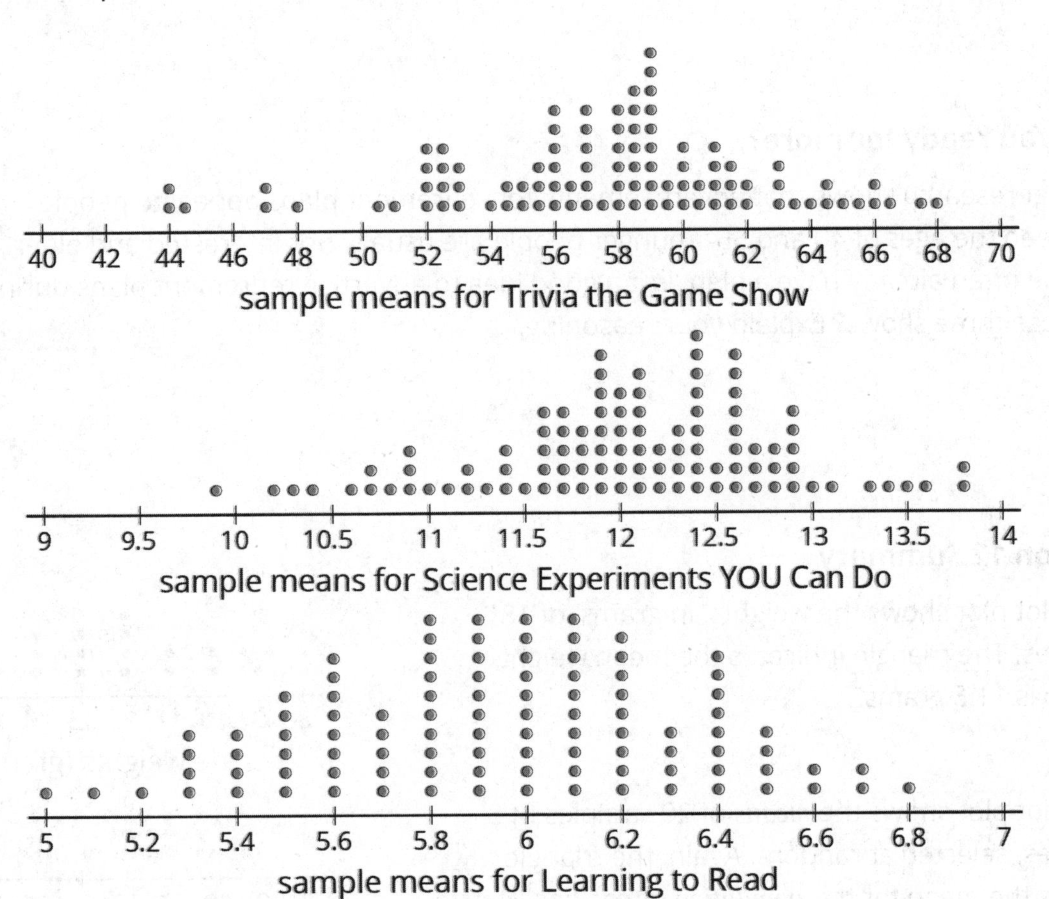


1. For each show, use the dot plot to estimate the *population* mean.

    a. Trivia the Game Show

    b. Science Experiments YOU Can Do

    c. Learning to Read

2. For each show, are most of the sample means within 1 year of your estimated population mean?

3. Suppose you take a new random sample of 10 viewers for each of the 3 shows. Which show do you expect to have the new sample mean closest to the population mean? Explain or show your reasoning.

### Are you ready for more?

Market research shows that advertisements for retirement plans appeal to people between the ages of 40 and 55. Younger people are usually not interested and older people often already have a plan. Is it a good idea to advertise retirement plans during any of these three shows? Explain your reasoning.

## Lesson 12 Summary

This dot plot shows the weights, in grams, of 18 cookies. The triangle indicates the mean weight, which is 11.6 grams.

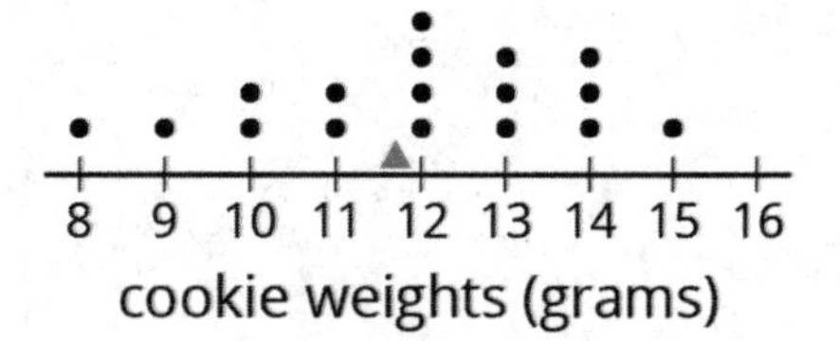


This dot plot shows the *means* of 20 samples of 5 cookies, selected at random. Again, the triangle shows the mean for the *population* of cookies. Notice that most of the sample means are fairly close to the mean of the entire population.

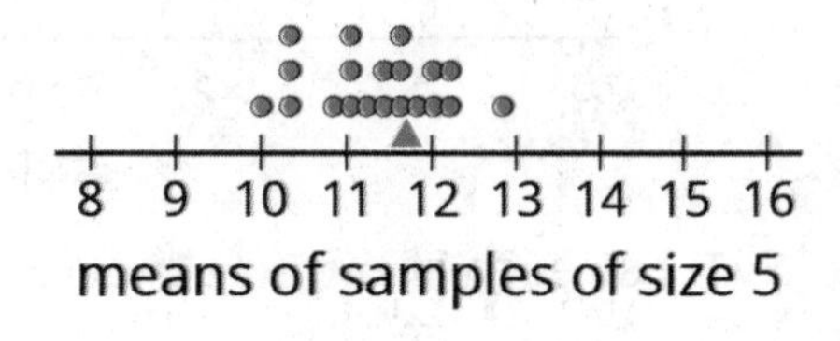


This dot plot shows the means of 20 samples of 10 cookies, selected at random. Notice that the means for these samples are even closer to the mean for the entire population.

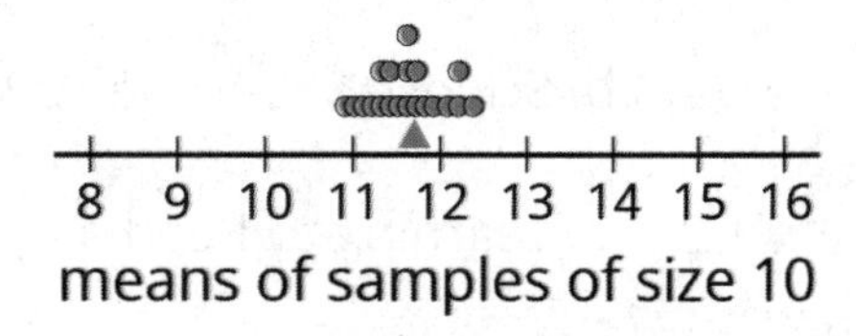


In general, as the sample size gets bigger, the mean of a sample is more likely to be closer to the mean of the population.

## Lesson 12 Practice Problems

1. One thousand baseball fans were asked how far they would be willing to travel to watch a professional baseball game. From this population, 100 different samples of size 40 were selected. Here is a dot plot showing the mean of each sample.

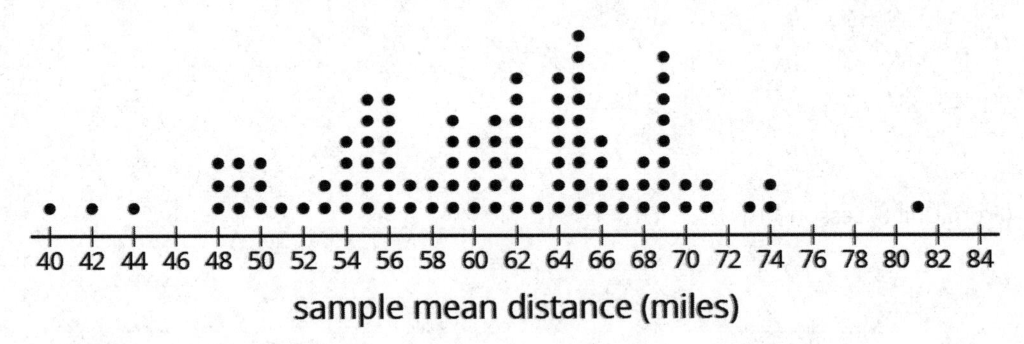


   Based on the distribution of sample means, what do you think is a reasonable estimate for the mean of the population?

2. Last night, everyone at the school music concert wrote their age on a slip of paper and placed it in a box. Today, each of the students in a math class selected a random sample of size 10 from the box of papers. Here is a dot plot showing their sample means, rounded to the nearest year.

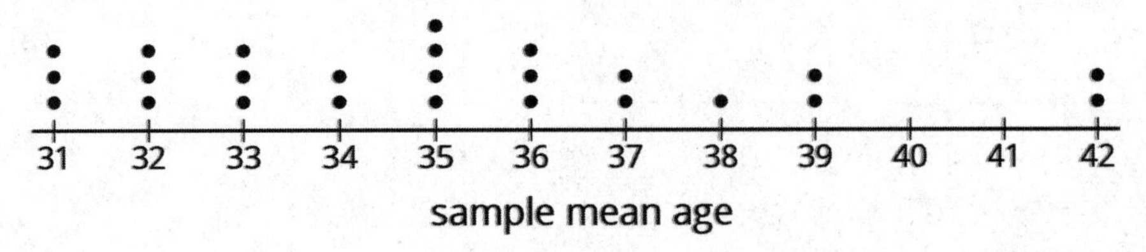


   a. Does the number of dots on the dot plot tell you how many people were at the concert or how many students are in the math class?

   b. The mean age for the population was 35 years. If Elena picks a new sample of size 10 from this population, should she expect her sample mean to be within 1 year of the population mean? Explain your reasoning.

   c. What could Elena do to select a random sample that is more likely to have a sample mean within 1 year of the population mean?

3. Andre would like to estimate the mean number of books the students at his school read over the summer break. He has a list of the names of all the students at the school, but he doesn't have time to ask every student how many books they read.

   What should Andre do to estimate the mean number of books?

   (From Unit 8, Lesson 11.)

# Lesson 13: What Are Probabilities?

Let's find out what's possible.

## 13.1: Which Game Would You Choose?

Which game would you choose to play? Explain your reasoning.

Game 1: You flip a coin and win if it lands showing heads.

Game 2: You roll a standard number cube and win if it lands showing a number that is divisible by 3.

## 13.2: What's Possible?

1. For each situation, list the **sample space** and tell how many outcomes there are.

    a. Han rolls a standard number cube once.

    b. Clare spins this spinner once.

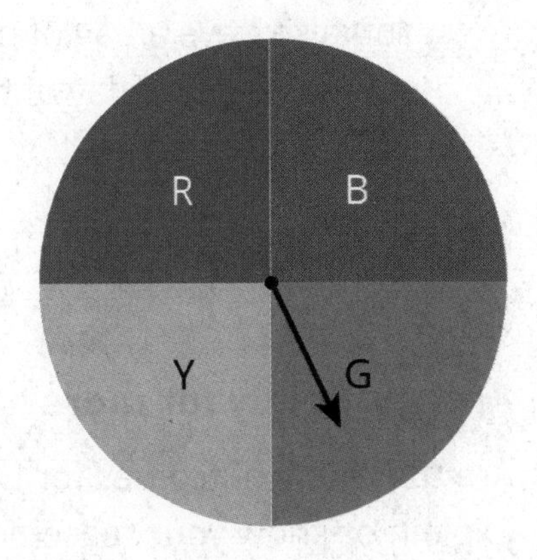


    c. Kiran selects a letter at **random** from the word "MATH."

d. Mai selects a letter at random from the alphabet.

e. Noah picks a card at random from a stack that has cards numbered 5 through 20.

2. Next, compare the likelihood of these outcomes. Be prepared to explain your reasoning.

    a. Is Clare more likely to have the spinner stop on the red or blue section?

    b. Is Kiran or Mai more likely to get the letter T?

    c. Is Han or Noah more likely to get a number that is greater than 5?

3. Suppose you have a spinner that is evenly divided showing all the days of the week. You also have a bag of papers that list the months of the year. Are you more likely to spin the current day of the week or pull out the paper with the current month?

## Are you ready for more?

Are there any outcomes for two people in this activity that have the same likelihood? Explain or show your reasoning.

## 13.3: What's in the Bag?

Your teacher will give your group a bag of paper slips with something printed on them. Repeat these steps until everyone in your group has had a turn.

- As a group, guess what is printed on the papers in the bag and record your guess in the table.
- Without looking in the bag, one person takes out one of the papers and shows it to the group.
- Everyone in the group records what is printed on the paper.
- The person who took out the paper puts it back into the bag, shakes the bag to mix up the papers, and passes the bag to the next person in the group.

| | Guess the sample space. | What is printed on the paper? |
|---|---|---|
| person 1 | | |
| person 2 | | |
| person 3 | | |
| person 4 | | |

1. How was guessing the sample space the fourth time different from the first?

2. What could you do to get a better guess of the sample space?

3. Look at all the papers in the bag. Were any of your guesses correct?

4. Are all of the possible outcomes equally likely? Explain.

5. Use the sample space to determine the **probability** that a fifth person would get the same outcome as person 1.

## Lesson 13 Summary

The **probability** of an event is a measure of the likelihood that the event will occur. Probabilities are expressed using numbers from 0 to 1.

- If the probability is 0, that means the event is impossible. For example, when you flip a coin, the probability that it will turn into a bottle of ketchup is 0. The closer the probability of some event is to 0, the less likely it is.
- If the probability is 1, that means the event is certain. For example, when you flip a coin, the probability that it will land somewhere is 1. The closer the probability of some event is to 1, the more likely it is.

If we list all of the possible outcomes for a chance experiment, we get the **sample space** for that experiment. For example, the sample space for rolling a standard number cube includes six outcomes: 1, 2, 3, 4, 5, and 6. The probability that the number cube will land showing the number 4 is $\frac{1}{6}$. In general, if all outcomes in an experiment are equally likely and there are $n$ possible outcomes, then the probability of a single outcome is $\frac{1}{n}$.

Sometimes we have a set of possible outcomes and we want one of them to be selected at **random**. That means that we want to select an outcome in a way that each of the outcomes is *equally likely*. For example, if two people both want to read the same book, we could flip a coin to see who gets to read the book first.

## Glossary

- probability
- random
- sample space

## Lesson 13 Practice Problems

1. List the *sample space* for each chance experiment.

    a. Flipping a coin

    b. Selecting a random season of the year

    c. Selecting a random day of the week

2. A computer randomly selects a letter from the alphabet.

    a. How many different outcomes are in the sample space?

    b. What is the probability the computer produces the first letter of your first name?

3. What is the probability of selecting a random month of the year and getting a month that starts with the letter "J?" If you get stuck, consider listing the sample space.

4. $E$ represents an object's weight on Earth and $M$ represents that same object's weight on the Moon. The equation $M = \frac{1}{6}E$ represents the relationship between these quantities.

   a. What does the $\frac{1}{6}$ represent in this situation?

   b. Give an example of what a person might weigh on Earth and on the Moon.

(From Unit 5, Lesson 1.)

# Lesson 14: Estimating Probabilities Through Repeated Experiments

Let's do some experimenting.

## 14.1: Decimals on the Number Line

1. Locate and label these numbers on the number line.

2. Choose one of the numbers from the previous question. Describe a game in which that number represents your probability of winning.

## 14.2: In the Long Run

Mai plays a game in which she only wins if she rolls a 1 or a 2 with a standard number cube.

1. List the outcomes in the sample space for rolling the number cube.

2. What is the probability Mai will win the game? Explain your reasoning.

3. If Mai is given the option to flip a coin and win if it comes up heads, is that a better option for her to win?

4. With your group, follow these instructions 10 times to create the graph.

    - One person rolls the number cube. Everyone records the outcome.
    - Calculate the fraction of rolls that are a win for Mai so far. Approximate the fraction with a decimal value rounded to the hundredths place. Record both the fraction and the decimal in the last column of the table.
    - On the graph, plot the number of rolls and the fraction that were wins.
    - Pass the number cube to the next person in the group.

| roll | outcome | total number of wins for Mai | fraction of games played that are wins |
|---|---|---|---|
| 1 | | | |
| 2 | | | |
| 3 | | | |
| 4 | | | |
| 5 | | | |
| 6 | | | |
| 7 | | | |
| 8 | | | |
| 9 | | | |
| 10 | | | |

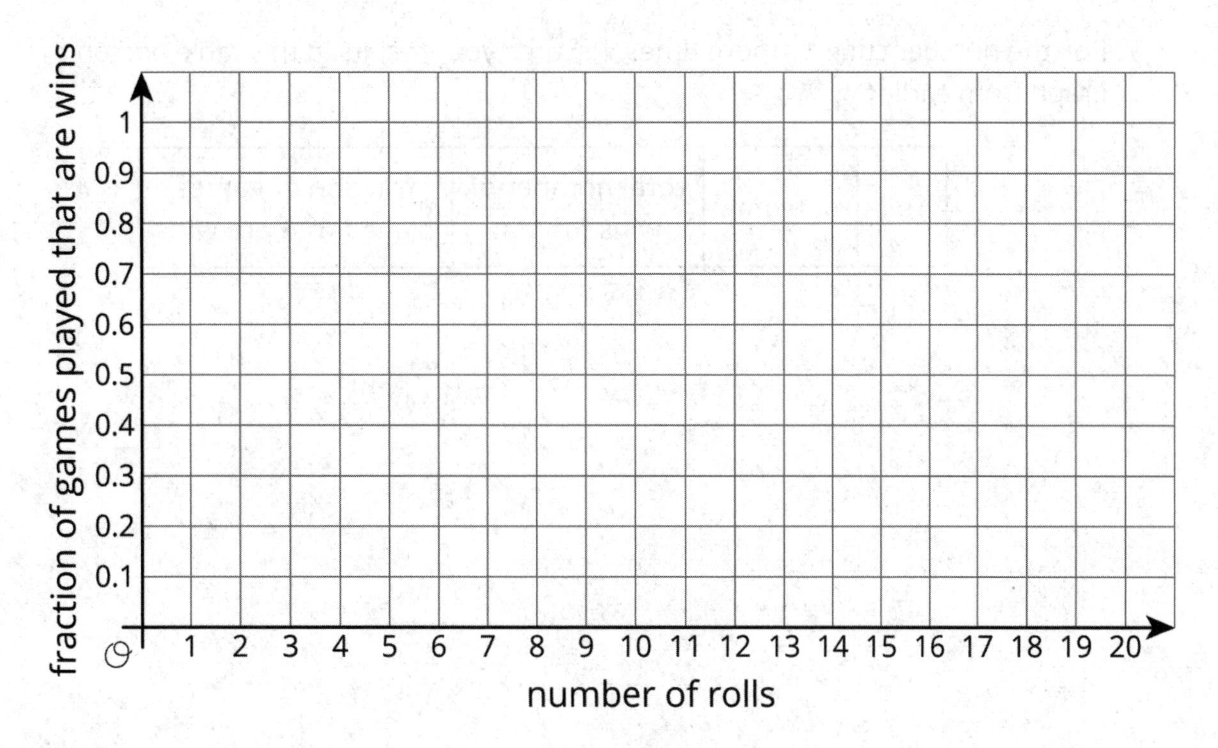


5. What appears to be happening with the points on the graph?

6. a. After 10 rolls, what fraction of the total rolls were a win?

b. How close is this fraction to the probability that Mai will win?

7. Roll the number cube 10 more times. Records your results in this table and on the graph from earlier.

| roll | outcome | total number of wins for Mai | fraction of games played that are wins |
|---|---|---|---|
| 11 | | | |
| 12 | | | |
| 13 | | | |
| 14 | | | |
| 15 | | | |
| 16 | | | |
| 17 | | | |
| 18 | | | |
| 19 | | | |
| 20 | | | |

8. a. After 20 rolls, what fraction of the total rolls were a win?

   b. How close is this fraction to the probability that Mai will win?

## 14.3: Due For a Win

1. For each situation, do you think the result is surprising or not? Is it possible? Be prepared to explain your reasoning.

    a. You flip the coin once, and it lands heads up.

    b. You flip the coin twice, and it lands heads up both times.

    c. You flip the coin 100 times, and it lands heads up all 100 times.

2. If you flip the coin 100 times, how many times would you expect the coin to land heads up? Explain your reasoning.

3. If you flip the coin 100 times, what are some other results that would not be surprising?

4. You've flipped the coin 3 times, and it has come up heads once. The cumulative fraction of heads is currently $\frac{1}{3}$. If you flip the coin one more time, will it land heads up to make the cumulative fraction $\frac{2}{4}$?

### Lesson 14 Summary

A probabilityfor an event represents the proportion of the time we expect that event to occur in the long run. For example, the probability of a coin landing heads up after a flip is $\frac{1}{2}$, which means that if we flip a coin many times, we expect that it will land heads up about half of the time.

Even though the probability tells us what we should expect if we flip a coin many times, that doesn't mean we are more likely to get heads if we just got three tails in a row. The chances of getting heads are the same every time we flip the coin, no matter what the outcome was for past flips.

# Lesson 14 Practice Problems

1. A carnival game has 160 rubber ducks floating in a pool. The person playing the game takes out one duck and looks at it.

    - If there's a red mark on the bottom of the duck, the person wins a small prize.
    - If there's a blue mark on the bottom of the duck, the person wins a large prize.
    - Many ducks do not have a mark.

    After 50 people have played the game, only 3 of them have won a small prize, and none of them have won a large prize.

    Estimate the number of the 160 ducks that you think have red marks on the bottom. Then estimate the number of ducks you think have blue marks. Explain your reasoning.

2. Lin wants to know if flipping a quarter really does have a probability of $\frac{1}{2}$ of landing heads up, so she flips a quarter 10 times. It lands heads up 3 times and tails up 7 times. Has she proven that the probability is not $\frac{1}{2}$? Explain your reasoning.

3. A spinner has four equal sections, with one letter from the word "MATH" in each section.

    a. You spin the spinner 20 times. About how many times do you expect it will land on A?

    b. You spin the spinner 80 times. About how many times do you expect it will land on something other than A? Explain your reasoning.

4. A spinner is spun 40 times for a game. Here is a graph showing the fraction of games that are wins under some conditions.

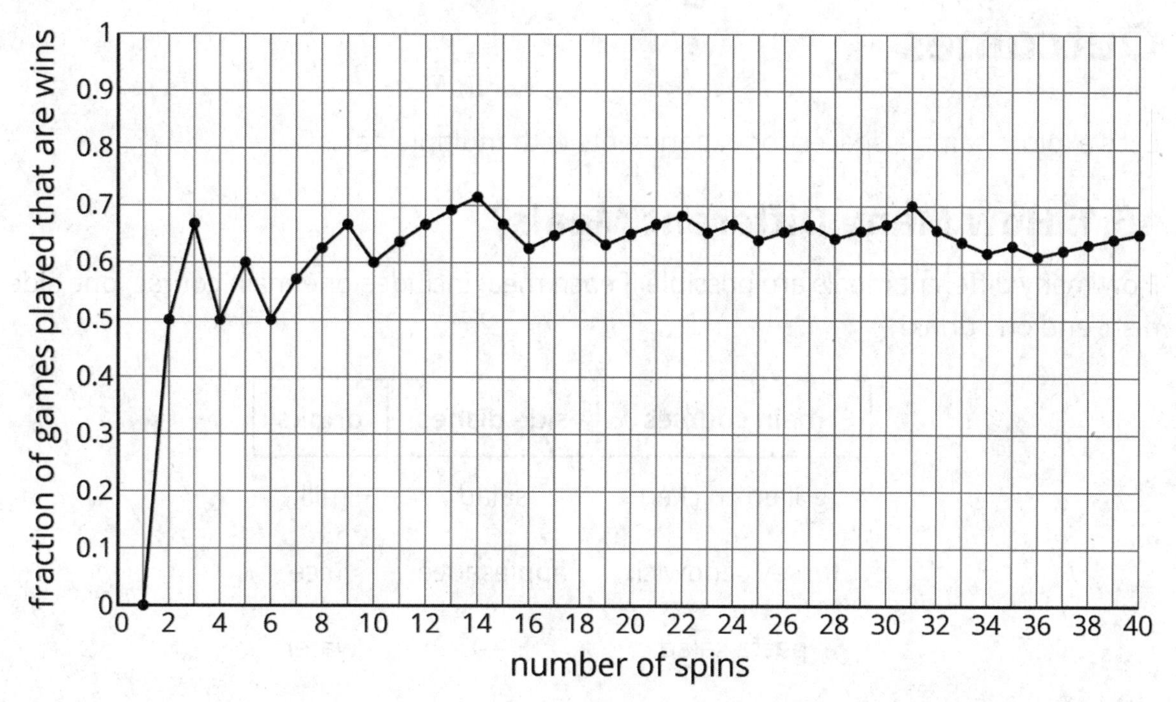


Estimate the probability of a spin winning this game based on the graph.

5. Which event is more likely: rolling a standard number cube and getting an even number, or flipping a coin and having it land heads up?

This problem is from an earlier lesson

6. Noah will select a letter at random from the word "FLUTE." Lin will select a letter at random from the word "CLARINET."

Which person is more likely to pick the letter "E?" Explain your reasoning.

(From Unit 8, Lesson 13.)

# Lesson 15: Keeping Track of All Possible Outcomes

Let's explore sample spaces for experiments with multiple parts.

## 15.1: How Many Different Meals?

How many different meals are possible if each meal includes one main course, one side dish, and one drink?

| main courses | side dishes | drinks |
|---|---|---|
| grilled chicken | salad | milk |
| turkey sandwich | applesauce | juice |
| pasta salad | — | water |

## 15.2: Lists, Tables, and Trees

Consider the experiment: Flip a coin, and then roll a number cube.

Elena, Kiran, and Priya each use a different method for finding the sample space of this experiment.

- Elena carefully writes a list of all the options: Heads 1, Heads 2, Heads 3, Heads 4, Heads 5, Heads 6, Tails 1, Tails 2, Tails 3, Tails 4, Tails 5, Tails 6.
- Kiran makes a table:

| | 1 | 2 | 3 | 4 | 5 | 6 |
|---|---|---|---|---|---|---|
| H | H1 | H2 | H3 | H4 | H5 | H6 |
| T | T1 | T2 | T3 | T4 | T5 | T6 |

- Priya draws a tree with branches in which each pathway represents a different outcome:

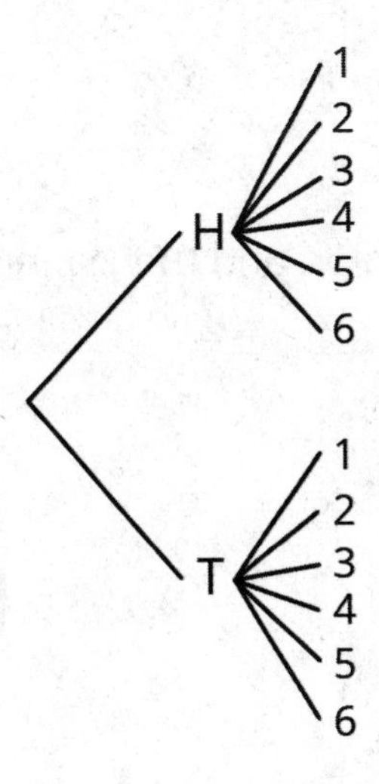


1. Compare the three methods. What is the same about each method? What is different? Be prepared to explain why each method produces all the different outcomes without repeating any.

2. Which method do you prefer for this situation?

   Pause here so your teacher can review your work.

3. Find the sample space for each of these experiments using any method. Make sure you list every possible outcome without repeating any.

   a. Flip a dime, then flip a nickel, and then flip a penny. Record whether each lands heads or tails up.

   b. Han's closet has: a blue shirt, a gray shirt, a white shirt, blue pants, khaki pants, and black pants. He must select one shirt and one pair of pants to wear for the day.

   c. Spin a color, and then spin a number.

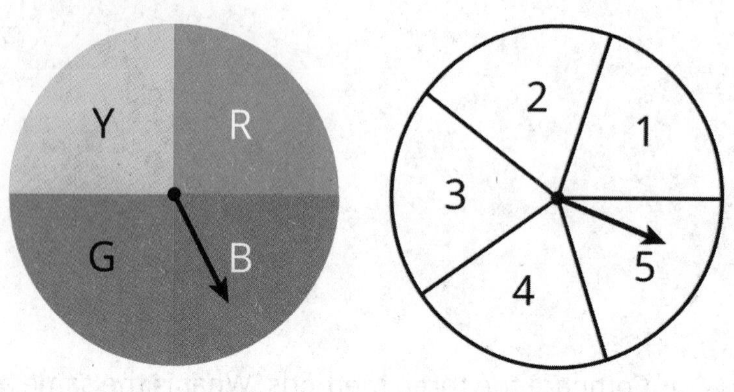


   d. Spin the hour hand on an analog clock, and then choose a.m. or p.m.

## 15.3: How Many Sandwiches?

1. A submarine sandwich shop makes sandwiches with one kind of bread, one protein, one choice of cheese, and *two* vegetables. How many different sandwiches are possible? Explain your reasoning. You do not need to write out the sample space.

    - Breads: Italian, white, wheat
    - Proteins: Tuna, ham, turkey, beans
    - Cheese: Provolone, Swiss, American, none
    - Vegetables: Lettuce, tomatoes, peppers, onions, pickles

2. Andre knows he wants a sandwich that has ham, lettuce, and tomatoes on it. He doesn't care about the type of bread or cheese. How many of the different sandwiches would make Andre happy?

3. If a sandwich is made by randomly choosing each of the options, what is the probability it will be a sandwich that Andre would be happy with?

### Are you ready for more?

Describe a situation that involves three parts and has a total of 24 outcomes in the sample space.

## Lesson 15 Summary

Sometimes we need a systematic way to count the number of outcomes that are possible in a given situation. For example, suppose there are 3 people (A, B, and C) who want to run for the president of a club and 4 different people (1, 2, 3, and 4) who want to run for vice president of the club. We can use a *tree*, a *table*, or an *ordered list* to count how many different combinations are possible for a president to be paired with a vice president.

With a tree, we can start with a branch for each of the people who want to be president. Then for each possible president, we add a branch for each possible vice president, for a total of $3 \cdot 4 = 12$ possible pairs. We can also start by counting vice presidents first and then adding a branch for each possible president, for a total of $3 \cdot 4 = 12$ possible pairs.

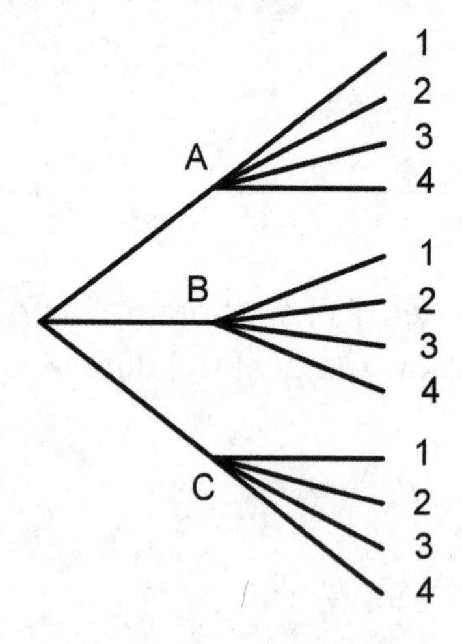


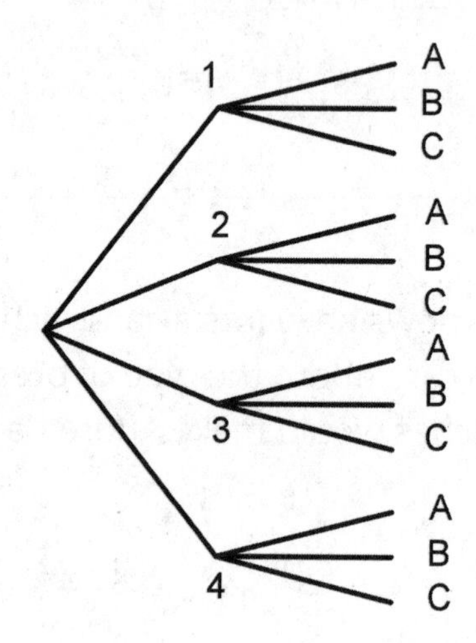


A table can show the same result:

| | 1 | 2 | 3 | 4 |
|---|---|---|---|---|
| A | A, 1 | A, 2 | A, 3 | A, 4 |
| B | B, 1 | B, 2 | B, 3 | B, 4 |
| C | C, 1 | C, 2 | C, 3 | C, 4 |

So does this ordered list:

A1, A2, A3, A4, B1, B2, B3, B4, C1, C2, C3, C4

## Lesson 15 Practice Problems

1. Noah is planning his birthday party. Here is a tree showing all of the possible themes, locations, and days of the week that Noah is considering.

    a. How many themes is Noah considering?

    b. How many locations is Noah considering?

    c. How many days of the week is Noah considering?

    d. One possibility that Noah is considering is a party with a space theme at the skating rink on Sunday. Write two other possible parties Noah is considering.

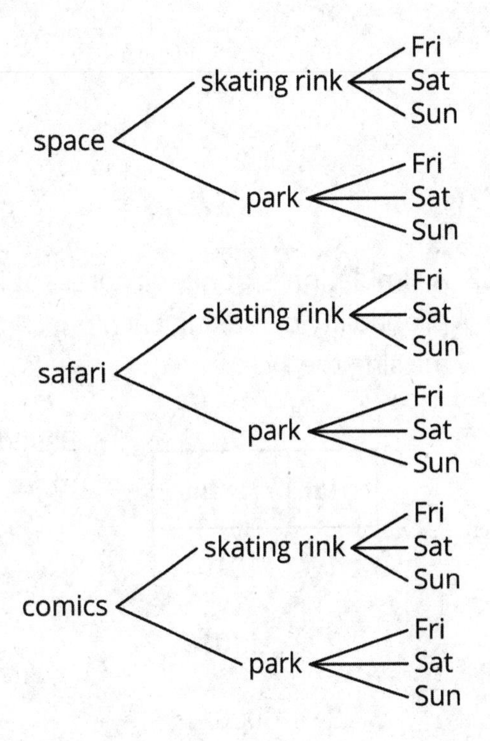


    e. How many different possible outcomes are in the sample space?

2. For each event, write the sample space and tell how many outcomes there are.

    a. Lin selects one type of lettuce and one dressing to make a salad.

        Lettuce types: iceberg, romaine
        Dressings: ranch, Italian, French

    b. Diego chooses rock, paper, or scissors, and Jada chooses rock, paper, or scissors.

c. Spin these 3 spinners.

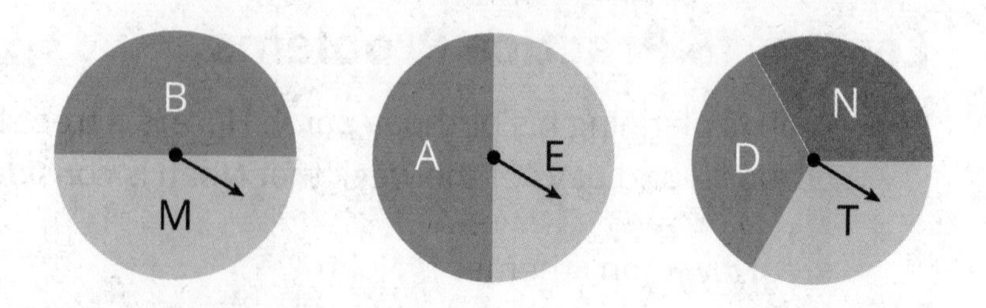


3. A simulation is done to represent kicking 5 field goals in a single game with a 72% probability of making each one. A 1 represents making the kick and a 0 represents missing the kick.

| trial | result |
|---|---|
| 1 | 10101 |
| 2 | 11010 |
| 3 | 00011 |
| 4 | 11111 |
| 5 | 10011 |

Based on these results, estimate the probability that 3 or more kicks are made.

This problem is from an earlier lesson

4. There is a bag of 50 marbles.

   - Andre takes out a marble, records its color, and puts it back in. In 4 trials, he gets a green marble 1 time.
   - Jada takes out a marble, records its color, and puts it back in. In 12 trials, she gets a green marble 5 times.
   - Noah takes out a marble, records its color, and puts it back in. In 9 trials, he gets a green marble 3 times.

   Estimate the probability of getting a green marble from this bag. Explain your reasoning.

   (From Unit 8, Lesson 14.)

# Lesson 16: Multi-step Experiments

Let's look at probabilities of experiments that have multiple steps.

## 16.1: True or False?

Is each equation true or false? Explain your reasoning.

$8 = (8 + 8 + 8 + 8) \div 3$

$(10 + 10 + 10 + 10 + 10) \div 5 = 10$

$(6 + 4 + 6 + 4 + 6 + 4) \div 6 = 5$

## 16.2: Spinning a Color and Number

The other day, you wrote the sample space for spinning each of these spinners once.

What is the probability of getting:

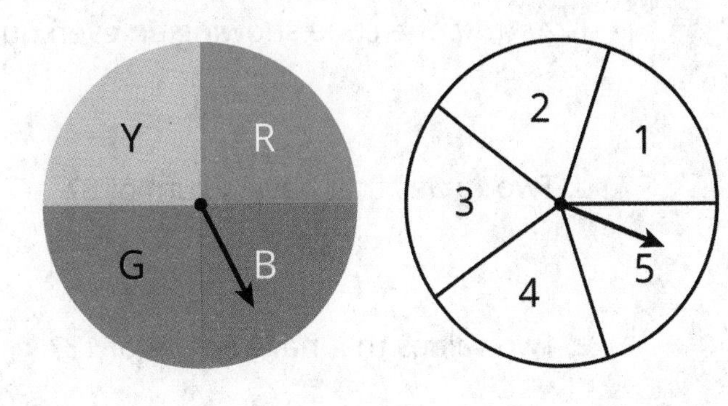


1. Green and 3?

2. Blue and any odd number?

3. Any color other than red and any number other than 2?

## 16.3: Cubes and Coins

The other day you looked at a list, a table, and a tree that showed the sample space for rolling a number cube and flipping a coin.

1. Your teacher will assign you one of these three structures to use to answer these questions. Be prepared to explain your reasoning.

   a. What is the probability of getting tails and a 6?

   b. What is the probability of getting heads and an odd number?

   Pause here so your teacher can review your work.

2. Suppose you roll two number cubes. What is the probability of getting:

   a. Both cubes showing the same number?

   b. *Exactly* one cube showing an even number?

   c. *At least* one cube showing an even number?

   d. Two values that have a sum of 8?

   e. Two values that have a sum of 13?

3. Jada flips three quarters. What is the probability that all three will land showing the same side?

## 16.4: Pick a Card

Imagine there are 5 cards. They are colored red, yellow, green, white, and black. You mix up the cards and select one of them without looking. Then, without putting that card back, you mix up the remaining cards and select another one.

1. Write the sample space and tell how many possible outcomes there are.

2. What structure did you use to write all of the outcomes (list, table, tree, something else)? Explain why you chose that structure.

3. What is the probability that:

    a. You get a white card and a red card (in either order)?

    b. You get a black card (either time)?

    c. You do not get a black card (either time)?

    d. You get a blue card?

    e. You get 2 cards of the same color?

    f. You get 2 cards of different colors?

## Are you ready for more?

In a game using five cards numbered 1, 2, 3, 4, and 5, you take two cards and add the values together. If the sum is 8, you win. Would you rather pick a card and put it back before picking the second card, or keep the card in your hand while you pick the second card? Explain your reasoning.

## Lesson 16 Summary

Suppose we have two bags. One contains 1 star block and 4 moon blocks. The other contains 3 star blocks and 1 moon block.

If we select one block at random from each, what is the probability that we will get two star blocks or two moon blocks?

To answer this question, we can draw a tree diagram to see all of the possible outcomes.

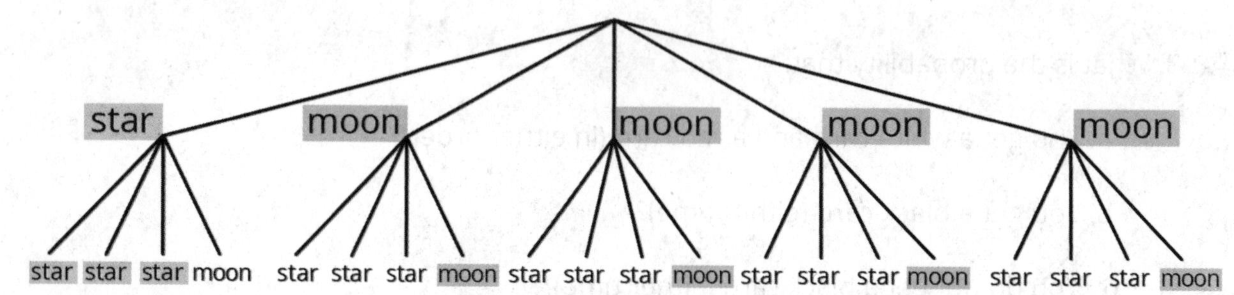


There are $5 \cdot 4 = 20$ possible outcomes. Of these, 3 of them are both stars, and 4 are both moons. So the probability of getting 2 star blocks or 2 moon blocks is $\frac{7}{20}$.

In general, if all outcomes in an experiment are equally likely, then the probability of an event is the fraction of outcomes in the sample space for which the event occurs.

## Lesson 16 Practice Problems

1. A vending machine has 5 colors (white, red, green, blue, and yellow) of gumballs and an equal chance of dispensing each. A second machine has 4 different animal-shaped rubber bands (lion, elephant, horse, and alligator) and an equal chance of dispensing each. If you buy one item from each machine, what is the probability of getting a yellow gumball and a lion band?

2. The numbers 1 through 10 are put in one bag. The numbers 5 through 14 are put in another bag. When you pick one number from each bag, what is the probability you get the same number?

3. When rolling 3 standard number cubes, the probability of getting all three numbers to match is $\frac{6}{216}$. What is the probability that the three numbers *do not* all match? Explain your reasoning.

4. For each event, write the sample space and tell how many outcomes there are.

a. Roll a standard number cube. Then flip a quarter.

b. Select a month. Then select 2020 or 2025.

(From Unit 8, Lesson 15.)

5. On a graph of the area of a square vs. its perimeter, a few points are plotted.

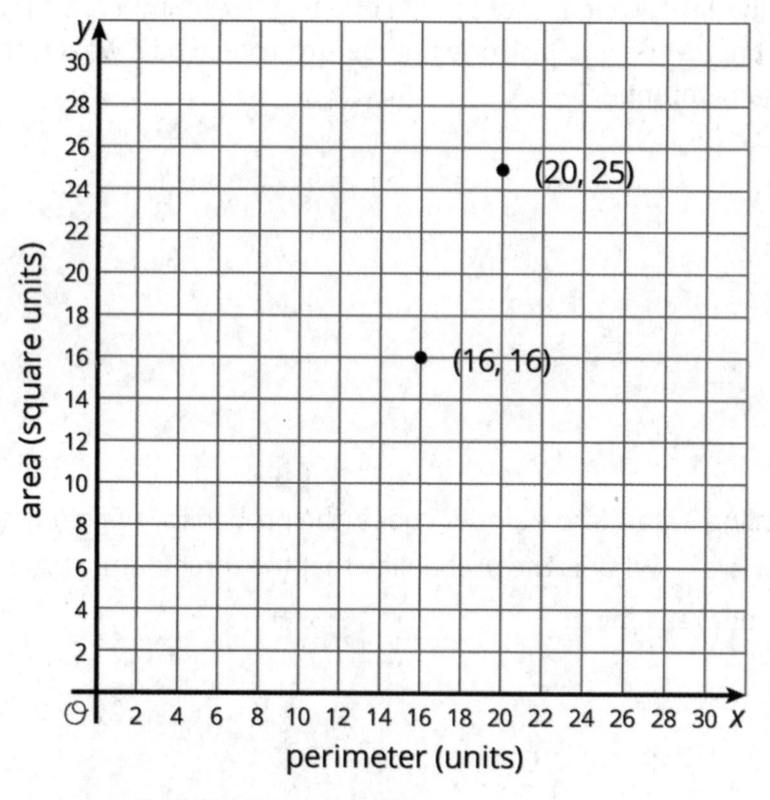


a. Add some more ordered pairs to the graph.

b. Is there a proportional relationship between the area and perimeter of a square? Explain how you know.

(From Unit 5, Lesson 7.)

# Lesson 17: Designing Simulations

Let's simulate some real-life scenarios.

## 17.1: Number Talk: Division

Find the value of each expression mentally.

$(4.2 + 3) \div 2$

$(4.2 + 2.6 + 4) \div 3$

$(4.2 + 2.6 + 4 + 3.6) \div 4$

$(4.2 + 2.6 + 4 + 3.6 + 3.6) \div 5$

## 17.2: Breeding Mice

A scientist is studying the genes that determine the color of a mouse's fur. When two mice with brown fur breed, there is a 25% chance that each baby will have white fur. For the experiment to continue, the scientist needs at least 2 out of 5 baby mice to have white fur.

To simulate this situation, you can flip a coin twice for each baby mouse.

- If the coin lands heads up both times, it represents a mouse with white fur.
- Any other result represents a mouse with brown fur.

1. Simulate 3 litters of 5 baby mice and record your results in the table.

| | mouse 1 | mouse 2 | mouse 3 | mouse 4 | mouse 5 | Do at least 2 have white fur? |
|---|---|---|---|---|---|---|
| simulation 1 | | | | | | |
| simulation 2 | | | | | | |
| simulation 3 | | | | | | |

2. Based on the results from everyone in your group, estimate the probability that the scientist's experiment will be able to continue.

3. How could you improve your estimate?

**Are you ready for more?**

For a certain pair of mice, the genetics show that each offspring has a probability of $\frac{1}{16}$ that they will be albino. Describe a simulation you could use that would estimate the probability that at least 2 of the 5 offspring are albino.

## 17.3: Designing Simulations

Your teacher will give your group a paper describing a situation.

1. Design a simulation that you could use to estimate a probability. Show your thinking. Organize it so it can be followed by others.

2. Explain how you used the simulation to answer the questions posed in the situation.

## Lesson 17 Summary

Many real-world situations are difficult to repeat enough times to get an estimate for a probability. If we can find probabilities for parts of the situation, we may be able to simulate the situation using a process that is easier to repeat.

For example, if we know that each egg of a fish in a science experiment has a 13% chance of having a mutation, how many eggs do we need to collect to make sure we have 10 mutated eggs? If getting these eggs is difficult or expensive, it might be helpful to have an idea about how many eggs we need before trying to collect them.

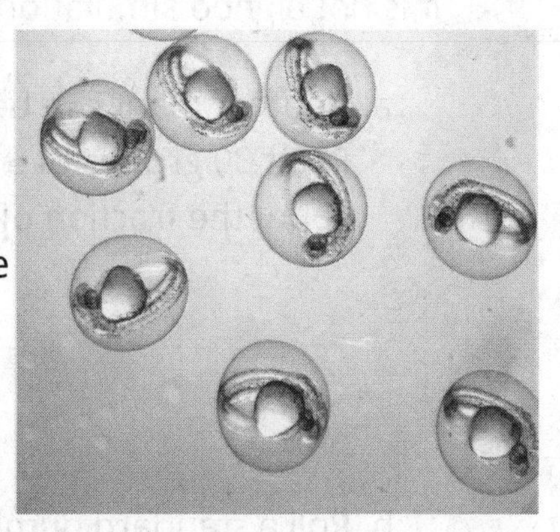

We could simulate this situation by having a computer select random numbers between 1 and 100. If the number is between 1 and 13, it counts as a mutated egg. Any other number would represent a normal egg. This matches the 13% chance of each fish egg having a mutation.

We could continue asking the computer for random numbers until we get 10 numbers that are between 1 and 13. How many times we asked the computer for a random number would give us an estimate of the number of fish eggs we would need to collect.

To improve the estimate, this entire process should be repeated many times. Because computers can perform simulations quickly, we could simulate the situation 1,000 times or more.

## Lesson 17 Practice Problems

1. A rare and delicate plant will only produce flowers from 10% of the seeds planted. To see if it is worth planting 5 seeds to see any flowers, the situation is going to be simulated. Which of these options is the best simulation? For the others, explain why it is not a good simulation.

   a. Another plant can be genetically modified to produce flowers 10% of the time. Plant 30 groups of 5 seeds each and wait 6 months for the plants to grow and count the fraction of groups that produce flowers.

   b. Roll a standard number cube 5 times. Each time a 6 appears, it represents a plant producing flowers. Repeat this process 30 times and count the fraction of times at least one number 6 appears.

   c. Have a computer produce 5 random digits (0 through 9). If a 9 appears in the list of digits, it represents a plant producing flowers. Repeat this process 300 times and count the fraction of times at least one number 9 appears.

   d. Create a spinner with 10 equal sections and mark one of them "flowers." Spin the spinner 5 times to represent the 5 seeds. Repeat this process 30 times and count the fraction of times that at least 1 "flower" was spun.

2. Jada and Elena learned that 8% of students have asthma. They want to know the probability that in a team of 4 students, at least one of them has asthma. To simulate this, they put 25 slips of paper in a bag. Two of the slips say "asthma." Next, they take four papers out of the bag and record whether at least one of them says "asthma." They repeat this process 15 times.

    - Jada says they could improve the accuracy of their simulation by using 100 slips of paper and marking 8 of them.
    - Elena says they could improve the accuracy of their simulation by conducting 30 trials instead of 15.

    a. Do you agree with either of them? Explain your reasoning.

    b. Describe another method of simulating the same scenario.

3. Match each expression in the first list with an equivalent expression from the second list.

A. $(8x + 6y) + (2x + 4y)$

B. $(8x + 6y) - (2x + 4y)$

C. $(8x + 6y) - (2x - 4y)$

D. $8x - 6y - 2x + 4y$

E. $8x - 6y + 2x - 4y$

F. $8x - (-6y - 2x + 4y)$

1. $10(x + y)$

2. $10(x - y)$

3. $6(x - \frac{1}{3}y)$

4. $8x + 6y + 2x - 4y$

5. $8x + 6y - 2x + 4y$

6. $8x - 2x + 6y - 4y$

(From Unit 4, Lesson 11.)

# Learning Targets

### Lesson 1: Representing Data Graphically

- I can describe the information presented in tables, dot plots, and bar graphs.
- I can use tables, dot plots, and bar graphs to represent distributions of data.

### Lesson 2: Using Dot Plots to Answer Statistical Questions

- I can use a dot plot to represent the distribution of a data set and answer questions about the real-world situation.
- I can use center and spread to describe data sets, including what is typical in a data set.

### Lesson 3: Interpreting Histograms

- I can use a histogram to describe the distribution of data and determine a typical value for the data.
- I can use a histogram to get information about the distribution of data and explain what it means in a real-world situation.

### Lesson 4: The Mean

- I can describe what the mean tells us in the context of the data.
- I can find the mean for a numerical data set.

### Lesson 5: Variability and MAD

- I can use means and MADs to compare groups.
- I know what the mean absolute deviation (MAD) measures and what information it provides.

### Lesson 6: The Median

- I can determine when the mean or the median is more appropriate to describe the center of data.
- I can find the median for a set of data.

### Lesson 7: Box Plots and Interquartile Range

- I can use IQR to describe the spread of data.
- I know what information a box plot shows and how it is constructed.

### Lesson 8: Larger Populations

- I can explain why it may be useful to gather data on a sample of a population.
- When I read or hear a statistical question, I can name the population of interest and give an example of a sample for that population.

### Lesson 9: What Makes a Good Sample?

- I can determine whether a sample is representative of a population by considering the shape, center, and spread of each of them.
- I know that some samples may represent the population better than others.
- I remember that when a distribution is not symmetric, the median is a better estimate of a typical value than the mean.

### Lesson 10: Sampling in a Fair Way

- I can describe ways to get a random sample from a population.
- I know that selecting a sample at random is usually a good way to get a representative sample.

### Lesson 11: Estimating Population Measures of Center

- I can consider the variability of a sample to get an idea for how accurate my estimate is.
- I can estimate the mean or median of a population based on a sample of the population.

### Lesson 12: More about Sampling Variability

- I can use the means from many samples to judge how accurate an estimate for the population mean is.
- I know that as the sample size gets bigger, the sample mean is more likely to be close to the population mean.

### Lesson 13: What Are Probabilities?

- I can use the sample space to calculate the probability of an event when all outcomes are equally likely.
- I can write out the sample space for a simple chance experiment.

### Lesson 14: Estimating Probabilities Through Repeated Experiments

- I can estimate the probability of an event based on the results from repeating an experiment.
- I can explain whether certain results from repeated experiments would be surprising or not.

### Lesson 15: Keeping Track of All Possible Outcomes

- I can write out the sample space for a multi-step experiment, using a list, table, or tree diagram.

### Lesson 16: Multi-step Experiments

- I can use the sample space to calculate the probability of an event in a multi-step experiment.

### Lesson 17: Designing Simulations

- I can design a simulation to estimate the probability of a multi-step real-world situation.

# Illustrative Mathematics

## Unit 9

STUDENT EDITION

Book 2

# Lesson 1: Fermi Problems

Let's make some estimates.

## 1.1: Ant Trek

How long would it take an ant to run from Los Angeles to New York City?

## 1.2: Stacks and Stacks of Cereal Boxes

Imagine a warehouse that has a rectangular floor and that contains all of the boxes of breakfast cereal bought in the United States in one year.

If the warehouse is 10 feet tall, what could the side lengths of the floor be?

## 1.3: Covering the Washington Monument

How many tiles would it take to cover the Washington Monument?

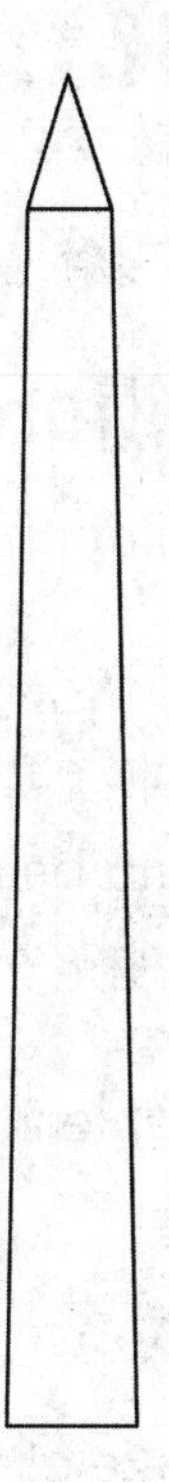

# Lesson 2: If Our Class Were the World

Let's use math to better understand our world.

## 2.1: All 7.4 Billion of Us

There are 7.4 billion people in the world. If the whole world were represented by a 30-person class:

- 14 people would eat rice as their main food.
- 12 people would be under the age of 20.
- 5 people would be from Africa.

1. How many people in the class would *not* eat rice as their main food?

2. What percentage of the people in the class would be under the age of 20?

3. Based on the number of people in the class representing people from Africa, how many people live in Africa?

## 2.2: About the People in the World

With the members of your group, write a list of questions about the people in the world. Your questions should begin with "How many people in the world. . ." Then, choose several questions on the list that you find most interesting.

## 2.3: If Our Class Were the World

Suppose your class represents all the people in the world.

Choose several characteristics about the world's population that you have investigated. Find the number of students in *your* class that would have the same characteristics.

Create a visual display that includes a diagram that represents this information. Give your display the title "If Our Class Were the World."

# Lesson 3: Rectangle Madness

Let's cut up rectangles.

## 3.1: Squares in Rectangles

1. Rectangle $ABCD$ is not a square. Rectangle $ABEF$ is a square.

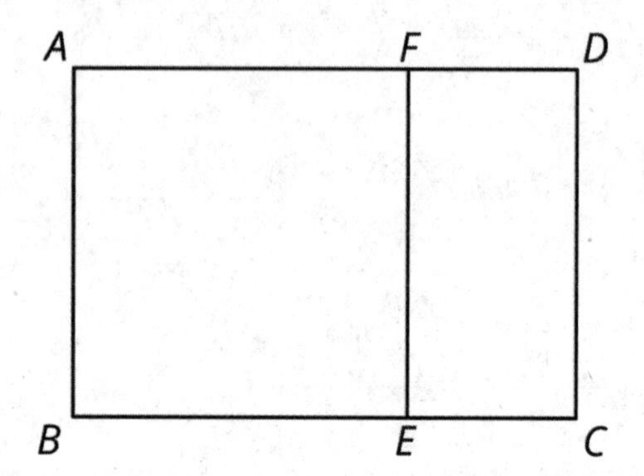


   a. Suppose segment $AF$ were 5 units long and segment $FD$ were 2 units long. How long would segment $AD$ be?

   b. Suppose segment $BC$ were 10 units long and segment $BE$ were 6 units long. How long would segment $EC$ be?

   c. Suppose segment $AF$ were 12 units long and segment $FD$ were 5 units long. How long would segment $FE$ be?

   d. Suppose segment $AD$ were 9 units long and segment $AB$ were 5 units long. How long would segment $FD$ be?

2. Rectangle $JKXW$ has been decomposed into squares.

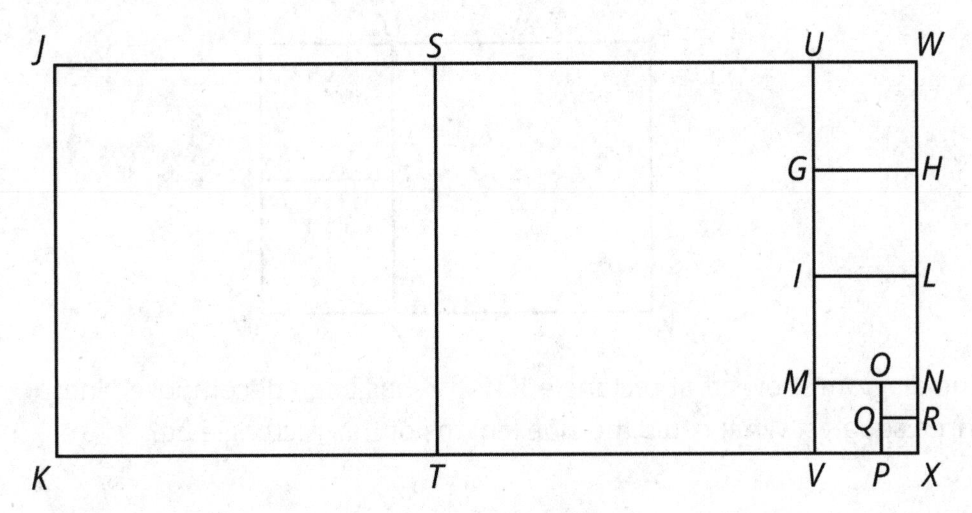


Segment $JK$ is 33 units long and segment $JW$ is 75 units long. Find the areas of all of the squares in the diagram.

3. Rectangle $ABCD$ is 16 units by 5 units.

a. In the diagram, draw a line segment that decomposes $ABCD$ into two regions: a square that is the largest possible and a new rectangle.

b. Draw another line segment that decomposes the *new* rectangle into two regions: a square that is the largest possible and another new rectangle.

c. Keep going until rectangle $ABCD$ is entirely decomposed into squares.

d. List the side lengths of all the squares in your diagram.

**Are you ready for more?**

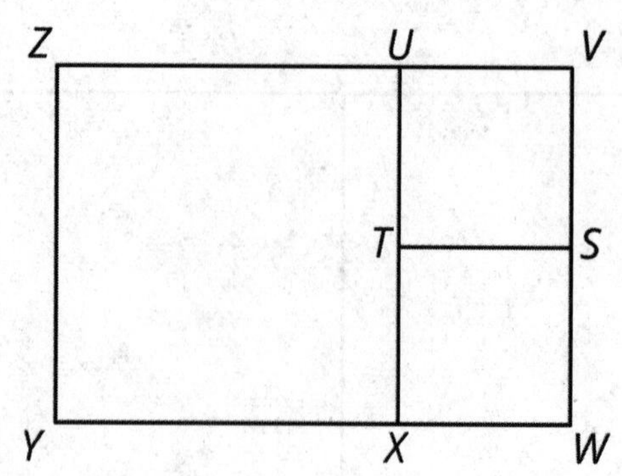


1. The diagram shows that rectangle $VWYZ$ has been decomposed into three squares. What could the side lengths of this rectangle be?

2. How many different side lengths can you find for rectangle $VWYZ$?

3. What are some rules for possible side lengths of rectangle $VWYZ$?

## 3.2: More Rectangles, More Squares

1. Draw a rectangle that is 21 units by 6 units.

   a. In your rectangle, draw a line segment that decomposes the rectangle into a new rectangle and a square that is as large as possible. Continue until the diagram shows that your original rectangle has been entirely decomposed into squares.

   b. How many squares of each size are in your diagram?

   c. What is the side length of the smallest square?

2. Draw a rectangle that is 28 units by 12 units.

   a. In your rectangle, draw a line segment that decomposes the rectangle into a new rectangle and a square that is as large as possible. Continue until the diagram shows that your original rectangle has been decomposed into squares.

   b. How many squares of each size are in your diagram?

   c. What is the side length of the smallest square?

3. Write each of these fractions as a mixed number with the smallest possible numerator and denominator:

   a. $\frac{16}{5}$

   b. $\frac{21}{6}$

   c. $\frac{28}{12}$

4. What do the fraction problems have to do with the previous rectangle decomposition problems?

## 3.3: Finding Equivalent Fractions

1. Accurately draw a rectangle that is 9 units by 4 units.

   a. In your rectangle, draw a line segment that decomposes the rectangle into a new rectangle and a square that is as large as possible. Continue until your original rectangle has been entirely decomposed into squares.

   b. How many squares of each size are there?

   c. What are the side lengths of the last square you drew?

   d. Write $\frac{9}{4}$ as a mixed number.

2. Accurately draw a rectangle that is 27 units by 12 units.

a. In your rectangle, draw a line segment that decomposes the rectangle into a new rectangle and a square that is as large as possible. Continue until your original rectangle has been entirely decomposed into squares.

b. How many squares of each size are there?

c. What are the side lengths of the last square you drew?

d. Write $\frac{27}{12}$ as a mixed number.

e. Compare the diagram you drew for this problem and the one for the previous problem. How are they the same? How are they different?

3. What is the greatest common factor of 9 and 4? What is the greatest common factor of 27 and 12? What does this have to do with your diagrams of decomposed rectangles?

## Are you ready for more?

We have seen some examples of rectangle tilings. A *tiling* means a way to completely cover a shape with other shapes, without any gaps or overlaps. For example, here is a tiling of rectangle $KXWJ$ with 2 large squares, 3 medium squares, 1 small square, and 2 tiny squares.

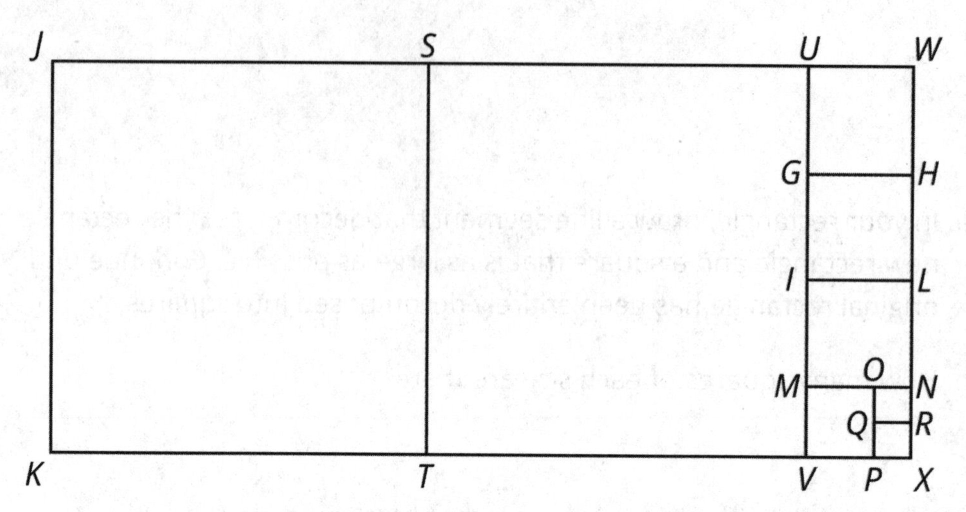


Some of the squares used to tile this rectangle have the same size.

Might it be possible to tile a rectangle with squares where the squares are *all different sizes*?

If you think it is possible, find such a rectangle and such a tiling. If you think it is not possible, explain why it is not possible.

## 3.4: It's All About Fractions

1. Accurately draw a 37-by-16 rectangle. (Use graph paper, if possible.)

   a. In your rectangle, draw a line segment that decomposes the rectangle into a new rectangle and a square that is as large as possible. Continue until your original rectangle has been entirely decomposed into squares.

   b. How many squares of each size are there?

   c. What are the dimensions of the last square you drew?

   d. What does this have to do with $2 + \frac{1}{3+\frac{1}{5}}$?

2. Consider a 52-by-15 rectangle.

a. In your rectangle, draw a line segment that decomposes the rectangle into a new rectangle and a square that is as large as possible. Continue until your original rectangle has been entirely decomposed into squares.

b. Write a fraction equal to this expression: $3 + \frac{1}{2+\frac{1}{7}}$.

c. Notice some connections between the rectangle and the fraction.

d. What is the greatest common factor of 52 and 15?

3. Consider a 98-by-21 rectangle.

a. In your rectangle, draw a line segment that decomposes the rectangle into a new rectangle and a square that is as large as possible. Continue until your original rectangle has been entirely decomposed into squares.

b. Write a fraction equal to this expression: $4 + \frac{1}{1+\frac{7}{14}}$.

c. Notice some connections between the rectangle and the fraction.

d. What is the greatest common factor of 98 and 21?

4. Consider a 121-by-38 rectangle.

a. Use the decomposition-into-squares process to write a continued fraction for $\frac{121}{38}$. Verify that it works.

b. What is the greatest common factor of 121 and 38?

# Lesson 4: Measurement Error (Part 1)

Let's check how accurate our measurements are.

## 4.1: How Long Are These Pencils?

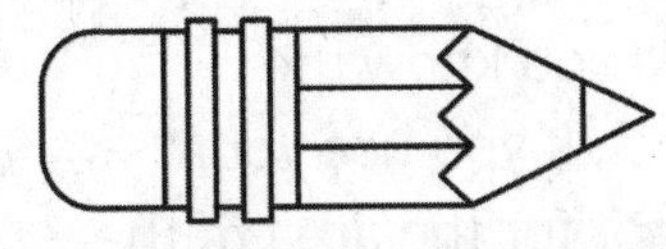

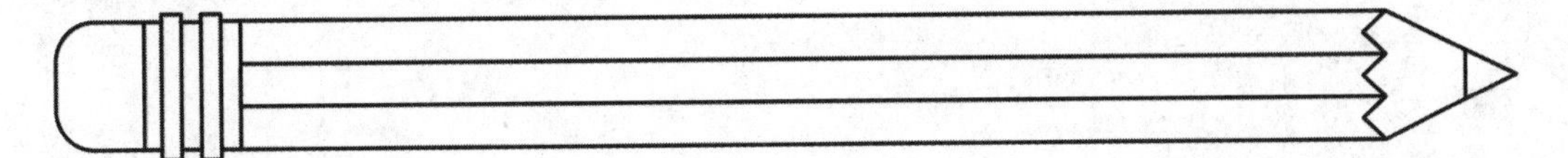

1. Estimate the length of each pencil.

2. How accurate are your estimates?

3. For each estimate, what is the largest possible percent error?

## 4.2: How Long Are These Floor Boards?

A wood floor is made by laying multiple boards end to end. Each board is measured with a maximum percent error of 5%. What is the maximum percent error for the total length of the floor?

# Lesson 5: Measurement Error (Part 2)

Let's check how accurate our calculations are.

## 5.1: Measurement Error for Area

Imagine that you measure the length and width of a rectangle and you know the measurements are accurate within 5% of the actual measurements. If you use your measurements to find the area, what is the maximum percent error for the area of the rectangle?

## 5.2: Measurement Error for Volume

1. The length, width, and height of a rectangular prism were measured to be 10 cm, 12 cm, and 25 cm. Assuming that these measurements are accurate to the nearest cm, what is the largest percent error possible for:

    a. each of the dimensions?

    b. the volume of the prism?

2. If the length, width, and height of a right rectangular prism have a maximum percent error of 1%, what is the largest percent error possible for the volume of the prism?

# Lesson 6: How Do We Choose?

Let's vote and choose a winner!

## 6.1: Which Was "Yessier"?

Two sixth-grade classes, A and B, voted on whether to give the answers to their math problems in poetry. The "yes" choice was more popular in both classes.

| | yes | no |
|---|---|---|
| class A | 24 | 16 |
| class B | 18 | 9 |

Was one class more in favor of math poetry, or were they equally in favor? Find three or more ways to answer the question.

## 6.2: Which Class Voted Purpler?

The school will be painted over the summer. Students get to vote on whether to change the color to purple (a "yes" vote), or keep it a beige color (a "no" vote).

The principal of the school decided to analyze voting results by class. The table shows some results.

In both classes, a majority voted for changing the paint color to purple. Which class was more in favor of changing?

| | yes | no |
|---|---|---|
| class A | 26 | 14 |
| class B | 31 | 19 |

## 6.3: Supermajorities

1. Another school is also voting on whether to change their school's color to purple. Their rules require a $\frac{2}{3}$ supermajority to change the colors. A total of 240 people voted, and 153 voted to change to purple. Were there enough votes to make the change?

2. This school also is thinking of changing their mascot to an armadillo. To change mascots, a 55% supermajority is needed. How many of the 240 students need to vote "yes" for the mascot to change?

3. At this school, which requires more votes to pass: a change of mascot or a change of color?

## 6.4: Best Restaurant

A town's newspaper held a contest to decide the best restaurant in town. Only people who subscribe to the newspaper can vote. 25% of the people in town subscribe to the newspaper. 20% of the subscribers voted. 80% of the people who voted liked Darnell's BBQ Pit best.

Darnell put a big sign in his restaurant's window that said, "80% say Darnell's is the best!"

Do you think Darnell's sign is making an accurate statement? Support your answer with:

- Some calculations
- An explanation in words
- A diagram that accurately represents the people in town, the newspaper subscribers, the voters, and the people who liked Darnell's best

# Lesson 7: More than Two Choices

Let's explore different ways to determine a winner.

## 7.1: Field Day

Students in a sixth-grade class were asked, "What activity would you most like to do for field day?" The results are shown in the table.

| activity | number of votes |
|---|---|
| softball game | 16 |
| scavenger hunt | 10 |
| dancing talent show | 8 |
| marshmallow throw | 4 |
| no preference | 2 |

1. What percentage of the class voted for softball?

2. What percentage did not vote for softball as their first choice?

## 7.2: School Lunches (Part 1)

Suppose students at our school are voting for the lunch menu over the course of one week. The following is a list of options provided by the caterer.

Menu 1: Meat Lovers

- Meat loaf
- Hot dogs
- Pork cutlets
- Beef stew
- Liver and onions

Menu 2: Vegetarian

- Vegetable soup and peanut butter sandwich
- Hummus, pita, and veggie sticks
- Veggie burgers and fries
- Chef's salad
- Cheese pizza every day
- Double desserts every day

Menu 3: Something for Everyone

- Chicken nuggets
- Burgers and fries
- Pizza
- Tacos
- Leftover day (all the week's leftovers made into a casserole)
- Bonus side dish: pea jello (green gelatin with canned peas)

Menu 4: Concession Stand

- Choice of hamburger or hot dog, with fries, every day

To vote, draw one of the following symbols next to each menu option to show your first, second, third, and last choices. If you use the slips of paper from your teacher, use only the column that says "symbol."

1st choice

2nd choice

3rd choice

4th choice

1. Meat Lovers _________
2. Vegetarian _________
3. Something for Everyone _________
4. Concession Stand _________

Here are two voting systems that can be used to determine the winner.

- Voting System #1. *Plurality*: The option with the most first-choice votes (stars) wins.
- Voting System #2. *Runoff*: If no choice received a majority of the votes, leave out the choice that received the fewest first-choice votes (stars). Then have another vote.

  If your first vote is still a choice, vote for that. If not, vote for your second choice that you wrote down.

  If there is still no majority, leave out the choice that got the fewest votes, and then vote again. Vote for your first choice if it's still in, and if not, vote for your second choice. If your second choice is also out, vote for your third choice.

1. How many people in our class are voting? How many votes does it take to win a majority?

2. How many votes did the top option receive? Was this a majority of the votes?

3. People tend to be more satisfied with election results if their top choices win. For how many, and what percentage, of people was the winning option:

   a. their first choice?

   b. their second choice?

   c. their third choice?

   d. their last choice?

4. After the second round of voting, did any choice get a majority? If so, is it the same choice that got a plurality in Voting System #1?

5. Which choice won?

6. How satisfied were the voters by the election results? For how many, and what percentage, of people was the winning option:

   a. their first choice?

   b. their second choice?

   c. their third choice?

   d. their last choice?

7. Compare the satisfaction results for the plurality voting rule and the runoff rule. Did one produce satisfactory results for more people than the other?

## 7.3: School Lunch (Part 2)

Let's analyze a different election.

In another class, there are four clubs. Everyone in each club agrees to vote for the lunch menu exactly the same way, as shown in this table.

| | Barbecue Club (21 members) | Garden Club (13 members) | Sports Boosters (7 members) | Film Club (9 members) |
|---|---|---|---|---|
| A. Meat Lovers | ☆ | ✕ | ✕ | ✕ |
| B. Vegetarian | ☺ | ☆ | 😐 | 😐 |
| C. Something for Everyone | 😐 | 😐 | ☺ | ☆ |
| D. Concession Stand | ✕ | ☺ | ☆ | ☺ |

1. Figure out which option won the election by answering these questions.

    a. On the first vote, when everyone voted for their first choice, how many votes did each option get? Did any choice get a majority?

    b. Which option is removed from the next vote?

    c. On the second vote, how many votes did each of the remaining three menu options get? Did any option get a majority?

    d. Which menu option is removed from the next vote?

    e. On the third vote, how many votes did each of the remaining two options get? Which option won?

2. Estimate how satisfied all the voters were.

    a. For how many people was the winner their first choice?

    b. For how many people was the winner their second choice?

    c. For how many people was the winner their third choice?

    d. For how many people was the winner their last choice?

3. Compare the satisfaction results for the plurality voting rule and the runoff rule. Did one produce satisfactory results for more people than the other?

## 7.4: Just Vote Once

Your class just voted using the *instant runoff* system. Use the class data for following questions.

1. For our class, which choice received the most points?
2. Does this result agree with that from the runoff election in an earlier activity?
3. For the other class, which choice received the most points?
4. Does this result agree with that from the runoff election in an earlier activity?

5. The runoff method uses information about people's first, second, third, and last choices when it is not clear that there is a winner from everyone's first choices. How does the instant runoff method include the same information?

6. After comparing the results for the three voting rules (plurality, runoff, instant runoff) and the satisfaction surveys, which method do you think is fairest? Explain.

**Are you ready for more?**

Numbering your choices 0 through 3 might not really describe your opinions. For example, what if you really liked A and C a lot, and you really hated B and D? You might want to give A and C both a 3, and B and D both a 0.

1. Design a numbering system where the size of the number accurately shows how much you like a choice. Some ideas:
    - The same 0 to 3 scale, but you can choose more than one of each number, or even decimals between 0 and 3.
    - A scale of 1 to 10, with 10 for the best and 1 for the worst.

2. Try out your system with the people in your group, using the same school lunch options for the election.

3. Do you think your system gives a more fair way to make choices? Explain your reasoning.

## 7.5: Weekend Choices

Clare, Han, Mai, Tyler, and Noah are deciding what to do on the weekend. Their options are cooking, hiking, and bowling. Here are the points for their instant runoff vote. Each first choice gets 2 points, the second choice gets 1 point, and the last choice gets 0 points.

| | cooking | hiking | bowling |
|---|---|---|---|
| Clare | 2 | 1 | 0 |
| Han | 2 | 1 | 0 |
| Mai | 2 | 1 | 0 |
| Tyler | 0 | 2 | 1 |
| Noah | 0 | 2 | 1 |

1. Which activity won using the instant runoff method? Show your calculations and use expressions or equations.

2. Which activity would have won if there was just a vote for their top choice, with a majority or plurality winning?

3. Which activity would have won if there was a runoff election?

4. Explain why this happened.

# Lesson 8: How Crowded Is this Neighborhood?

Let's see how proportional relationships apply to where people live.

## 8.1: Dot Density

The figure shows four squares. Each square encloses an array of dots. Squares A and B have side length 2 inches. Squares C and D have side length 1 inch.

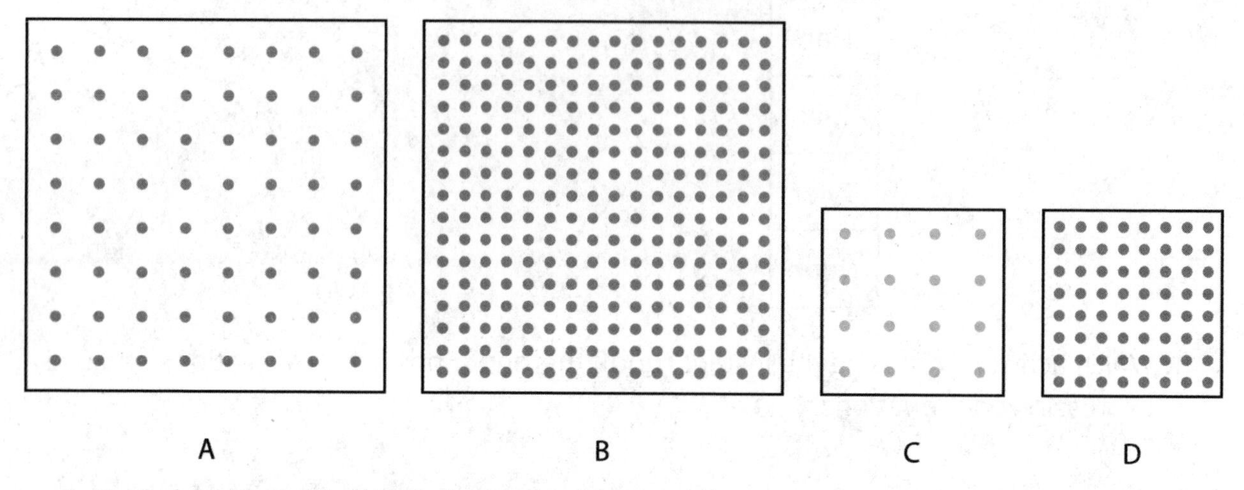


1. Complete the table with information about each square.

| square | area of the square in square inches | number of dots | number of dots per square inch |
|---|---|---|---|
| A | | | |
| B | | | |
| C | | | |
| D | | | |

2. Compare each square to the others. What is the same and what is different?

## 8.2: Dot Density with a Twist

The figure shows two arrays, each enclosed by a square that is 2 inches wide.

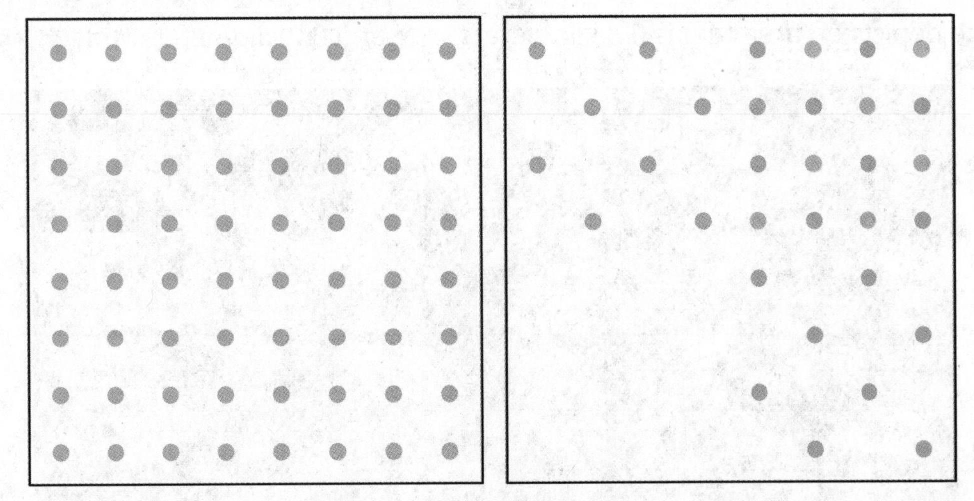

1. Let $a$ be the area of the square and $d$ be the number of dots enclosed by the square. For each square, plot a point that represents its values of $a$ and $d$.

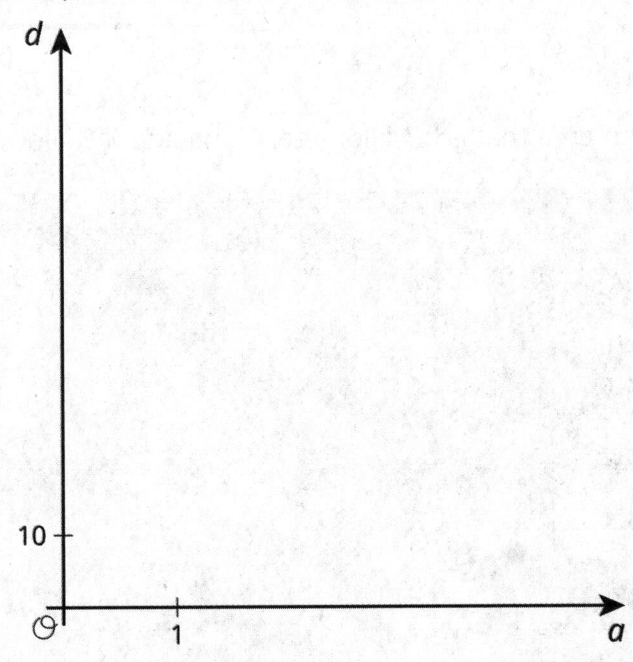


2. Draw lines from $(0, 0)$ to each point. For each line, write an equation that represents the proportional relationship.

3. What is the constant of proportionality for each relationship? What do the constants of proportionality tell us about the dots and squares?

## 8.3: Housing Density

Here are pictures of two different neighborhoods.

This image depicts an area that is 0.3 kilometers long and 0.2 kilometers wide.

**0.1 km**

This image depicts an area that is 0.4 kilometers long and 0.2 kilometers wide.

**0.1 km**

For each neighborhood, find the number of houses per square kilometer.

## 8.4: Population Density

- New York City has a population of 8,406 thousand people and covers an area of 1,214 square kilometers.
- Los Angeles has a population of 3,884 thousand people and covers an area of 1,302 square kilometers.

1. The points labeled $A$ and $B$ each correspond to one of the two cities. Which is which? Label them on the graph.

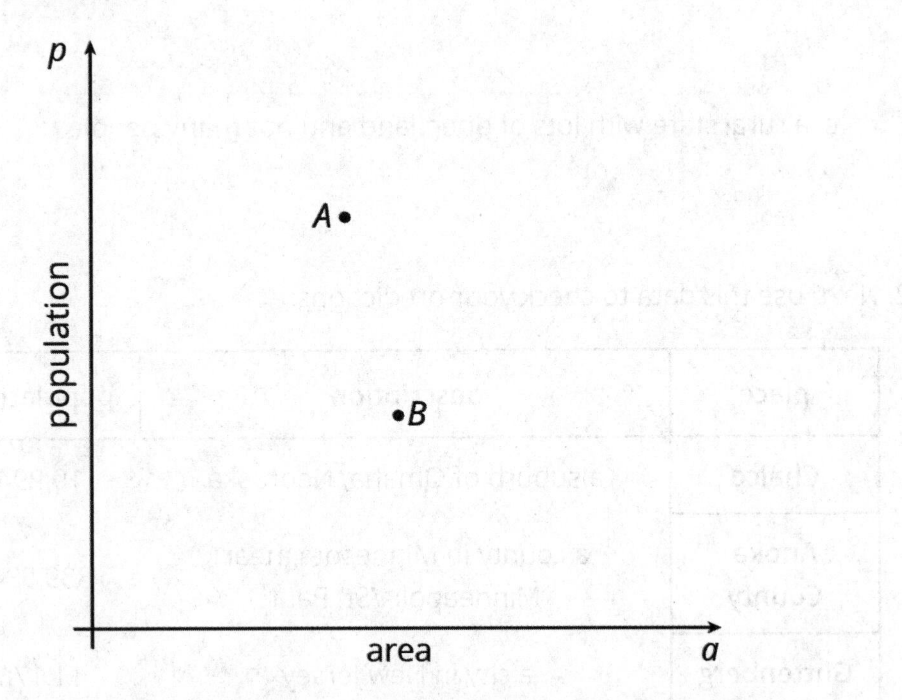


2. Write an equation for the line that passes through $(0, 0)$ and $A$. What is the constant of proportionality?

3. Write an equation for the line that passes through $(0, 0)$ and $B$. What is the constant of proportionality?

4. What do the constants of proportionality tell you about the crowdedness of these two cities?

## Are you ready for more?

1. Predict where these types of regions would be shown on the graph:

    a. a suburban region where houses are far apart, with big yards

    b. a neighborhood in an urban area with many high-rise apartment buildings

    c. a rural state with lots of open land and not many people

2. Next, use this data to check your predictions:

| place | description | population | area ($km^2$) |
|---|---|---|---|
| **Chalco** | a suburb of Omaha, Nebraska | 10,994 | 7.5 |
| **Anoka County** | a county in Minnesota, near Minneapolis/St. Paul | 339,534 | 1,155 |
| **Guttenberg** | a city in New Jersey | 11,176 | 0.49 |
| **New York** | a state | 19,746,227 | 141,300 |
| **Rhode Island** | a state | 1,055,173 | 3,140 |
| **Alaska** | a state | 736,732 | 1,717,856 |
| **Tok** | a community in Alaska | 1,258 | 342.7 |

# Lesson 9: Picking Representatives

Let's think about fair representation.

## 9.1: Computers for Kids

A program gives computers to families with school-aged children. They have a certain number of computers to distribute fairly between several families. How many computers should each family get?

1. One month the program has 8 computers. The families have these numbers of school-aged children: 4, 2, 6, 2, 2.

   a. How many children are there in all?

   b. Counting all the children in all the families, how many children would use each computer? This is the number of children per computer. Call this number $A$.

   c. Fill in the third column of the table. Decide how many computers to give to each family if we use $A$ as the basis for distributing the computers.

| family | number of children | number of computers, using $A$ |
|---|---|---|
| Baum | 4 | |
| Chu | 2 | |
| Davila | 6 | |
| Eno | 2 | |
| Farouz | 2 | |

   d. Check that 8 computers have been given out in all.

2. The next month they again have 8 computers. There are different families with these numbers of children: 3, 1, 2, 5, 1, 8.

   a. How many children are there in all?

   b. Counting all the children in all the families, how many children would use each computer? This is the number of children per computer. Call this number $B$.

   c. Does it make sense that $B$ is not a whole number? Why?

   d. Fill in the third column of the table. Decide how many computers to give to each family if we use $B$ as the basis for distributing the computers.

| family | number of children | number of computers, using $B$ | number of computers, your way | children per computer, your way |
|---|---|---|---|---|
| Gray | 3 | | | |
| Hernandez | 1 | | | |
| Ito | 2 | | | |
| Jones | 5 | | | |
| Krantz | 1 | | | |
| Lo | 8 | | | |

   e. Check that 8 computers have been given out in all.

   f. Does it make sense that the number of computers for one family is not a whole number? Explain your reasoning.

   g. Find and describe a way to distribute computers to the families so that each family gets a whole number of computers. Fill in the fourth column of the table.

h. Compute the number of children per computer in each family and fill in the last column of the table.

i. Do you think your way of distributing the computers is fair? Explain your reasoning.

## 9.2: School Mascot (Part 1)

A school is deciding on a school mascot. They have narrowed the choices down to the Banana Slugs or the Sea Lions.

The principal decided that each class gets one vote. Each class held an election, and the winning choice was the one vote for the whole class. The table shows how three classes voted.

| | banana slugs | sea lions | class vote |
|---|---|---|---|
| class A | 9 | 3 | banana slug |
| class B | 14 | 10 | |
| class C | 6 | 30 | |

1. Which mascot won, according to the principal's plan? What percentage of the votes did the winner get under this plan?

2. Which mascot received the most student votes in all? What percentage of the votes did this mascot receive?

3. The students thought this plan was not very fair. They suggested that bigger classes should have more votes to send to the principal. Make up a proposal for the principal where there are as few votes as possible, but the votes proportionally represent the number of students in each class.

4. Decide how to assign the votes for the results in the class. (Do they all go to the winner? Or should the loser still get some votes?)

5. In your system, which mascot is the winner?

6. In your system, how many representative votes are there? How many students does each vote represent?

## 9.3: Advising the School Board

1. In a very small school district, there are four schools, D, E, F, and G. The district wants a total of 10 advisors for the students. Each school should have at least one advisor.

| school | number of students | number of advisors, using $A$ |
|---|---|---|
| D | 48 | |
| E | 12 | |
| F | 24 | |
| G | 36 | |

   a. How many students are in this district in all?

   b. If the advisors could represent students at different schools, how many students per advisor should there be? Call this number $A$.

   c. Using $A$ students per advisor, how many advisors should each school have? Complete the table with this information for schools D, E, F, and G.

2. Another district has four schools; some are large, others are small. The district wants 10 advisors in all. Each school should have at least one advisor.

| school | number of students | number of advisors, using $B$ | number of advisors, your way | students per advisor, your way |
|---|---|---|---|---|
| Dr. King School | 500 | | | |
| O'Connor School | 200 | | | |
| Science Magnet School | 140 | | | |
| Trombone Academy | 10 | | | |

a. How many students are in this district in all?

b. If the advisors didn't have to represent students at the same school, how many students per advisor should there be? Call this number $B$.

c. Using $B$ students per advisor, how many advisors should each school have? Give your quotients to the tenths place. Fill in the first "number of advisors" column of the table. Does it make sense to have a tenth of an advisor?

d. Decide on a consistent way to assign advisors to schools so that there are only whole numbers of advisors for each school, and there is a total of 10 advisors among the schools. Fill in the "your way" column of the table.

e. How many students per advisor are there at each school? Fill in the last row of the table.

f. Do you think this is a fair way to assign advisors? Explain your reasoning.

## 9.4: School Mascot (Part 2)

The whole town gets interested in choosing a mascot. The mayor of the town decides to choose representatives to vote.

There are 50 blocks in the town, and the people on each block tend to have the same opinion about which mascot is best. Green blocks like sea lions, and gold blocks like banana slugs. The mayor decides to have 5 representatives, each representing a district of 10 blocks.

Here is a map of the town, with preferences shown.

1. Suppose there were an election with each block getting one vote. How many votes would be for banana slugs? For sea lions? What percentage of the vote would be for banana slugs?

2. Suppose the districts are shown in the next map. What did the people in each district prefer? What did their representative vote? Which mascot would win the election?

Complete the table with this election's results.

| district | number of blocks for banana slugs | number of blocks for sea lions | percentage of blocks for banana slugs | representative's vote |
|---|---|---|---|---|
| 1 | 10 | 0 | | banana slugs |
| 2 | | | | |
| 3 | | | | |
| 4 | | | | |
| 5 | | | | |

3. Suppose, instead, that the districts are shown in the new map below. What did the people in each district prefer? What did their representative vote? Which mascot would win the election?

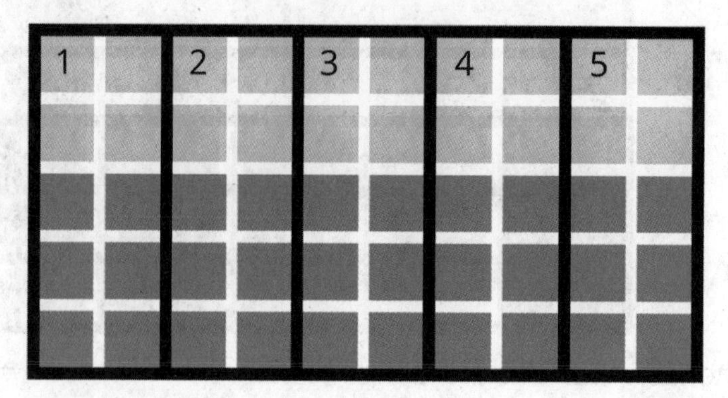


Complete the table with this election's results.

| district | number of blocks for banana slugs | number of blocks for sea lions | percentage of blocks for banana slugs | representative's vote |
|---|---|---|---|---|
| 1 | | | | |
| 2 | | | | |
| 3 | | | | |
| 4 | | | | |
| 5 | | | | |

4. Suppose the districts are designed in yet another way, as shown in the next map. What did the people in each district prefer? What did their representative vote? Which mascot would win the election?

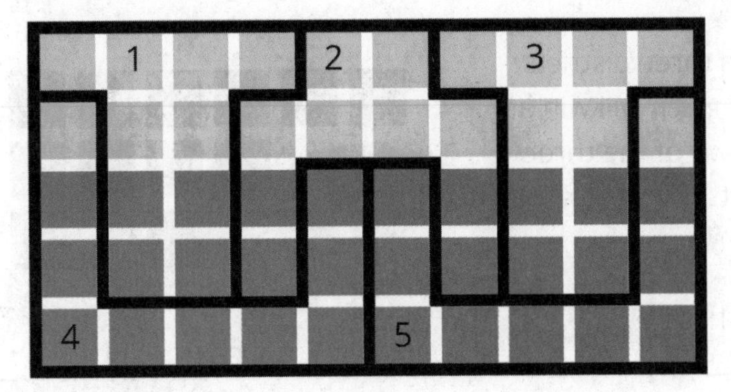


Complete the table with this election's results.

| district | number of blocks for banana slugs | number of blocks for sea lions | percentage of blocks for banana slugs | representative's vote |
|---|---|---|---|---|
| 1 | | | | |
| 2 | | | | |
| 3 | | | | |
| 4 | | | | |
| 5 | | | | |

5. Write a headline for the local newspaper for each of the ways of splitting the town into districts.

6. Which systems on the three maps of districts do you think are more fair? Are any totally unfair?

## 9.5: Fair and Unfair Districts

1. Smallville's map is shown, with opinions shown by block in green and gold. Decompose the map to create three connected, equal-area districts in two ways:

   a. Design three districts where *green* will win at least two of the three districts. Record results in Table 1.

Table 1:

| district | number of blocks for green | number of blocks for gold | percentage of blocks for green | representative's vote |
|---|---|---|---|---|
| 1 | | | | |
| 2 | | | | |
| 3 | | | | |

   b. Design three districts where *gold* will win at least two of the three districts. Record results in Table 2.

Table 2:

| district | number of blocks for green | number of blocks for gold | percentage of blocks for green | representative's vote |
|---|---|---|---|---|
| 1 | | | | |
| 2 | | | | |
| 3 | | | | |

2. Squaretown's map is shown, with opinions by block shown in green and gold. Decompose the map to create five connected, equal-area districts in two ways:

   a. Design five districts where *green* will win at least three of the five districts. Record the results in Table 3.

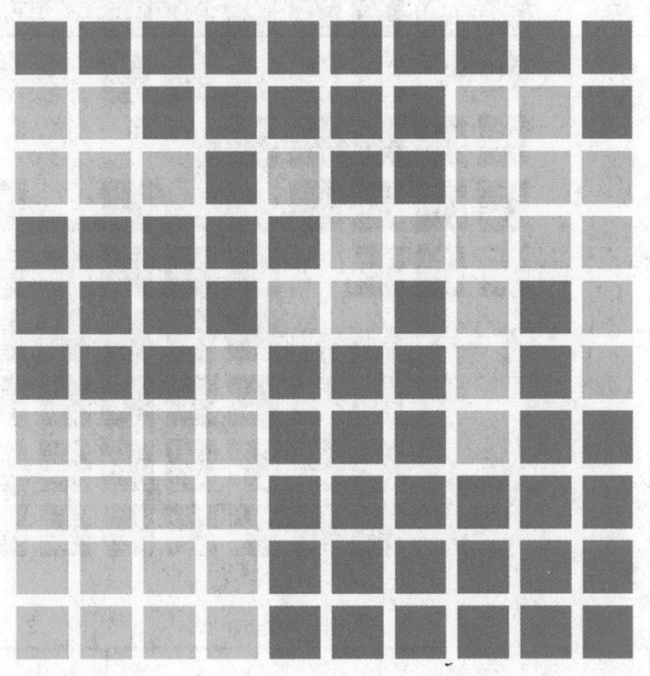

Table 3:

| district | number of blocks for green | number of blocks for gold | percentage of blocks for green | representative's vote |
|---|---|---|---|---|
| 1 | | | | |
| 2 | | | | |
| 3 | | | | |
| 4 | | | | |
| 5 | | | | |

b. Design five districts where *gold* will win at least three of the five districts. Record the results in Table 4.

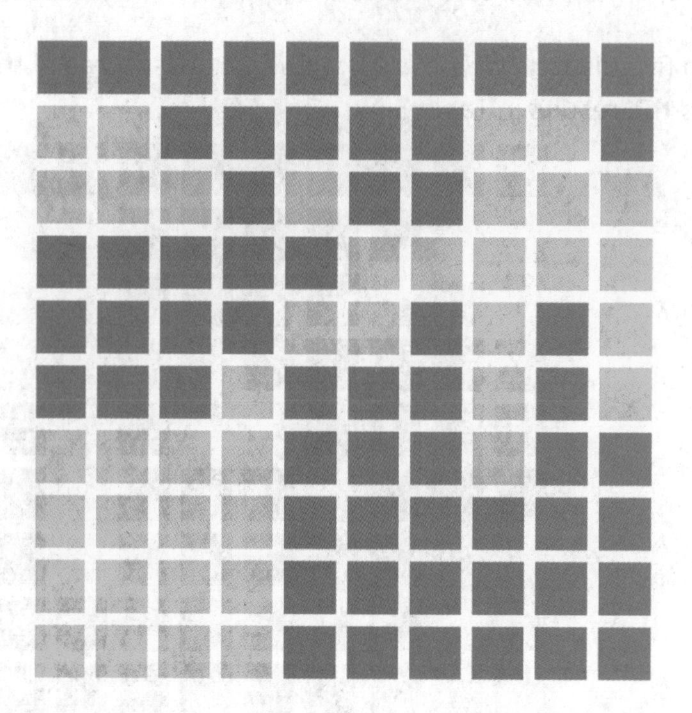

Table 4:

| district | number of blocks for green | number of blocks for gold | percentage of blocks for green | representative's vote |
|---|---|---|---|---|
| 1 | | | | |
| 2 | | | | |
| 3 | | | | |
| 4 | | | | |
| 5 | | | | |

3. Mountain Valley's map is shown, with opinions by block shown in green and gold. (This is a town in a narrow valley in the mountains.) Can you decompose the map to create three connected, equal-area districts in the two ways described here?

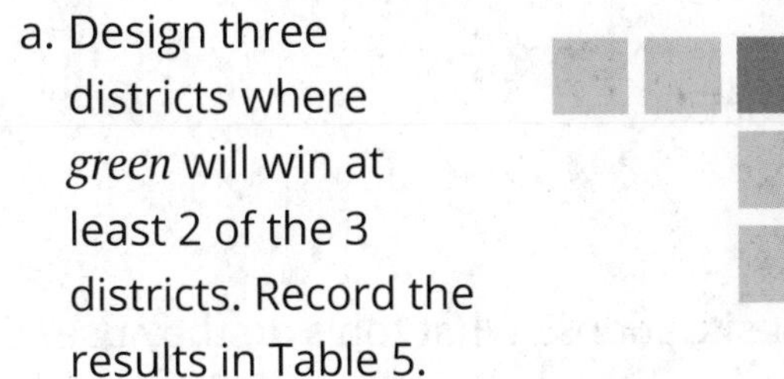

a. Design three districts where *green* will win at least 2 of the 3 districts. Record the results in Table 5.

Table 5:

| district | number of blocks for green | number of blocks for gold | percentage of blocks for green | representative's vote |
|---|---|---|---|---|
| 1 | | | | |
| 2 | | | | |
| 3 | | | | |

b. Design three districts where *gold* will win at least 2 of the 3 districts. Record the results in Table 6.

Table 6:

| district | number of blocks for green | number of blocks for gold | percentage of blocks for green | representative's vote |
|---|---|---|---|---|
| 1 | | | | |
| 2 | | | | |
| 3 | | | | |

# Lesson 10: Measuring Long Distances Over Uneven Terrain

Let's measure long distances over uneven terrain.

## 10.1: How Far Is It?

How do people measure distances in different situations? What tools do they use? Come up with at least three different methods and situations where those methods are used.

## 10.2: Planning a 5K Course

The school is considering holding a 5K fundraising walk on the school grounds. Your class is supposed to design the course for the walk.

1. What will you need to do to design the course for the walk?

2. Come up with a method to measure the course. Pause here so your teacher can review your plan.

## 10.3: Comparing Methods

Let's see how close different measuring methods are to each other. Your teacher will show you a path to measure.

1. Use your method to measure the length of the path at least two times.

2. Decide what distance you will report to the class.

3. Compare your results with those of two other groups. Express the differences between the measurements in terms of percentages.

4. Discuss the advantages and disadvantages of each group's method.

# Lesson 11: Building a Trundle Wheel

Let's build a trundle wheel.

## 11.1: What Is a Trundle Wheel?

A tool that surveyors use to measure distances is called a trundle wheel.

1. How does a trundle wheel measure distance?

2. Why is this method of measuring distances better than the methods we used in the previous lesson?

3. How could we construct a simple trundle wheel? What materials would we need?

## 11.2: Building a Trundle Wheel

Your teacher will give you some supplies. Construct a trundle wheel and use it to measure the length of the classroom. Record:

1. the diameter of your trundle wheel

2. the number of clicks across classroom

3. the length of the classroom (Be prepared to explain your reasoning.)

# Lesson 12: Using a Trundle Wheel to Measure Distances

Let's use our trundle wheels.

## 12.1: Measuring Distances with the Trundle Wheel

Earlier you made trundle wheels so that you can measure long distances. Your teacher will show you a path to measure.

1. Measure the path with your trundle wheel three times and calculate the distance. Record your results in the table.

| trial number | number of clicks | computation | distance |
|---|---|---|---|
| 1 | | | |
| 2 | | | |
| 3 | | | |

2. Decide what distance you will report to the class. Be prepared to explain your reasoning.

3. Compare this distance with the distance you measured the other day for this same path.

4. Compare your results with the results of two other groups. Express the differences between the measurements in terms of percentages.

# Lesson 13: Designing a 5K Course

Let's map out the 5K course.

## 13.1: Make a Proposal

Your teacher will give you a map of the school grounds.

1. On the map, draw in the path you measured earlier with your trundle wheel and label its length.

2. Invent another route for a walking course and draw it on your map. Estimate the length of the course you drew.

3. How many laps around your course must someone complete to walk 5 km?

## 13.2: Measuring and Finalizing the Course

1. Measure your proposed race course with your trundle wheel at least two times. Decide what distance you will report to the class.

2. Revise your course, if needed.

3. Create a visual display that includes:

    - A map of your final course
    - The starting and ending locations
    - The number of laps needed to walk 5 km
    - Any other information you think would be helpful to the race organizers

### Are you ready for more?

The map your teacher gave you didn't include a scale. Create one.

# Glossary

**absolute value**
The absolute value of a number is its distance from 0 on the number line.

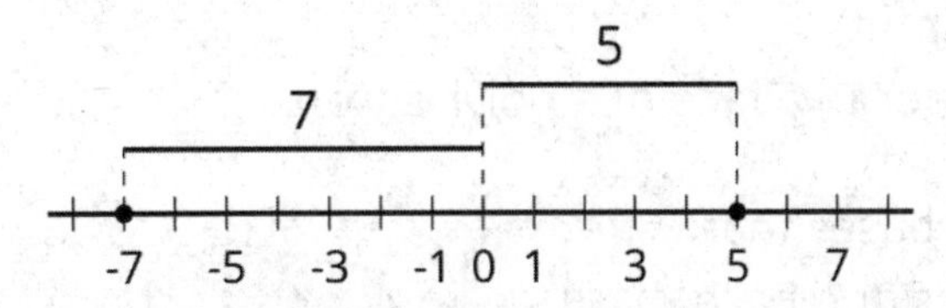


The absolute value of -7 is 7, because it is 7 units away from 0. The absolute value of 5 is 5, because it is 5 units away from 0.

**area**
Area is the number of square units that cover a two-dimensional region, without any gaps or overlaps.

For example, the area of region A is 8 square units. The area of the shaded region of B is $\frac{1}{2}$ square unit.

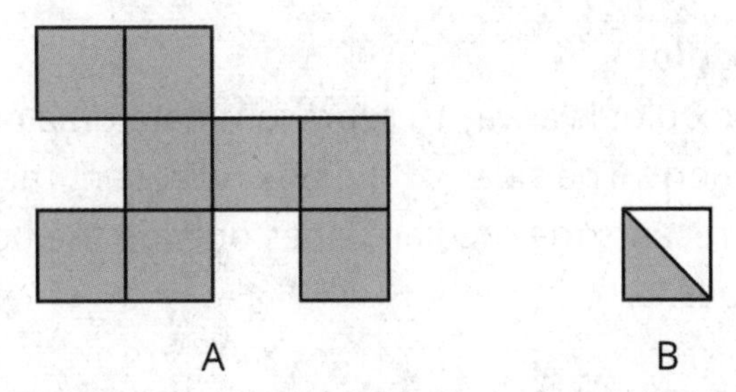


**average**
The average is another name for the mean of a data set.

For the data set 3, 5, 6, 8, 11, 12, the average is 7.5.

$3 + 5 + 6 + 8 + 11 + 12 = 45$

$45 \div 6 = 7.5$

**base (of a parallelogram or triangle)**
We can choose any side of a parallelogram or triangle to be the shape's base. Sometimes we use the word *base* to refer to the length of this side.

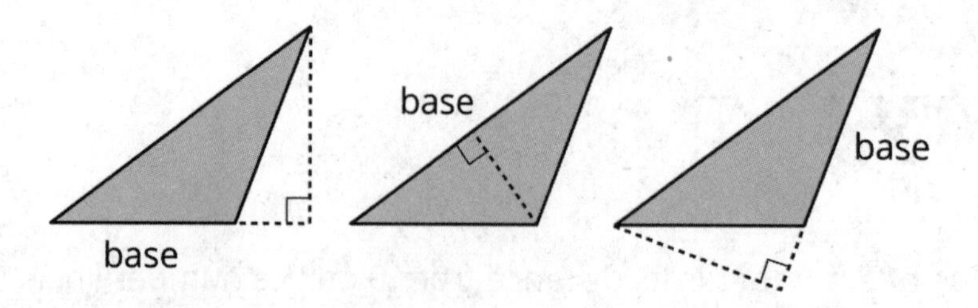


**base (of a prism or pyramid)**
The word *base* can also refer to a face of a polyhedron.

A prism has two identical bases that are parallel. A pyramid has one base.

A prism or pyramid is named for the shape of its base.

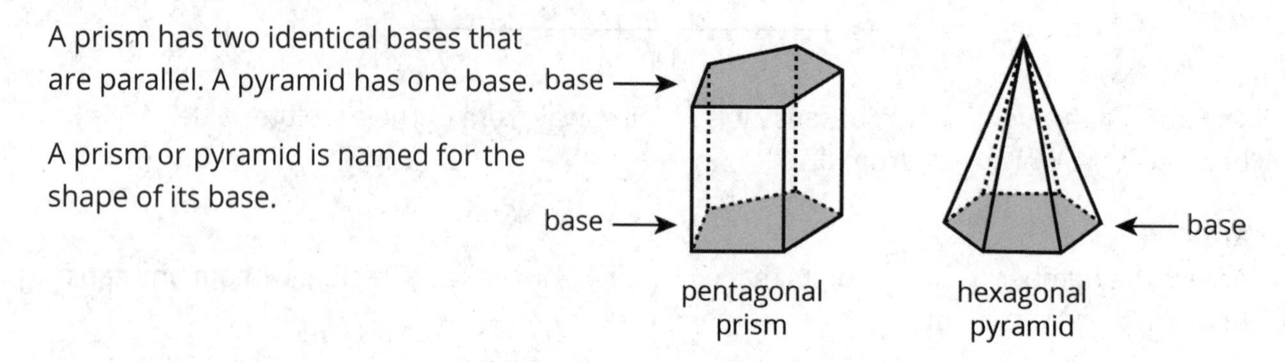


**box plot**
A box plot is a way to represent data on a number line. The data is divided into four sections. The sides of the box represent the first and third quartiles. A line inside the box represents the median. Lines outside the box connect to the minimum and maximum values.

For example, this box plot shows a data set with a minimum of 2 and a maximum of 15. The median is 6, the first quartile is 5, and the third quartile is 10.

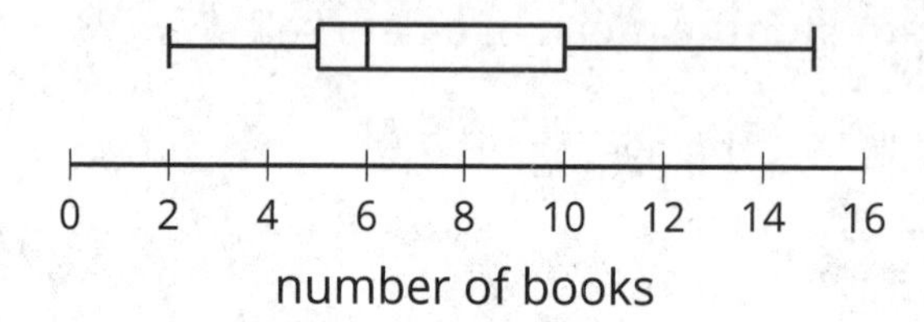


**center**
The center of a set of numerical data is a value in the middle of the distribution. It represents a typical value for the data set.

For example, the center of this distribution of cat weights is between 4.5 and 5 kilograms.

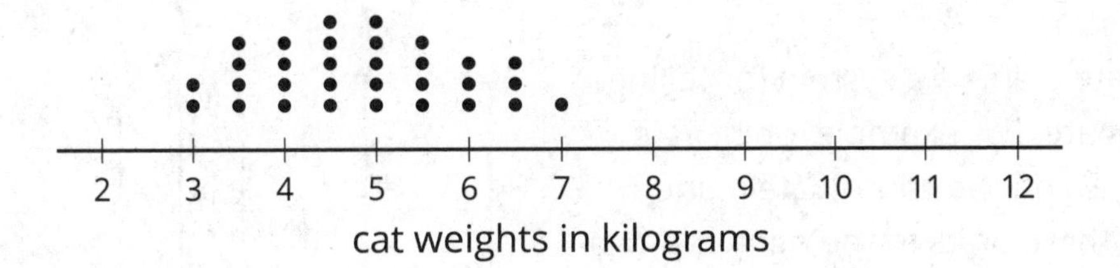


**circle**

A circle is made out of all the points that are the same distance from a given point.

For example, every point on this circle is 5 cm away from point $A$, which is the center of the circle.

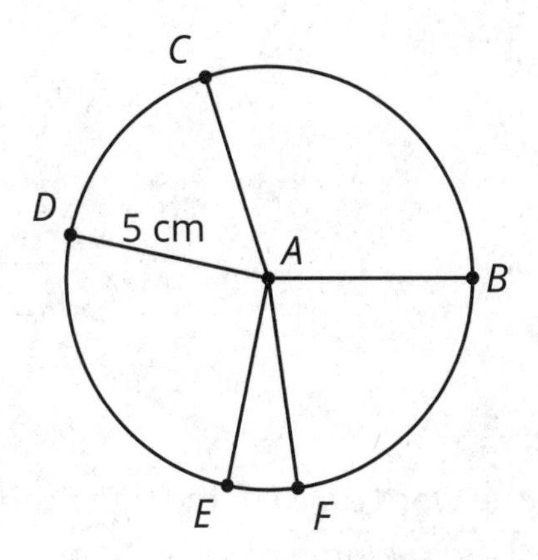


**circumference**

The circumference of a circle is the distance around the circle. If you imagine the circle as a piece of string, it is the length of the string. If the circle has radius $r$ then the circumference is $2\pi r$.

The circumference of a circle of radius 3 is $2 \cdot \pi \cdot 3$, which is $6\pi$, or about 18.85.

**coefficient**

A coefficient is a number that is multiplied by a variable.

For example, in the expression $3x + 5$, the coefficient of $x$ is 3. In the expression $y + 5$, the coefficient of $y$ is 1, because $y = 1 \cdot y$.

**compose**

Compose means "put together." We use the word *compose* to describe putting more than one figure together to make a new shape.

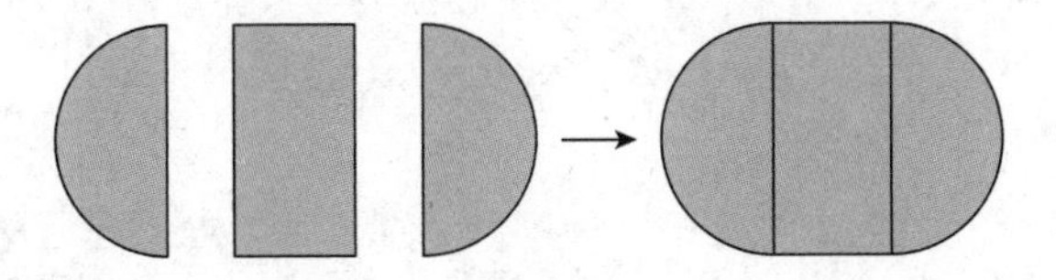

**coordinate plane**

The coordinate plane is a system for telling where points are. For example. point $R$ is located at $(3, 2)$ on the coordinate plane, because it is three units to the right and two units up.

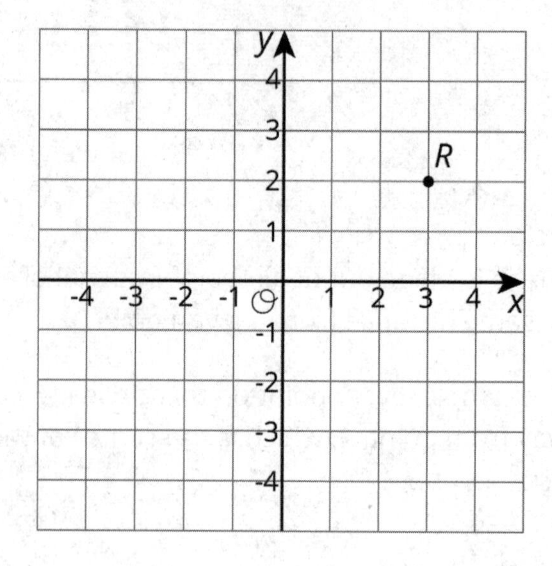


**cubed**

We use the word *cubed* to mean "to the third power." This is because a cube with side length $s$ has a volume of $s \cdot s \cdot s$, or $s^3$.

**decompose**

Decompose means "take apart." We use the word *decompose* to describe taking a figure apart to make more than one new shape.

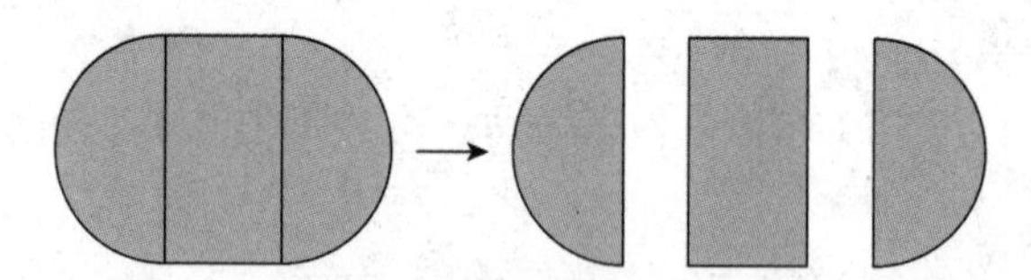

**dependent variable**

The dependent variable is the result of a calculation.

For example, a boat travels at a constant speed of 25 miles per hour. The equation $d = 25t$ describes the relationship between the boat's distance and time. The dependent variable is the distance traveled, because $d$ is the result of multiplying 25 by $t$.

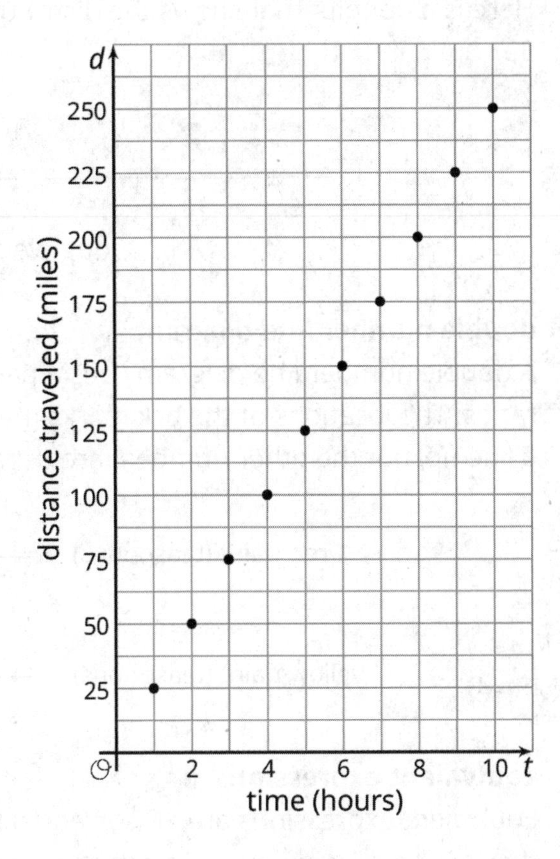


**deposit**

When you put money into an account, it is called a *deposit*.

For example, a person added $60 to their bank account. Before the deposit, they had $435. After the deposit, they had $495, because $435 + 60 = 495$.

**diameter**

A diameter is a line segment that goes from one edge of a circle to the other and passes through the center. A diameter can go in any direction. Every diameter of the circle is the same length. We also use the word *diameter* to mean the length of this segment.

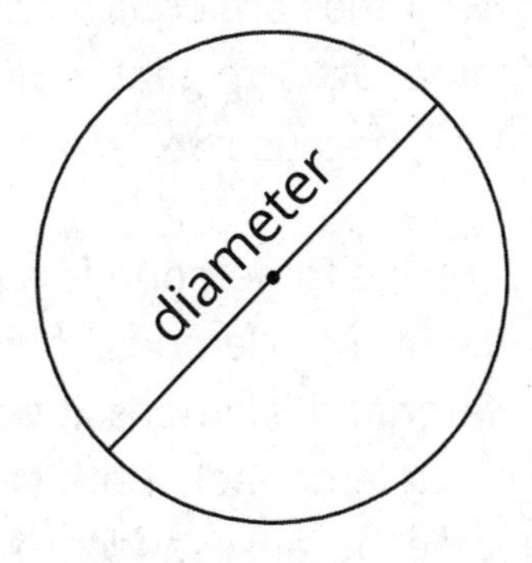


**distribution**

The distribution tells how many times each value occurs in a data set. For example, in the data set blue, blue, green, blue, orange, the distribution is 3 blues, 1 green, and 1 orange.

Here is a dot plot that shows the distribution for the data set 6, 10, 7, 35, 7, 36, 32, 10, 7, 35.

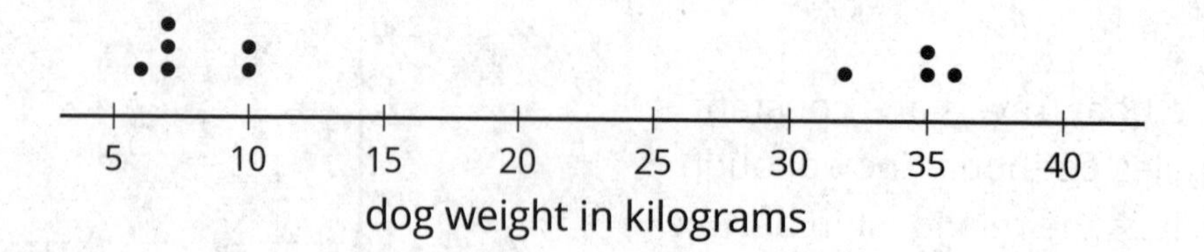


**double number line diagram**

A double number line diagram uses a pair of parallel number lines to represent equivalent ratios. The locations of the tick marks match on both number lines. The tick marks labeled 0 line up, but the other numbers are usually different.

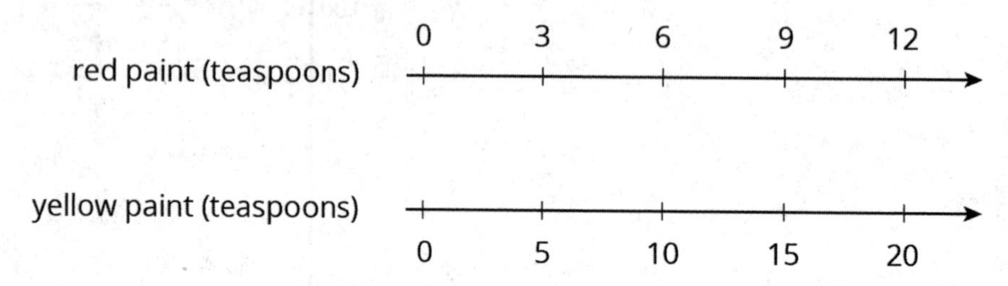


**equivalent expressions**

Equivalent expressions are always equal to each other. If the expressions have variables, they are equal whenever the same value is used for the variable in each expression.

For example, $3x + 4x$ is equivalent to $5x + 2x$. No matter what value we use for $x$, these expressions are always equal. When $x$ is 3, both expressions equal 21. When $x$ is 10, both expressions equal 70.

**equivalent ratios**

Two ratios are equivalent if you can multiply each of the numbers in the first ratio by the same factor to get the numbers in the second ratio. For example, $8 : 6$ is equivalent to $4 : 3$, because $8 \cdot \frac{1}{2} = 4$ and $6 \cdot \frac{1}{2} = 3$.

A recipe for lemonade says to use 8 cups of water and 6 lemons. If we use 4 cups of water and 3 lemons, it will make half as much lemonade. Both recipes taste the same, because  and  are equivalent ratios.

| cups of water | number of lemons |
|---|---|
| 8 | 6 |
| 4 | 3 |

**exponent**

In expressions like $5^3$ and $8^2$, the 3 and the 2 are called exponents. They tell you how many factors to multiply. For example, $5^3 = 5 \cdot 5 \cdot 5$, and $8^2 = 8 \cdot 8$.

**face**

Each flat side of a polyhedron is called a face. For example, a cube has 6 faces, and they are all squares.

**frequency**

The frequency of a data value is how many times it occurs in the data set.

For example, there were 20 dogs in a park. The table shows the frequency of each color.

| color | frequency |
|---|---|
| white | 4 |
| brown | 7 |
| black | 3 |
| multi-color | 6 |

**height (of a parallelogram or triangle)**

The height is the shortest distance from the base of the shape to the opposite side (for a parallelogram) or opposite vertex (for a triangle).

We can show the height in more than one place, but it will always be perpendicular to the chosen base.

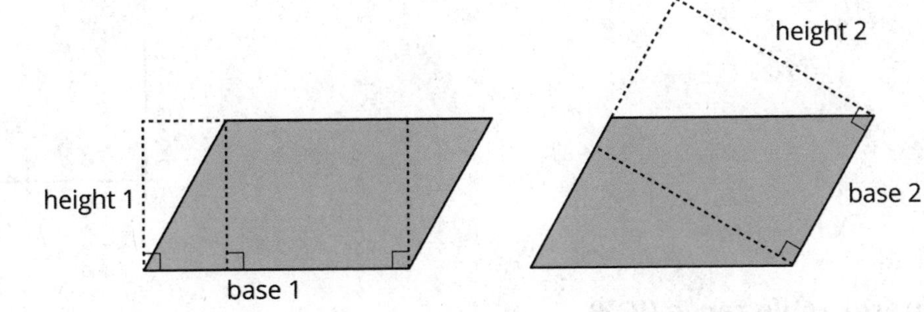


**histogram**

A histogram is a way to represent data on a number line. Data values are grouped by ranges. The height of the bar shows how many data values are in that group.

This histogram shows there were 10 people who earned 2 or 3 tickets. We can't tell how many of them earned 2 tickets or how many earned 3. Each bar includes the left-end value but not the right-end value. (There were 5 people who earned 0 or 1 tickets and 13 people who earned 6 or 7 tickets.)

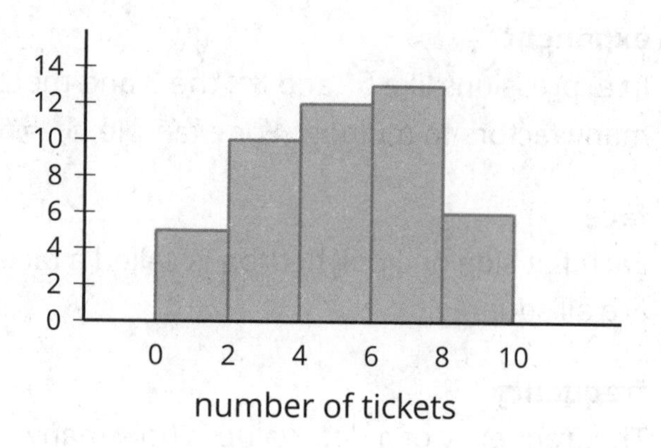


**independent variable**

The independent variable is used to calculate the value of another variable.

For example, a boat travels at a constant speed of 25 miles per hour. The equation $d = 25t$ describes the relationship between the boat's distance and time. The independent variable is time, because $t$ is multiplied by 25 to get $d$.

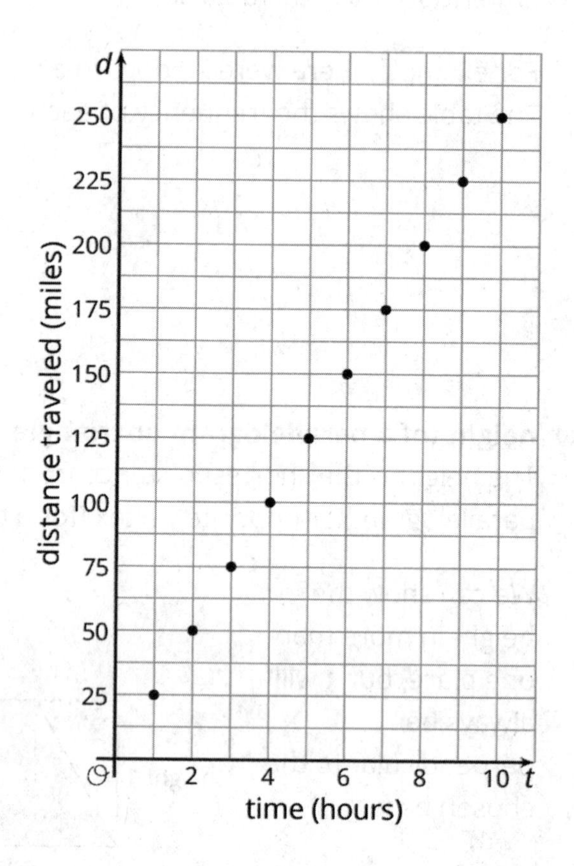


**interquartile range (IQR)**

The interquartile range is one way to measure how spread out a data set is. We sometimes call this the IQR. To find the interquartile range we subtract the first quartile from the third quartile.

For example, the IQR of this data set is 20 because $50 - 30 = 20$.

| 22 | 29 | 30 | 31 | 32 | 43 | 44 | 45 | 50 | 50 | 59 |
|---|---|---|---|---|---|---|---|---|---|---|
| | | Q1 | | | Q2 | | | Q3 | | |

**long division**

Long division is an algorithm for finding the quotient of two numbers expressed in decimal form. It works by building up the quotient one digit at a time, from left to right. Each time you get a new digit, you multiply the divisor by the corresponding base ten value and subtract that from the dividend.

$$
\begin{array}{r}
128 \\
4\overline{)513} \\
\underline{400} \\
113 \\
\underline{80} \\
33 \\
\underline{32} \\
1
\end{array}
$$

Using long division we see that $513 \div 4 = 128\frac{1}{4}$. We can also write this as $513 = 128 \times 4 + 1$.

**long division**

Long division is a way to show the steps for dividing numbers in decimal form. It finds the quotient one digit at a time, from left to right.

For example, here is the long division for $57 \div 4$.

$$
\begin{array}{r}
14.25 \\
4\overline{)57.00} \\
\underline{-4\phantom{0.00}} \\
17\phantom{.00} \\
\underline{-16\phantom{.00}} \\
10\phantom{0} \\
\underline{-8\phantom{0}} \\
20 \\
\underline{-20} \\
0
\end{array}
$$

**mean**

The mean is one way to measure the center of a data set. We can think of it as a balance point. For example, for the data set 7, 9, 12, 13, 14, the mean is 11.

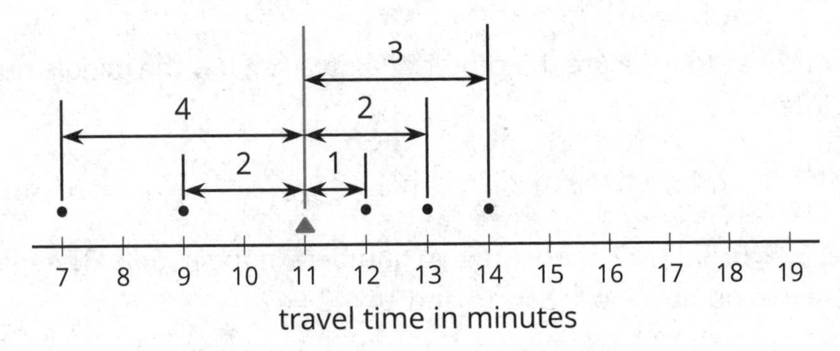


To find the mean, add up all the numbers in the data set. Then, divide by how many numbers there are. $7 + 9 + 12 + 13 + 14 = 55$ and $55 \div 5 = 11$.

**mean absolute deviation (MAD)**

The mean absolute deviation is one way to measure how spread out a data set is. Sometimes we call this the MAD. For example, for the data set 7, 9, 12, 13, 14, the MAD is

2.4. This tells us that these travel times are typically 2.4 minutes away from the mean, which is 11.

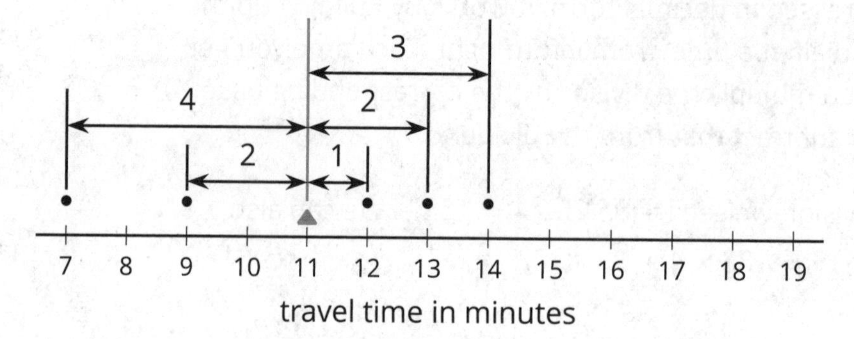


To find the MAD, add up the distance between each data point and the mean. Then, divide by how many numbers there are.

$4 + 2 + 1 + 2 + 3 = 12$ and $12 \div 5 = 2.4$

**measure of center**

A measure of center is a value that seems typical for a data distribution.

Mean and median are both measures of center.

**measurement error**

Measurement error is the positive difference between a measured amount and the actual amount.

For example, Diego measures a line segment and gets 5.3 cm. The actual length of the segment is really 5.32 cm. The measurement error is 0.02 cm, because $5.32 - 5.3 = 0.02$.

**median**

The median is one way to measure the center of a data set. It is the middle number when the data set is listed in order.

For the data set 7, 9, 12, 13, 14, the median is 12.

For the data set 3, 5, 6, 8, 11, 12, there are two numbers in the middle. The median is the average of these two numbers. $6 + 8 = 14$ and $14 \div 2 = 7$.

**negative number**

A negative number is a number that is less than zero. On a horizontal number line, negative numbers are usually shown to the left of 0.

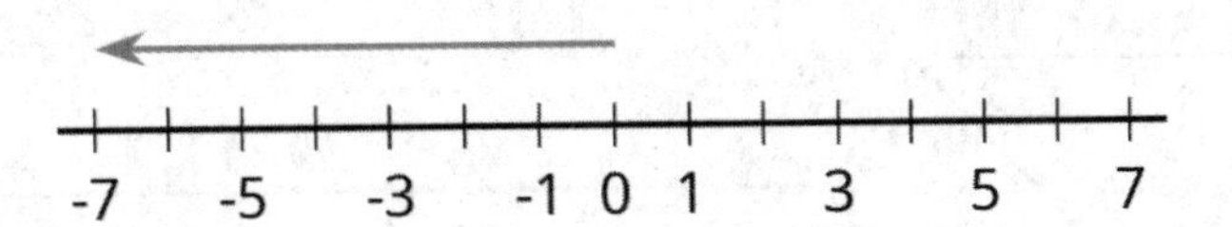


**net**
A net is a two-dimensional figure that can be folded to make a polyhedron.

Here is a net for a cube.

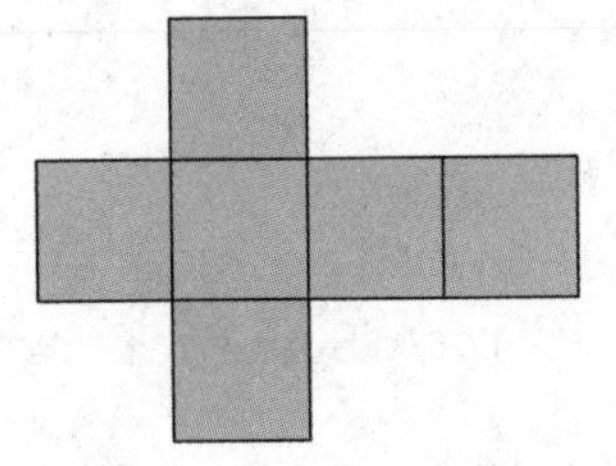

**opposite**
Two numbers are opposites if they are the same distance from 0 and on different sides of the number line.

For example, 4 is the opposite of -4, and -4 is the opposite of 4. They are both the same distance from 0. One is negative, and the other is positive.

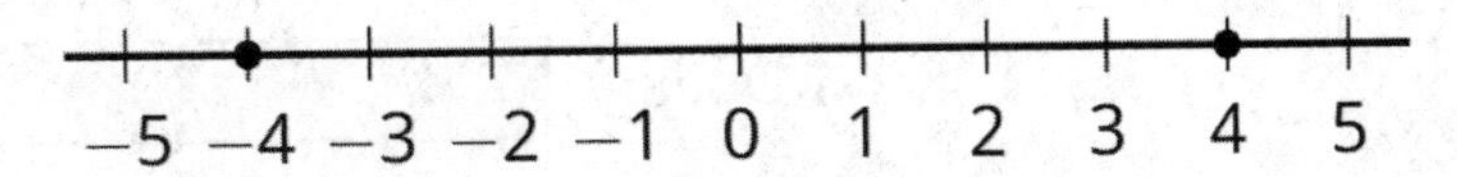


**opposite vertex**
For each side of a triangle, there is one vertex that is not on that side. This is the opposite vertex.

For example, point $A$ is the opposite vertex to side $BC$.

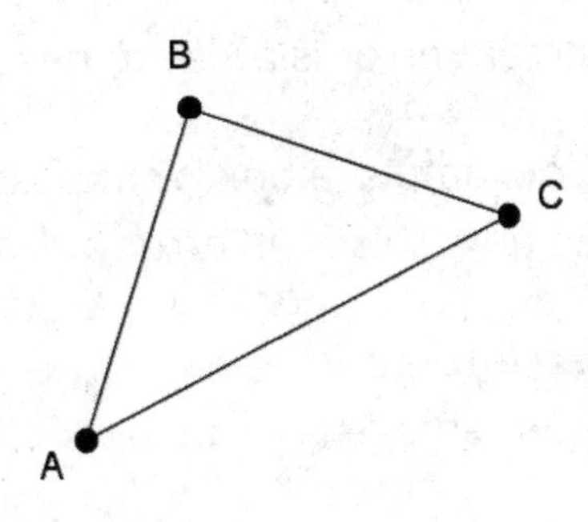


**parallelogram**
A parallelogram is a type of quadrilateral that has two pairs of parallel sides.

Here are two examples of parallelograms.

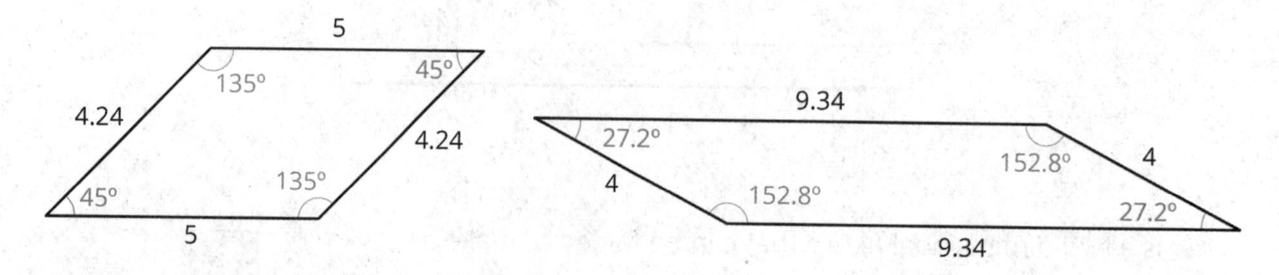


**per**

The word *per* means "for each." For example, if the price is $5 per ticket, that means you will pay $5 *for each* ticket. Buying 4 tickets would cost $20, because $4 \cdot 5 = 20$.

**percent**

The word *percent* means "for each 100." The symbol for percent is %.

For example, a quarter is worth 25 cents, and a dollar is worth 100 cents. We can say that a quarter is worth 25% of a dollar.

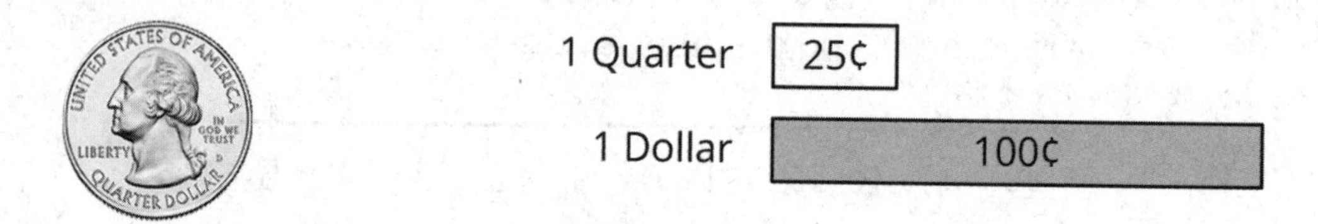


**percent error**

Percent error is a way to describe error, expressed as a percentage of the actual amount.

For example, a box is supposed to have 150 folders in it. Clare counts only 147 folders in the box. This is an error of 3 folders. The percent error is 2%, because 3 is 2% of 150.

**percentage**

A percentage is a rate per 100.

For example, a fish tank can hold 36 liters. Right now there is 27 liters of water in the tank. The percentage of the tank that is full is 75%.

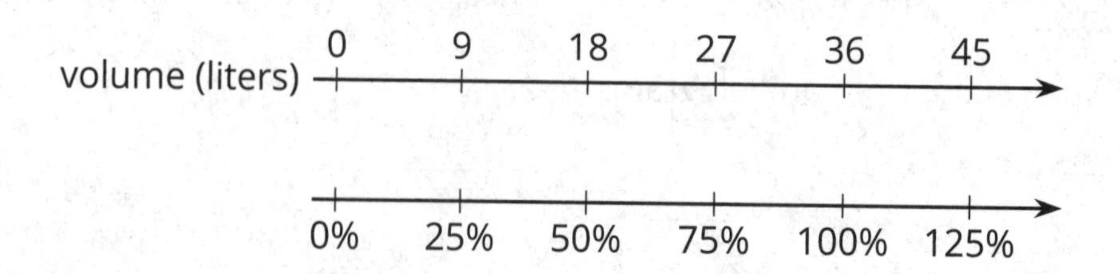

**percentage decrease**

A percentage decrease tells how much a quantity went down, expressed as a percentage of the starting amount.

For example, a store had 64 hats in stock on Friday. They had 48 hats left on Saturday. The amount went down by 16.

This was a 25% decrease, because 16 is 25% of 64.

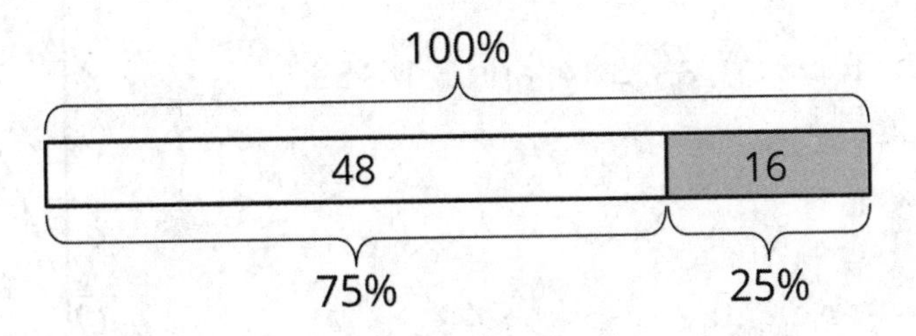


**percentage increase**

A percentage increase tell how much a quantity went up, expressed as a percentage of the starting amount.

For example, Elena had $50 in the bank on Monday. She had $56 on Tuesday. The amount went up by $6.

This was a 12% increase, because 6 is 12% of 50.

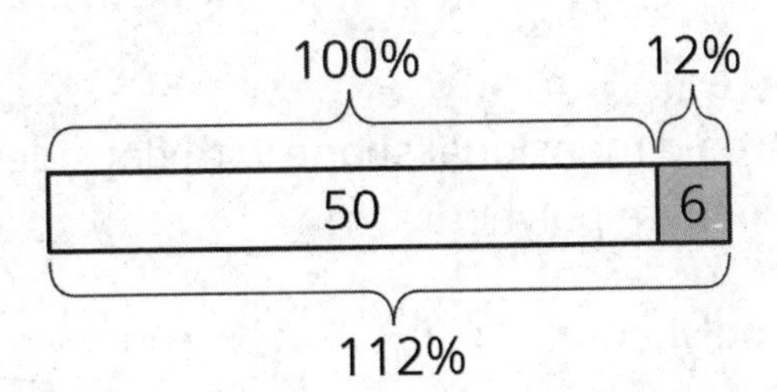

**pi ($\pi$)**

There is a proportional relationship between the diameter and circumference of any circle. The constant of proportionality is pi. The symbol for pi is $\pi$.

We can represent this relationship with the equation $C = \pi d$, where $C$ represents the circumference and $d$ represents the diameter.

Some approximations for $\pi$ are $\frac{22}{7}$, 3.14, and 3.14159.

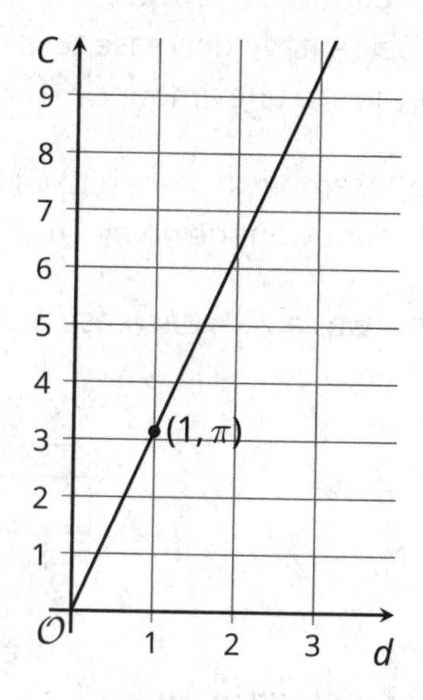


**polygon**

A polygon is a closed, two-dimensional shape with straight sides that do not cross each other.

Figure $ABCDE$ is an example of a polygon.

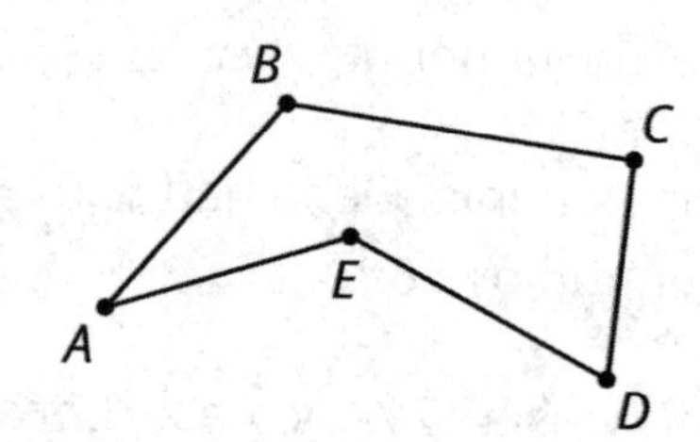


**polyhedron**

A polyhedron is a closed, three-dimensional shape with flat sides. When we have more than one polyhedron, we call them polyhedra.

Here are some drawings of polyhedra.

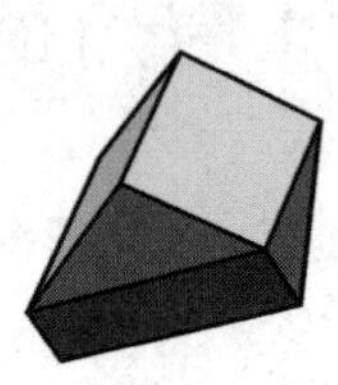

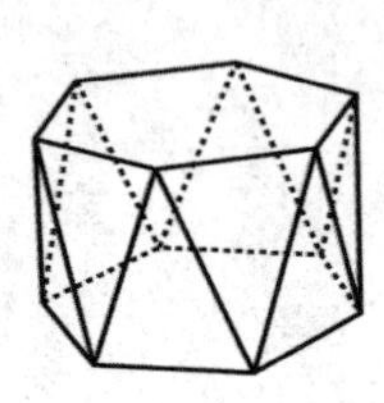

**population**

A population is a set of people or things that we want to study.

For example, if we want to study the heights of people on different sports teams, the population would be all the people on the teams.

**positive number**
A positive number is a number that is greater than zero. On a horizontal number line, positive numbers are usually shown to the right of 0.

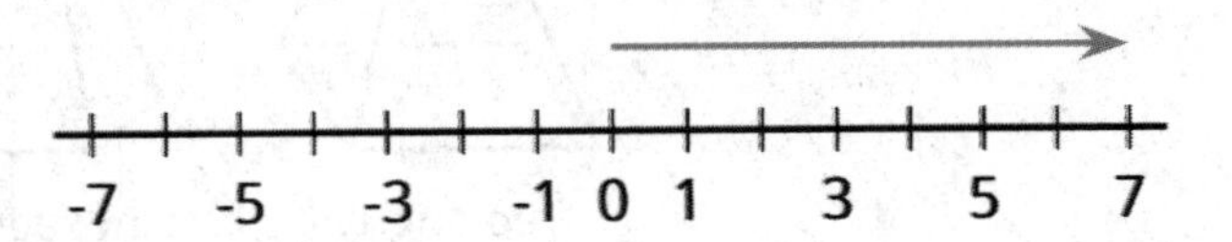


**prism**
A prism is a type of polyhedron that has two bases that are identical copies of each other. The bases are connected by rectangles or parallelograms.

Here are some drawings of prisms.

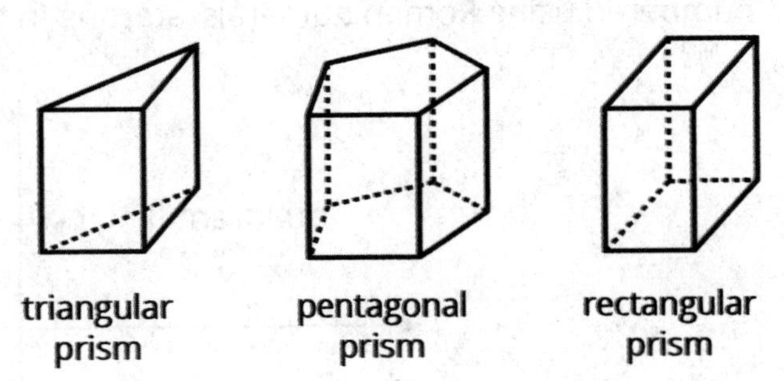


**probability**
The probability of an event is a number that tells how likely it is to happen. A probability of 1 means the event will always happen. A probability of 0 means the event will never happen.

For example, the probability of selecting a moon block at random from this bag is $\frac{4}{5}$.

**pyramid**
A pyramid is a type of polyhedron that has one base. All the other faces are triangles, and they all meet at a single vertex.

Here are some drawings of pyramids.

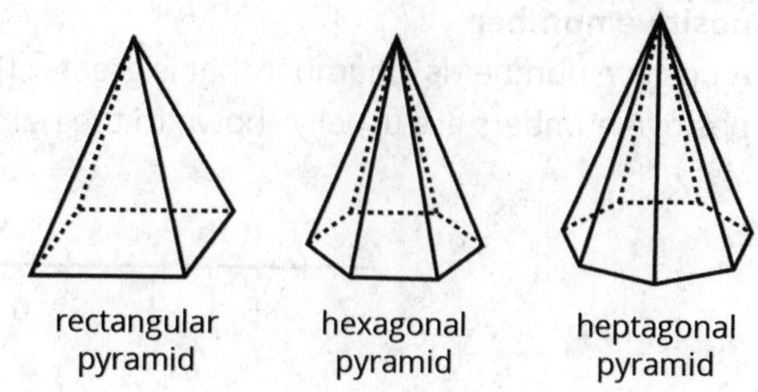


**quadrant**

The coordinate plane is divided into 4 regions called quadrants. The quadrants are numbered using Roman numerals, starting in the top right corner.

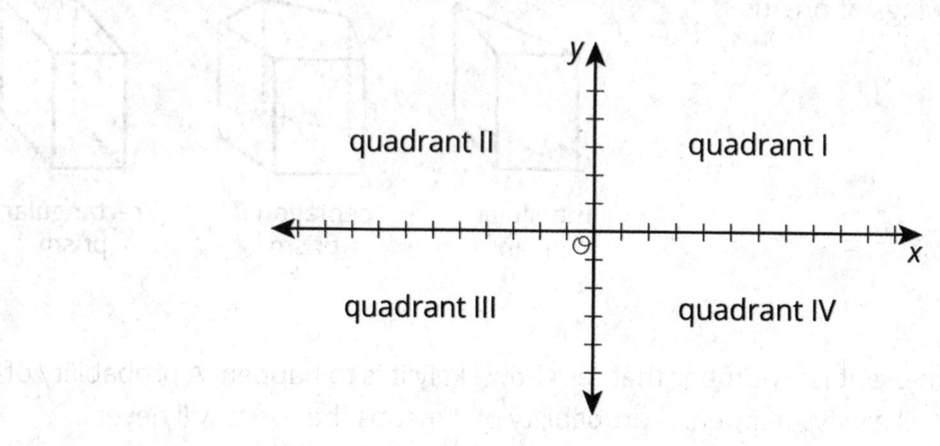


**quadrilateral**

A quadrilateral is a type of polygon that has 4 sides. A rectangle is an example of a quadrilateral. A pentagon is not a quadrilateral, because it has 5 sides.

**quartile**

Quartiles are the numbers that divide a data set into four sections that each have the same number of values.

For example, in this data set the first quartile is 30. The second quartile is the same thing as the median, which is 43. The third quartile is 50.

| 22 | 29 | 30 | 31 | 32 | 43 | 44 | 45 | 50 | 50 | 59 |
|---|---|---|---|---|---|---|---|---|---|---|
| | | Q1 | | | Q2 | | | Q3 | | |

**radius**

A radius is a line segment that goes from the center to the edge of a circle. A radius can go in any direction. Every radius of the circle is the same length. We also use the word *radius* to mean the length of this segment.

For example, $r$ is the radius of this circle with center $O$.

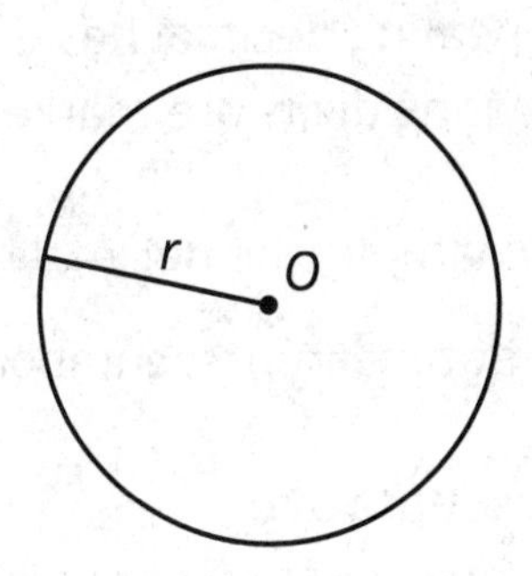


**random**

Outcomes of a chance experiment are random if they are all equally likely to happen.

**range**

The range is the distance between the smallest and largest values in a data set. For example, for the data set 3, 5, 6, 8, 11, 12, the range is 9, because $12 - 3 = 9$.

**ratio**

A ratio is an association between two or more quantities.

For example, the ratio $3 : 2$ could describe a recipe that uses 3 cups of flour for every 2 eggs, or a boat that moves 3 meters every 2 seconds. One way to represent the ratio $3 : 2$ is with a diagram that has 3 blue squares for every 2 green squares.

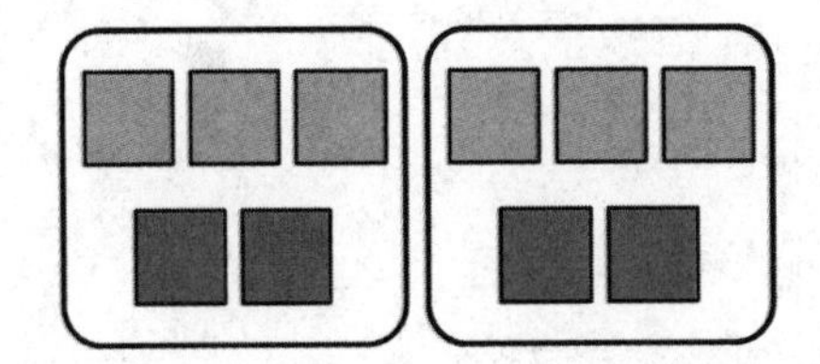

**rational number**

A rational number is a fraction or the opposite of a fraction.

For example, 8 and -8 are rational numbers because they can be written as $\frac{8}{1}$ and $-\frac{8}{1}$.

Also, 0.75 and -0.75 are rational numbers because they can be written as $\frac{75}{100}$ and $-\frac{75}{100}$.

**reciprocal**

Dividing 1 by a number gives the reciprocal of that number. For example, the reciprocal of 12 is $\frac{1}{12}$, and the reciprocal of $\frac{2}{5}$ is $\frac{5}{2}$.

**region**

A region is the space inside of a shape. Some examples of two-dimensional regions are inside a circle or inside a polygon. Some examples of three-dimensional regions are the inside of a cube or the inside of a sphere.

**repeating decimal**
A repeating decimal has digits that keep going in the same pattern over and over. The repeating digits are marked with a line above them.

For example, the decimal representation for $\frac{1}{3}$ is $0.\overline{3}$, which means 0.3333333 . . .
The decimal representation for $\frac{25}{22}$ is $1.1\overline{36}$ which means 1.136363636 . . .

**representative**
A sample is representative of a population if its distribution resembles the population's distribution in center, shape, and spread.

For example, this dot plot represents a population.

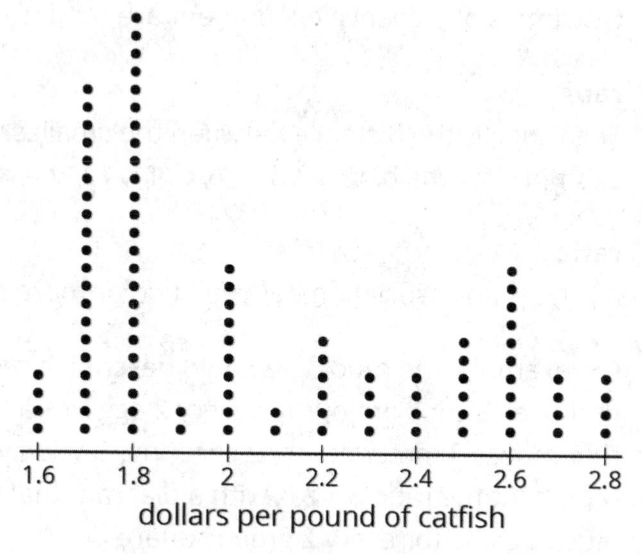


This dot plot shows a sample that is representative of the population.

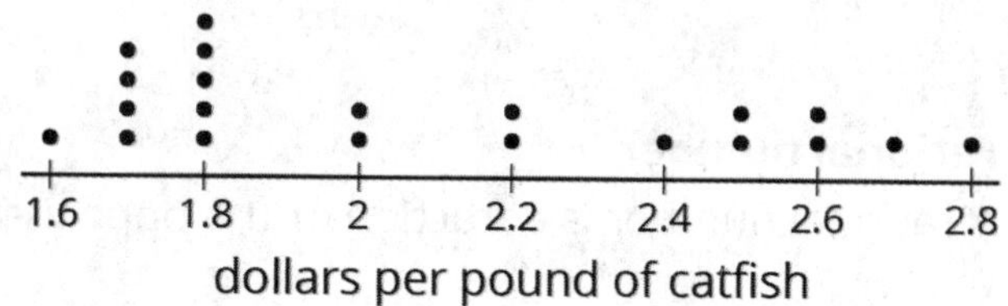


**same rate**
We use the words *same rate* to describe two situations that have equivalent ratios.

For example, a sink is filling with water at a rate of 2 gallons per minute. If a tub is also filling with water at a rate of 2 gallons per minute, then the sink and the tub are filling at the same rate.

**sample**
A sample is part of a population. For example, a population could be all the seventh grade students at one school. One sample of that population is all the seventh grade students who are in band.

**sample space**
The sample space is the list of every possible outcome for a chance experiment.

For example, the sample space for tossing two coins is:

| heads-heads | tails-heads |
|---|---|
| heads-tails | tails-tails |

**sign**
The sign of any number other than 0 is either positive or negative.

For example, the sign of 6 is positive. The sign of -6 is negative. Zero does not have a sign, because it is not positive or negative.

**solution to an equation**
A solution to an equation is a number that can be used in place of the variable to make the equation true.

For example, 7 is the solution to the equation $m + 1 = 8$, because it is true that $7 + 1 = 8$. The solution to $m + 1 = 8$ is not 9, because $9 + 1 \neq 8$.

**spread**
The spread of a set of numerical data tells how far apart the values are.

For example, the dot plots show that the travel times for students in South Africa are more spread out than for New Zealand.

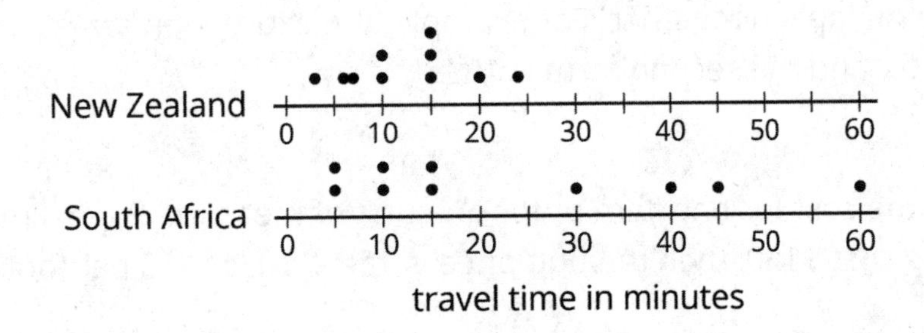


**squared**
We use the word *squared* to mean "to the second power." This is because a square with side length $s$ has an area of $s \cdot s$, or $s^2$.

**surface area**
The surface area of a polyhedron is the number of square units that covers all the faces of the polyhedron, without any gaps or overlaps.

For example, if the faces of a cube each have an area of 9 $cm^2$, then the surface area of the cube is $6 \cdot 9$, or 54 $cm^2$.

**table**

A table organizes information into horizontal *rows* and vertical *columns*. The first row or column usually tells what the numbers represent.

For example, here is a table showing the tail lengths of three different pets. This table has four rows and two columns.

| pet | tail length (inches) |
|---|---|
| dog | 22 |
| cat | 12 |
| mouse | 2 |

**tape diagram**

A tape diagram is a group of rectangles put together to represent a relationship between quantities.

For example, this tape diagram shows a ratio of 30 gallons of yellow paint to 50 gallons of blue paint.

| 10 | 10 | 10 | | |
|---|---|---|---|---|
| 10 | 10 | 10 | 10 | 10 |

If each rectangle were labeled 5, instead of 10, then the same picture could represent the equivalent ratio of 15 gallons of yellow paint to 25 gallons of blue paint.

**term**

A term is a part of an expression. It can be a single number, a variable, or a number and a variable that are multiplied together. For example, the expression $5x + 18$ has two terms. The first term is $5x$ and the second term is 18.

**unit price**

The unit price is the cost for one item or for one unit of measure. For example, if 10 feet of chain link fencing cost \$150, then the unit price is $150 \div 10$, or \$15 per foot.

**unit rate**

A unit rate is a rate per 1.

For example, 12 people share 2 pies equally. One unit rate is 6 people per pie, because $12 \div 2 = 6$. The other unit rate is $\frac{1}{6}$ of a pie per person, because $2 \div 12 = \frac{1}{6}$.

**variable**
A variable is a letter that represents a number. You can choose different numbers for the value of the variable.

For example, in the expression $10 - x$, the variable is $x$. If the value of $x$ is 3, then $10 - x = 7$, because $10 - 3 = 7$. If the value of $x$ is 6, then $10 - x = 4$, because $10 - 6 = 4$.

**withdrawal**
When you take money out of an account, it is called a withdrawal.

For example, a person removed $25 from their bank account. Before the withdrawal, they had $350. After the withdrawal, they had $325, because $350 - 25 = 325$.

# Attributions



## Image Attributions

By United States Census Bureau. Public Domain. American Fact Finder . http://factfinder.census.gov/faces/nav/jsf/pages/searchresults.xhtml?ref=geo&refresh=t&tab=map&src=bkmk.

Shrimp in aquarium, by uzilday. Public Domain. Pixabay. https://pixabay.com/en/shrimp-cherry-shrimp-aquarium-99521/.

Paper ream , by Bluesnap. Public Domain. Pixabay. https://pixabay.com/en/paper-ream-stack-tiered-white-224223/.

Stack of books, by Hermann. Public Domain. Pixabay. https://pixabay.com/en/books-education-school-literature-441866/.

"Steiger Ferris Wheel 1102009 1", by Zonk43. Public Domain. Wikimedia Commons. https://commons.wikimedia.org/wiki/File:Steiger_Ferris_Wheel_11102009_1.JPG.

By US Geological Survey. Public Domain. US Geological Survey. https://calval.cr.usgs.gov/rst-resources/sites_catalog/radiometric-sites/lake-tahoe/.

By United States Census Bureau. Public Domain. American Fact Finder. https://factfinder.census.gov/faces/nav/jsf/pages/.

Cheese Balls, by Andrew Stadel . CC BY-SA. Estimation 180. http://www.estimation180.com/.

By United States Census Bureau. Public Domain. American Fact Finder. http://factfinder.census.gov/faces/nav/jsf/pages/searchresults.xhtml?ref=geo&refresh=t&tab=map&src=bkmk.

By Dominique Feldwick-Davis . Public Domain. Pexels. https://www.pexels.com/photo/underwater-sea-turtle-turtle-sea-life-38452/.

Cars , by Pexels. Public Domain. Pixabay. https://pixabay.com/en/body-kit-car-car-wallpaper-custom-1869879/.

By makamuki0. Public Domain. Pixabay. https://pixabay.com/en/climber-scalar-rock-wall-escalation-1175425/.

Money Bag, by security_man. Public Domain. openclipart.org. https://openclipart.org/detail/245511/money-bag.

By Orca. Public Domain. Pixabay. https://pixabay.com/en/mountaineer-glacier-mountains-1008737/.

"US Navy 090226-N-9584H-018 Construction Electrician Constructionman Greg Langdon, assigned to Naval Mobile Construction Battalion (NMCB) 1, installs a new section of drill steel during a water well drilling operation", by U.S. Navy photo by Mass Communication Specialist Seaman Ernesto Hernandez Fonte. Public Domain. https://www.navy.mil/view_image.asp?id=69337.

Bathyscaphe Trieste, by U.S. Navy photo by Mass Communication Specialist Seaman Ernesto Hernandez Fonte. Public Domain. Wikimedia Commons. https://commons.wikimedia.org/wiki/File:Bathyscaphe_Trieste.jpg.

By Clker-Free-Vector-Images. Public Domain. Pixabay. https://pixabay.com/en/cat-cartoon-art-pet-cute-animal-304204/.

Submarine sandwich and chips, by jeffreyw. CC BY 2.0. Wikimedia Commons. https://commons.wikimedia.org/wiki/File:Submarine_sandwich_and_chips.jpg.

By Kapa65. Public Domain. Pixabay. https://pixabay.com/en/mice-mastomys-family-together-800875/.

Pawsox Mascot, by Paul Keleher. CC BY 2.0. Wikimedia Commons. https://commons.wikimedia.org/wiki/File:Pawsox_mascot.jpg.